AN INTRODUCTION
TO STATISTICS
FOR PSYCHOLOGY AND EDUCATION

AN INTRODUCTION TO STATISTICS
FOR PSYCHOLOGY AND EDUCATION

KEITH N. CLAYTON

Vanderbilt University

Charles E. Merrill Publishing Company
A Bell & Howell Company
Columbus   Toronto   London   Sydney

Published by Charles E. Merrill Publishing Co.
A Bell & Howell Company
Columbus, Ohio 43216

This book was set in Century Schoolbook
Production Editor: Rex Davidson
Text Designer: Cynthia Brunk
Developmental Editor: Annamaria Doney
Cover Design Coordination: Tony Faiola
Cover Photo: Melvin L. Prueitt, Los Alamos National Laboratory

Library of Congress Catalog Card Number: 83-62723
International Standard Book Number: 0-675-20154-3
Printed in the United States of America
1 2 3 4 5 6 7 8 9 10—88 87 86 85 84

Dedicated to
Norman James Clayton
Elizabeth May Clayton
Benjamin John Clayton

PREFACE

The principal features of this book were motivated by three considerations. First, since its major goal is to promote statistical understanding, a chapter is devoted to the concept of sampling distributions. This concept is often treated briefly, if at all, in introductory statistics texts. Yet it is central to statistical inference; so here students are encouraged to see that all descriptive and test statistics have sampling distributions and to understand their role in making claims about populations from samples.

A second consideration influencing this book is the increasing availability of powerful computer resources. Today, even inexpensive handheld calculators have considerable power. A major benefit is that instructors and texts can spend more time on comprehension and less time on the mechanics of computation. Regardless of computer resources, however, the student has to learn which statistic is appropriate and why. For instance, imagine a student approaching the computer center with data to be analyzed, having just read about t tests. This student might be surprised to learn that a computer package offers two or three t tests, but the choice among them is unclear. As computing becomes easier, more accessible, and less expensive, the skills appropriate for the statistics user will be to interact intelligently with powerful computing devices. These facts have had two effects on this book. First, a traditional topic has been excluded. Specifically, this book does not cover grouped data. Second, a chapter on computer statistics packages has been added. This chapter, Chapter 14, describes three widely available packages, MINITAB, SPSS, and BMDP. Chapter 14 also relates the packages to each of the statistics described in chapters 2–13. Thus, the student prepared by Chapter 7 to perform a t test will find in Chapter 14 the means to select from the statistical packages, and will find which programs correspond to the key equations in Chapter 7. On the other hand, neither computers nor Chapter 14 are essential to the use of this book.

A third consideration influencing this book is dissatisfaction with the available coverage of most introductory statistics textbooks. Specifically, these texts lack a treatment of within-subject analysis of variance. The purpose of an introductory course is to acquaint students with the techniques used in the research literature and to prepare them to perform their own analyses in subsequent courses, usually an experimental lab course. Within-subject designs are widely used and, in fact, are likely to be used by students in their own research. A comparison of within-subject and between-subject designs is covered in Chapter 9.

Acknowledgements. Several people have helped improve the quality of this book, or have facilitated my writing of it. First, I want to thank the students in statistics courses in classes at Eastern Michigan University and Vanderbilt University. They alerted me to ambiguities and computational errors. The instructors of these courses also provided important feedback. I especially thank Maureen Powers of Vanderbilt University for her very detailed reading and suggestions. Also appreciated are the comments and suggestions of Dave Chattin, Karen Mezynski, and for their assistance in reviewing the manuscript, I wish to thank Lyle Bourne, University of Colorado; Don Jackson, Eastern Michigan University; Thomas Nygren, Ohio State University; David Vaughn, University of Texas at Austin; Elliot Cramer, University of North Carolina; Edwin Shirkey, University of Central Florida; and David Dodd, University of Utah.

The writing of this book began during a semester leave of absence, for which I am grateful to Dean Jacque Voegeli, Vanderbilt University. For their assistance at Charles E. Merrill Publishing Company, I wish to thank Jane Sudbrink, Rex Davidson, and especially Annamaria Doney. I am grateful to the Literary Executor of the late Sir Ronald A. Fisher, F.R.S., to Dr. Frank Yates, F.R.S., and to Longman Group Ltd. London, for permission to reprint Tables III, IV, and VII from their book *Statistical tables for biological, agricultural and medical research* (6th edition, 1974). Finally, I am grateful to Christina Moser, who provided an essential supportive atmosphere in which to think and write.

KEITH NEIL CLAYTON

BRIEF CONTENTS

CONTENTS

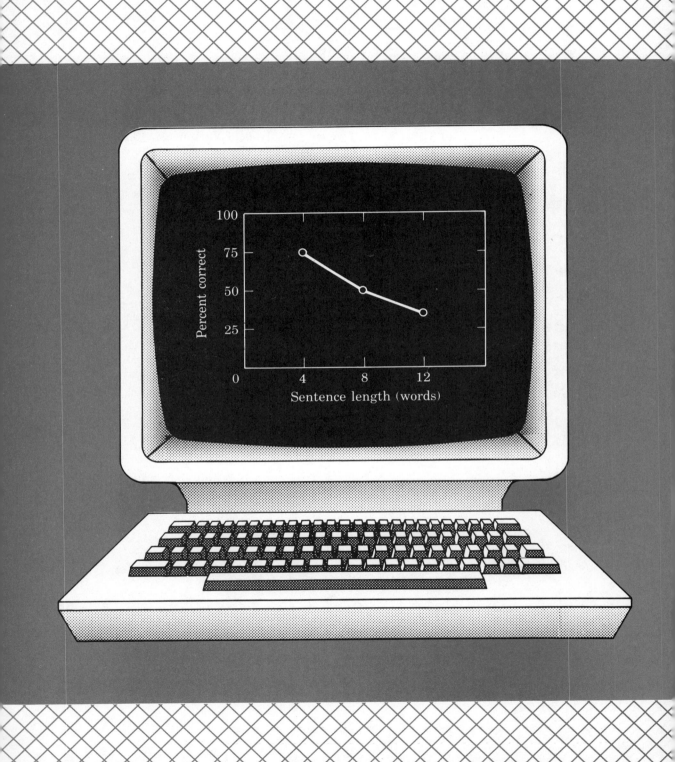

1 Introduction

T he various news media, television commercials, and advertisements are filled with examples of statistics:

- The average local rainfall in August is 3.2 inches.
- Auto production in the U.S. declined 15% in the first quarter of the year.
- Ninety-seven percent of all American households have at least one TV set.
- In one taste-test, 53% preferred one cola over another.

These examples illustrate the descriptive function of statistics; that is, the use of statistics to describe, or summarize, a set of observations. A **descriptive statistic**, then, is a number assigned to a collection of observations, or a set of data. The descriptive statistics in these examples are percentages (15%, 97%, 53%) and an average (3.2 inches). The *principal* function of statistics, however, is more than descriptive. Scientists make observations (collect data) in order to make generalizations about social, biological, or physical phenomena. The principal function of statistics is to help the scientist make decisions about the phenomenon under study. Statistical methods used in such decision making are called **inferential statistics**. The distinction between descriptive and inferential statistics, then, distinguishes between methods used to organize and summarize a set of observations and methods used to draw conclusions from that data.

In other words, social scientists rarely stop at description. In addition to calculating averages, percentages, and so forth, an investigator typically computes a *test* statistic as well. This test statistic is used by the scientist to decide how to generalize from the data. For example, in addition to calculating an average value, an experimenter may calculate a test statistic, called t. To make a conclusion about the obtained average, the value of t is compared to values expected when certain conditions are true. We show in later chapters exactly how test statistics such as t are used to draw inferences from data. But first, fundamental to understanding these test statistics are the concepts of *populations* and *samples*. These terms are defined in the next section, after which several other primary statistical concepts are introduced. Descriptive statistics are discussed in Chapter 2 and the remainder of the text is devoted to inferential statistics.

1.1

THE DISTINCTION BETWEEN POPULATIONS AND SAMPLES

A **population** is defined as a set of individuals or measurements having some common characteristic. A **sample** is defined as a subset of the population, and as such, is said to be "drawn" from the population. Public opinion polls nicely illustrate the distinction between samples and populations. Before beginning a poll, the pollster typically defines a target population, such as "all U.S. citizens of voting age" or "all members of labor unions." Then a number of such persons are selected for an interview. Recently a poll of 3000 female college students at six "highly selective" schools reported that 77% agreed with the statement "Mothers should not work at all, or work only part-time until their children are five years old." In this example, the sample is the set of 3000 students and the population would be all female students at "highly selective" schools. The pollster apparently wishes to generalize the results of the survey to all female students at such schools.

A population need not be a collection of people. Population has been defined as a *set of individuals* or *measurements*. In many applications of statistical inference, a population is thought of as a set of numbers assigned to observations. In such applications, repeated observations may be made on the same individuals. For example, a study of hearing may have subjects listen 100 times to an auditory signal. Each of the observations may be considered as a measure sampled from an infinitely large population of such measures that could have been taken. The sample in this case could have been 1000 instead of 100. As another example, consider an investigation of the effect of caffeine on reaction time. Suppose that subjects after having ingested caffeine are given 200 trials in which they are to respond to a signal as quickly as they can. The experimenter measures the latency of each response in milliseconds and computes an average reaction time. In this study, the average reaction time is the descriptive statistic and the sample is the set of 200 measures.

Some populations exist; others are hypothetical, or conceptual. For example, the population of the residents of the state of Texas is real and countable. Populations of measures are typically more theoretical. In the study on caffeine, the population could be defined as the set of all reaction times theoretically obtainable under the conditions of the experiment. Of course, to obtain all such measures would not be possible.

As we have seen, the principal function of statistics is not merely to provide descriptions of samples. Samples are used to draw inferences about populations. The major purpose of this text is to introduce the methods and concepts used to make such statistical inferences. For example, what conclusion can be drawn from a report that 53% of a number of people (a sample) preferred one brand of cola over another? The intent of the report, no doubt, is to suggest something about all cola drinkers (a population). But are we to conclude that *exactly* 53% of this population (all cola drinkers) would have the same preference? Are we even prepared to conclude that a majority (more than 50%) do?

How do we proceed to reach *any* conclusion from such data? Such questions constitute the major focus of this text.

1.2

THE DISTINCTION BETWEEN STATISTICS AND PARAMETERS

Corresponding to the distinction between samples and populations is a distinction between statistics and parameters. A **statistic** is an index associated with a sample. A **parameter** is an index associated with a population. In the poll of female students, 77% agreed with the mothers-working statement. The 77% is a statistic because it is an index on a sample. In the cola-preference example, 53% preferred a particular cola. The 53% is a statistic because it is an index on a sample of cola drinkers. However, the statement about TV sets ("97% of all American households have at least one TV set") is a statement about a parameter (97%), because it is a claim about a population (all American households). Any survey of all members of a population, such as taken by the U.S. Census Bureau every ten years, is an attempt to obtain values of parameters. Any attempt short of this, any attempt that samples a subset of the population, produces statistics.

We may now characterize inferential statistics using these new terms: *statistical inference* involves the use of the values of statistics from samples for the purpose of drawing inferences about the parameters of a population. For example, in the cola-preference report that 53% preferred a particular cola, the statistic (53%) is known but the parameter (the percent of all cola drinkers preferring that cola) is not known. Any claim made about the population is a claim about a parameter. As another example, it has been claimed that 10% of all males are color blind. Such a claim is a claim about the value of a parameter, the percent of all males that are color blind. However, the claim comes from a statistic, since all males have not been tested for color blindness (10% of some sample of males show color blindness). As a final example of the distinction, consider the experiment on caffeine and reaction time. Suppose it is found that the average reaction time is 425 milliseconds. This average value is a statistic. It may or may not have the same value as the parameter, the average reaction time for the population of such measures.

Again, notice that we have been describing two different activities. First, the investigator collects data (makes observations) and calculates statistics. This is the descriptive phase and the statistics are descriptive statistics. Second, the investigator operates on these statistics and makes decisions about population parameters. This second stage involves inferential statistics.

Statistics and parameters are symbolized by letters. To make clear whether a letter refers to a statistic or a parameter, <u>italic letters are traditionally used to refer to statistics and Greek letters are used to refer to parameters.</u> This tradition will be followed here. For instance, the symbol used to refer to the arithmetic average, or mean, of a sample will be M. The mean of a population, however, will be referred to by the Greek letter μ (mu). In the caffeine example, then, $M = 425$ milliseconds. In this example, the value of μ is not known. It might be larger or smaller than 425. The major purpose of obtaining M is to make inferences about μ.

1.3

VARIABLES

A **variable** is any characteristic that can take on different values. For example, age, weight, gender, and marital status are all variables. So too are response latency, number of correct responses, and so forth. There are two traditional ways to classify variables. One classification stems from descriptions of experimental designs. In discussions of experimental design, a distinction is made between independent and dependent variables. However, another important distinction is between discrete and continuous variables.

1.3.1 THE DISTINCTION BETWEEN DEPENDENT AND INDEPENDENT VARIABLES

The distinction made here is between a variable under the control of the experimenter and one that the experimenter measures. <u>The variable under the control of the experimenter, the one "manipulated" by the experimenter, is the **independent variable**</u>. The variable measured by the experimenter, the one that presumably "depends" on the <u>independent variable, is the **dependent variable**</u>. For example, a study of reading comprehension might have three different groups of students read paragraphs that average 4, 8, or 12 words per sentence, respectively. In this study the independent variable would be sentence length and the dependent variable would be some measure of comprehension.

The distinction between independent and dependent variables is particularly evident when experimental results are graphed. Traditionally, <u>the values of the dependent variable are plotted on the vertical, or *y*-axis and the levels of the independent variable are plotted on the horizontal, or *x*-axis.</u> The results are conceived, then, as showing the dependent variable "as a function of" the independent variable. An example is shown in Fig. 1.1, which plots the possible relation between reading comprehension and average sentence length. Fig. 1.1 plots a measure of comprehension "as a function of" sentence length.

FIGURE 1.1

An example of a possible result. The independent variable is sentence length. The dependent variable is percent correct, a measure of comprehension.

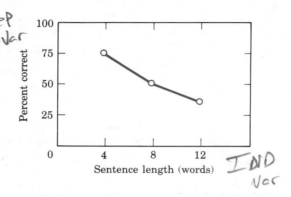

Other examples might be "learning errors as a function of reward magnitude," or "visual sensitivity as a function of wavelength." In each example, the dependent variable is listed first and the independent variable is listed second.

1.3.2 THE DISTINCTION BETWEEN DISCRETE AND CONTINUOUS VARIABLES

A variable has been defined as a characteristic that can take on several values. The distinction between discrete and continuous variables is one arising from limitations on the values the variable can take. A variable is called **continuous** if it can take on all possible real numbers over a particular range of values. More precisely, a variable that has no gaps or breaks between two points is said to be continuous between those two points. Height is one such variable because, in principle at least, it can be measured to any desired degree of precision. Response time (duration of time to respond) is another example of a continuous variable. If a variable is not continuous it is said to be discontinuous, or **discrete**. For example, consider a variable that can take on only the values of whole numbers, or integers. The variable "number of correct choices on the test" can take on only the values of 1, 2, 3, . . . , up to the number of items on the test. Other examples of discrete variables are: number of children, number of responses, number of trials to learn, etc.

1.4

THE DISTINCTION BETWEEN CATEGORICAL AND NUMERICAL DATA

Data (and variables) may also be classified as either categorical or numerical. Data are **categorical** (nonquantitative) when the observations are labelled, named, or categorized but not assigned a number. Examples of categories seen in psychological and educational research are: gender, classes of abnormal or deviant behavior, species, pass-fail, and marital status. **Numerical data** result from measurements that assign numbers to observations. Examples of numerical (quantitative) data are: numbers of correct responses, latency of response, age, and weight.

The distinction between categorical and numerical data is important because it determines the type of statistical analysis that is appropriate to the data. It should be obvious, for example, that the arithmetic average cannot meaningfully be calculated on categorical data. The distinction is even more important when we consider inferential statistics. Categorical data are analyzed differently from numerical data. Most of the analyses described in this book are based on the assumption that the data are numerical. Chapter 12, however, is devoted to analyses of categorical data. One of the first criteria for selecting among statistical procedures is to determine whether the data are categorical or numerical.

KEY TERMS

A **descriptive statistic** is a number assigned to a collection of observations or a set of data. It is used to describe the data and is to be distinguished from an **inferential statistic**, which is used to make inferences from the data. Inferential statistics involve the use of descriptive statistics on samples to make inferences about population parameters.

A **population** is defined as a complete set of individuals or measurements that have some common characteristic. It is distinguished from a **sample**, which is a subset of individuals or measurements "drawn" from the population for purposes of drawing inferences about the population.

A **statistic** is a characteristic of a sample and is usually symbolized by an italic letter. It is to be distinguished from a **parameter**, which is an index associated with a population. A parameter is traditionally symbolized by a Greek letter.

A **variable** is any characteristic that can take on different values.

An **independent variable** is one manipulated by the experimenter. It is to be distinguished from a **dependent variable**, which is measured by the experimenter.

A **continuous variable** is one that can take on all real numbers as values. It is distinguished from a **discrete variable**, which is limited to a finite set of values, usually whole integers.

Categorical data result when observations are labelled, named, or categorized. They are distinguished from **numerical** (quantitative) **data**, which result from measurements that assign numbers to observations.

QUESTIONS

1. A set of observations is a population or a sample depending on the purpose of the investigation. If the investigator is satisfied to describe the set without generalizing beyond it, then the set may be thought of as a population. If, however, the investigator wishes to draw some inference about observations beyond the set, then the set is a sample. From a description of a set you can usually tell which is intended. Indicate which of the following is a population and which is a sample.
   P (A) All the registered voters of Virginia.
   S (B) Every fifth car produced by Ford Motor Company.
   P (C) All the students in this statistics class.

2. Descriptive statistics describe a collection of observations. Inferential statistics are used to draw conclusions from a set of data. Consider the following experiment: Two groups of college students are presented with 20 three-letter trigrams. Group I receives meaningful trigrams like RED, HER, and DAD. Group II is given nonsense trigrams like DES, GUH, and XIB. Immediately after studying the list, the students were asked to write as many trigrams as they could recall. Indicate which of the following are descriptive (D) and which are inferential (I).
   D (A) The average number of words recalled for Group I is 15.
   D (B) Group II recalled 40% of the words.
   I (C) Group I recalled more words because there were brighter students in that group.
   I (D) The students in Group II do not like to be in experiments.
   D (E) The most words recalled in Group II was 20.

3. Identify each of the following statements as using inferential (I) or descriptive (D) statistics.
  (A) 58% of the custodial staff at State U. is Caucasian.
  (B) 9 out of 10 cats in America prefer Chowchow mix.
  (C) In a survey of 10,000 households, 28% of the families took a summer vacation of 10 days or longer.
  (D) The average time in last week's 10 kilometer race was 53:38.
  (E) There is a 40% chance of snow tomorrow.
  (F) The temperature was above 32°F for 79% of today.

For questions 4–8 answer the following:
(A) Which would be intended: inferential (I) or descriptive (D) statistics?
(B) Does the research involve a sample or a population? If a sample is used, what is the population that the researcher sampled from?
(C) What is the independent variable?
(D) What is the dependent variable?
(E) Is the dependent variable categorical or numerical?

4. Fifty freshmen participated in the testing of a new toothpaste, Extra Bright. Each participant was rated for whiteness of teeth (on a scale of 1 to 10) before and after a two-month period of using Extra Bright.

5. An educational psychologist wanted to know if computer-assisted instruction (CAI) helped fifth graders learn math better than traditional classroom instruction. Two hundred students from 10 public school classes participated in the CAI program; their scores on a standardized mathematics achievement test were compared with national fifth grade public school norms.

6. A demographer (one who is interested in the study of human populations) was interested in finding out if people who live in cities die at a younger age than people who live in rural areas. Using the 1980 census data, he chose data at random from several cities to compare with data from several rural counties.

7. The same demographer from the previous question also wanted to know if birthrates in cities differed from rural areas. He compared 1980 census data for number of births in all U.S. cities with all non-urban area births.

8. All incoming freshman at State U. were required to fill out a questionnaire regarding their academic interests and intended majors. The administration compared the responses of males and females with respect to preferred major: social sciences, natural sciences, humanities, or engineering.

9. Which is a parameter (P) and which is a statistic (S)?
  (A) μ
  (B) M
  (C) 83% of all voters in Virginia
  (D) 73% of cola drinkers sampled in a taste-test

(E) Average amount of food your dog eats in one day

(F) Public opinion poll results

10. Formulate your own questions that might be asked in a public opinion poll:

(A) One which would yield categorical data.

(B) One which would yield numerical data.

11. Suppose a questionnaire asks the following:

NAME

AGE

SEX

MARITAL STATUS

SOCIAL SECURITY NUMBER

CALORIC INTAKE PER DAY

CIGARETTES SMOKED PER DAY

WEIGHT

HEIGHT

Which are categorical and which are numerical?

12. Suppose the responses to the questionnaire in Question 11 are coded such that MALE = 1 and FEMALE = 2. Why would it not make sense to find the average for SEX?

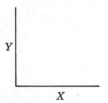

13. (A) Which axis on the above figure represents the independent variable of an experiment, and which represents the dependent variable?

(B) The ____ variable is the one the experimenter controls, such as "the amount of drug injected into a rat."

(C) The ____ variable is what the experimenter measures, such as "number of errors made by a rat in a maze."

For each of the following graphs, briefly summarize what the experiment is testing. For example, the graph below could be summarized as "the effects of print type size on reading speed."

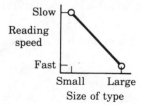

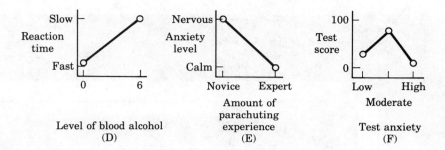

Level of blood alcohol
(D)

Amount of
parachuting
experience
(E)

Test anxiety
(F)

14. In the experiment described in Question 2, what is the independent variable? What is the dependent variable?

15. Consider the following experiment:
Several subjects are given different amounts of caffeine ranging from 0 grams for subject 1 to 15 grams for subject 15. The subjects are then asked to perform a reaction time task where their responses are measured to the nearest millisecond. What is the independent variable? What is the dependent variable?

16. The following represents the results of the experiment in Question 15:

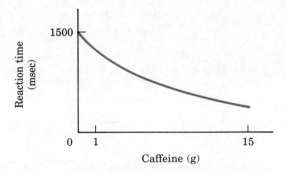

We could describe this result as ____ as a function of ____.

17. Define the term *variable*.

18. Distinguish between an independent variable and a dependent variable.

19. Distinguish between a discrete variable and a continuous variable.

20. Give an example of a continuous variable.

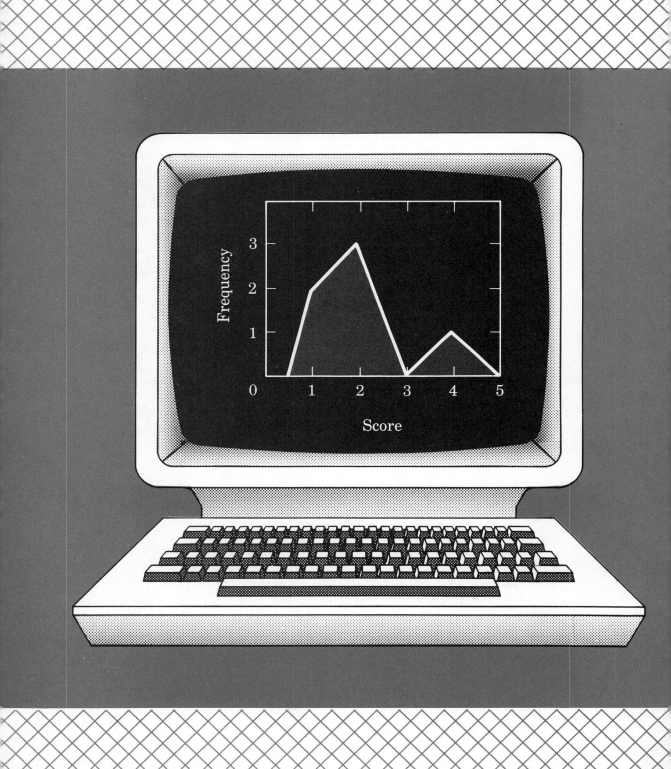

2 Descriptive Statistics

An initial way to organize data is to count the frequency that each category, or number, occurs. Consider, for example, the poll described in Chapter 1 on attitudes toward working mothers. Three thousand female college students were asked whether they agreed with the statement that mothers should not work until their children are at least five years old. The responses are placed into one of two categories, agree or disagree, and the frequencies in each category are tabulated. The resulting data may be displayed as in Table 2.1.

TABLE 2.1
Frequency distribution of categorical data.

Response to Question	Frequency
Agree	2310
Disagree	690
Total	3000

Data organized in this way are called **frequency distributions**, which summarize a collection of data by showing the frequency with which each category, or score value, occurs.

2.1

FREQUENCY DISTRIBUTIONS AND GRAPHS

In addition to tables, data may be presented in figures or graphs. We illustrate this by means of numerical data. Suppose, in a study of sharing, that six kindergartners are each given four opportunities to share prizes with another player. The score each child receives represents the number of times that he or she shares. Suppose the scores for each of the six children are 2, 1, 2, 4, 1, and 2. Table 2.2 organizes the six scores into a frequency distribution.

TABLE 2.2
Frequency distribution of quantitative data.

Score	Frequency
4	1
3	0
2	3
1	2
Total	6

Table 2.2 shows that one subject received a score of 4 (shared four times), three subjects received a score of 2, and two subjects received a score of 1. This frequency distribution is displayed graphically on the left side of Fig. 2.1.

FIGURE 2.1
Graphs of data in
Table 2.2.

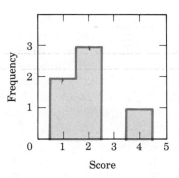 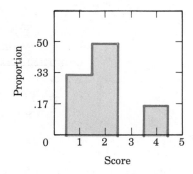

Another way to construct a distribution is to transform each frequency into a proportion of the total number of scores. To do so, for each value the frequency of scores associated with that value would be divided by the total number of scores. For example, in the data in Table 2.2 the value of 4 occurred one time; it has a frequency of 1. To transform this frequency into a proportion, we would divide 1 by the total number of scores, $1/6 = .17$. The value of 2 occurred 3 times; its proportion is $3/6 = .50$. Finally, the value of 1 has a frequency of 2. The proportion associated with 1, then, is $2/6 = .33$. The distribution is graphed on the right side of Fig. 2.1. <u>Why do we transform frequencies into proportions? This is valuable when comparisons are made between different distributions. If the two distributions have unequal sample sizes, transforming their frequencies into proportions simplifies the comparison.</u> This transformation is also valuable because of the important link between proportions and probabilities—a topic explored in the next chapter.

The graphs in Fig. 2.1 are both called **histograms.** Notice that in a histogram there are two axes: a horizontal baseline, or x-axis, and a vertical, or y-axis. The score values are represented along the x-axis, and the y-axis represents the frequency or proportion of each score value. For example, the fact that the value of 4 occurred one time (has a frequency of 1) is represented by the height of the bar above 4.

Another way to graph the data is shown in Fig. 2.2.

FIGURE 2.2
Frequency polygon of
data in Table 2.2.

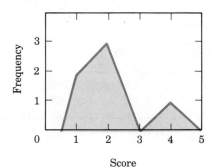

Fig. 2.2 is a *frequency polygon*. In a frequency polygon, each score value is represented by a single point. The height of the point represents the number of times that value was observed in the sample. The points for adjacent score values are connected, including those values with zero frequency. Thus, both ends of a frequency polygon drop down to the baseline at the value just below the lowest score and the value just above the highest value. The *area* under the frequency polygon is the same as that under the bar graph in Fig. 2.1. This assertion could be proved with elementary geometry, but proving such an assertion is not important here. On the other hand, the concept of area *is* important because, as we shall see, many developments in statistical inference make use of it.

2.2

CHARACTERISTICS OF DISTRIBUTIONS

The graphs in Fig. 2.3 illustrate the variety of ways distributions may differ.

FIGURE 2.3
Examples of distributions.

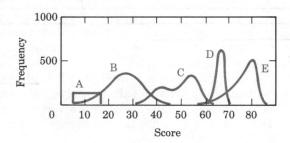

Distribution A is called a *rectangular*, or *uniform*, distribution; any scores that occur have the same frequency. Distribution B would be called *bell-shaped*. Distributions A and B also differ in that the scores in B are, in general, higher in value than those in A. Distribution E has the highest scores of all. Distribution C is unique in that it has two peaks, or humps, whereas the others have one. The five distributions also differ in the degree to which scores cluster, as in D, or are spread apart, as in B.

The purpose of descriptive statistics is to give us measures of these different characteristics. That is, the purpose of descriptive statistics is to provide numerical values that reflect properties of the distribution, such as its most typical score, where it is located on the *x*-axis, how spread out it is, how bell-shaped it is, and so forth. More formally, **descriptive statistics** are defined as numerical values calculated according to well-specified rules in order to provide measures of characteristics of distributions. For example, the **range** of a set of scores is a measure of the degree to which the scores either cluster or spread. The range is defined as the difference between the highest and lowest score in the set. The value of the range for the data in Table 2.2, for example, is 3 because the highest score is 4 and the lowest is 1 (4 − 1 = 3). The value of the range is large when the scores are widely scattered, as in distribution B of Fig. 2.3, and the value of the range is small when the scores are clustered tightly together, as in distribution D. Obviously, if all scores in a distribution have the identical value, the range would equal zero. The range and other statistics that reflect the degree to which scores are dissimilar, or vary, are called **measures of variability**.

The **mode** is the score that has the greatest frequency; that is, the score below the "hump" in the distribution. The value of the mode for the distribution in Table 2.2 is 2, because 2 has a greater frequency than any other score. The value of the mode is not to be confused with the frequency associated with the mode. In Table 2.2, 2 is the mode, which has a frequency of 3. Distribution B in Fig. 2.3 has a mode of 25; distribution D has a mode of 65. The mode may be thought of as the most typical score; it is one of several measures of central tendency. **Measures of central tendency** are also referred to as *measures of location*, because they generally reflect the location of the distribution on the *x*-axis.

The mode is also used to help describe the shape of distributions. Distribution C in Fig. 2.3 is called *bimodal* because it has two humps. The other distributions illustrate *unimodal* distributions. Another characteristic of the shape of a distribution is its *skewness*. Quantitative measures of skewness are available, but for our purposes skewness falls into one of three categories: positive, negative, or none. These are illustrated in Fig. 2.4.

FIGURE 2.4
Examples of positive, zero, and negative skew.

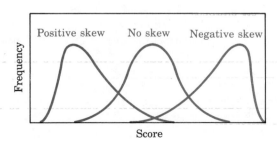

A distribution has *zero* or no skew when it is perfectly symmetrical; that is, it could be folded over at its center and both sides would coincide. When a distribution is **positively skewed**, the scores extend farther from the mode in the direction of large values rather than in the direction of small values. When a distribution is **negatively skewed**, the scores extend from the mode farther in the direction of small rather than large values. Stated another way, skewed (nonsymmetrical) distributions seem to have one long tail (or "skewer"). Whether the skew is positive or negative depends on whether the tail is pointing to the right or left. If the skewer is pointing to the right (toward increasingly positive scores), the distribution is positively skewed. If the skewer is pointed to the left (toward increasingly negative scores), the distribution is negatively skewed.

Two classes of descriptive statistics are taken up in the next two sections. Section 2.3 covers measures of central tendency, which we have seen to be measures of the most typical score, or measures of the location of the distribution on the *x*-axis. We have already defined the mode as one measure of central tendency. Section 2.4 covers measures of variability; in other words, measures of the degree to which the distribution is spread out or clustered. Range is an example of a measure of variability.

2.3

OTHER MEASURES OF CENTRAL TENDENCY

2.3.1 DEFINITIONS

The **mean** of a set of scores is defined as the sum of all scores in the set divided by the total number of scores. Thus, the mean of the scores in the set (3, 1, 8, 1, 7) is 4.0, because there are 5 scores that sum to 20 (3 + 1 + 8 + 1 + 7 = 20). Notice that the mean and mode do not have the same value for this distribution because the mode, the most frequent score, is 1.0.

The **median** of a set of scores is best understood after the scores have been rank-ordered; that is, arranged according to numerical value. Thus, the scores (3, 1, 8, 1, 7) would be rearranged (1, 1, 3, 7, 8). The median is in the middle of the group of scores. When there is an odd number of scores, the median is the middle score in the rank-order. For the example (1, 1, 3, 7, 8), the median is equal to 3. When there is an even number of scores, the median is halfway between the two middle scores. For example, the median of the scores (1, 3, 4, 8) is 3.5, and the median of the scores (1, 3, 4, 6, 8, 10) is 5.0.

2.3.2   FORMULA FOR THE MEAN

The formula for the mean is written

$$M = \frac{\Sigma X}{N} \tag{2.1}$$

In this formula the Greek capital letter *sigma*, Σ, means "the sum of." The variable X stands for any score. Thus, ΣX means literally to sum all the scores. The denominator of Equation 2.1, N, stands for the total number of scores. For example, given the set of scores: (1, 3, 6, 6),

$$\Sigma X = 1 + 3 + 6 + 6 = 16$$
$$N = 4$$
$$M = \frac{\Sigma X}{N} = \frac{16}{4} = 4.0$$

As another example, given the set of scores (2, 4, 6, 3, 0)

$$\Sigma X = 2 + 4 + 6 + 3 + 0 = 15$$
$$N = 5$$
$$M = \frac{\Sigma X}{N} = \frac{15}{5} = 3.0$$

Note that even though a score is zero and does not contribute to the *sum* of scores, its existence still contributes to N.

2.3.3   MORE ON THE SUMMATION SIGN

Sometimes the summation symbol is written with variables that are given subscripts, as in

$$\Sigma(X_i - k)$$

In this example, the subscript on X, i, means that the summation applies over all values of X and not over different values of k. In fact, in this example, k is a constant (does not vary), and the expression requires that each value of X be reduced by the magnitude of k and that the resulting differences be totalled. Suppose, for instance, that there are 5 scores ($N = 5$). The first score is labelled X_1, the second is labelled X_2, the third X_3, etc., to the fifth, which is X_5. For illustrative purposes, let $X_1 = 10, X_2 = 15, X_3 = 3, X_4 = 7$, and $X_5 = -1$. Finally, let $k = 4$. Then, in this case

$$\Sigma(X_i - k) = (X_1 - k) + (X_2 - k) + (X_3 - k) + (X_4 - k) + (X_5 - k)$$
$$= (10 - 4) + (15 - 4) + (3 - 4) + (7 - 4) + (-1 - 4)$$
$$= 6 + 11 + (-1) + 3 + (-5)$$
$$= 6 + 11 - 1 + 3 - 5 = 14$$

You may also see summation used when each score is multiplied by a constant, as in

$$\Sigma k X_i$$

Again, the subscript on X means that although X varies, k is a constant. So, for N scores,

$$\Sigma k X_i = k X_1 + k X_2 + \ldots + k X_N$$

Notice in this case k could be factored out, so that

$$\Sigma k X_i = \Sigma k(X_1 + X_2 + \ldots + X_N)$$
$$\Sigma k X_i = k\Sigma(X_1 + X_2 + \ldots + X_N)$$
$$\Sigma k X_i = k\Sigma X_i$$

2.3.4 GRAPHIC REPRESENTATION OF THE MEAN

So far the mean has been treated algebraically. But it has geometric meaning, too; that is, it makes sense to think of scores as located in space and to consider the mean as located somewhere in that space. To see this, examine the set of scores (1, 3, 6, 6) shown in Fig. 2.5.

FIGURE 2.5
The mean as center of gravity.

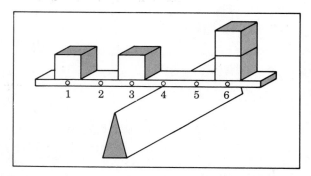

Fig. 2.5 depicts each score as an equal unit of mass residing on a weightless board. Here the mean is revealed as a center of gravity, a location on the board (x-axis), which could be used as the point of balance. As long as a sharp support is located as shown under the board at 4.0, the board will remain balanced. If one of the units

residing at "6" were shifted to 10, the center of gravity would shift to the right. The balance point would shift to 5, as we note from $\Sigma X/N =$ $(1 + 3 + 6 + 10)/4 = 20/4 = 5$.

2.3.5 COMPARISONS AMONG CENTRAL TENDENCY MEASURES

Unlike the median and mode, the mean is always influenced by the exact value of every score in the distribution. This may be appreciated algebraically or geometrically. Algebraically, the numerator of the mean is ΣX, the sum of all scores. It follows that if every score influences ΣX, changing any one would alter ΣX and hence the mean. Geometrically, think of the scores as balanced in Fig. 2.5. Moving any score to the right or left will tip the balance. In contrast, the median and mode are relatively unaffected by changes in individual scores. For example, the mode of the set of scores (1, 3, 6, 6) remains 6, even if 1 or 3 are permitted to range over all possible values. Similarly, the median of the set of scores (11, 13, 14, 15) would remain 13.5, even if the smallest score became zero or the largest score became infinity.

The mean is said to be relatively "sensitive" to extreme scores. This is seen when we examine the influence of skewness on the three measures of central tendency. This influence is illustrated in Fig. 2.6.

FIGURE 2.6

Influence of skewness on M, median, and mode.

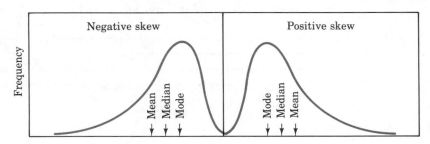

When there is no skew, all measures of central tendency have the identical value. For example, distribution B in Fig. 2.3 is symmetrical; its mean, median, and mode all equal 25. However, when skew is positive, the measures become ordered in increasing magnitude: mode, median, and mean. When the skew is negative, the reverse order is seen. When the skew is severe, the mean becomes a misleading measure of central tendency. For example, the set of scores (1, 1, 1, 2, 50) is severely positively skewed; it has a mode and median equal to 1, but a mean equal to 11. Clearly, the mean is a relatively poor index of the "typical" score in this set. The distribution of family incomes is a classic example of a positively skewed distribution for which the mean is a poor index of location. The distribution of annual family income has a vast majority of scores well below $100,000 with very few families receiving greater than that. For this reason, the median is typically chosen as the measure of central tendency of incomes. (Although in news media accounts it is often difficult to tell which measure is being reported.)

So far we have compared the mean, median, and mode in terms of how differently they are influenced by extreme scores. We have seen that for severely skewed distributions, the median is preferred when the distribution is to be described by a single measure of central tendency. The measures differ in another important characteristic, one we refer to as "stability." To understand stability as used here, it is necessary to imagine a situation in which several samples are drawn from the same population and the mean, median, and mode are calculated for each sample. What we would likely find is that the sample means would tend to cluster together, the medians would be somewhat more variable, and the modes would be most widely scattered. That is, successive means would tend to be somewhat more in agreement with one another than would be successive medians and modes. Thus, relative to the median and mode, the mean is said to be more "stable"—it varies the least from sample to sample. Given our interest in drawing inferences about parameters from statistics, this property of stability is very important. For these and other reasons, then, the mean is the more widely used in inferential statistics.

We now make further comparisons among measures of central tendency, but without considering the issue of which to select. Rather, these comparisons are useful devices for introducing new concepts. We begin with a discussion of *deviation*. In general, deviation refers to the distance between scores. The magnitude of the deviation of a score from M is the size of the difference when M is subtracted from the score. That is,

$$\text{deviation of } X \text{ from the mean} = X - M$$

If X is 3 and M is 4, the deviation is -1. For any set of scores the sum of the signed values of the deviation of each score from M is equal to zero. That is,

$$\Sigma(X_i - M) = 0 \qquad\qquad (2.2)$$

This is illustrated in Table 2.3. The table also shows $\Sigma(X_i - M)^2$, which will be dealt with shortly.

As seen in Table 2.3, the sum of the deviations from the mean equals zero. This point may also be appreciated from Fig. 2.5, which illustrates M as the center of gravity. The center of gravity is the point at which the sum of the distances (deviations) to all scores on its right

TABLE 2.3
Sum of deviations
from the mean.

Score	$(X_i - M)$	$(X_i - M)^2$
1	-3	9
3	-1	1
6	2	4
6	2	4
Sum 16	0	18

$$M = \frac{\Sigma X}{N} = \frac{16}{4} = 4$$

$$\Sigma(X_i - M) = 0$$

$$\Sigma(X_i - M)^2 = 18$$

(positive deviations) is the same as the sum of the distance to all scores on its left (negative deviations).

Unlike the mean, the median and mode do not have this property relative to summed deviations. In other words, only the mean has the property that for all sets of scores, the sum of the deviations from it will equal zero.

Deviations of scores from the mean play an important role in *standard scores*, defined in Section 2.5. Another powerful concept is that of *squared deviation*. As might be expected, squared deviation is defined as the square of the deviation of a score from the mean. That is,

$$\text{squared deviation of } X \text{ from the mean} = (X - M)^2$$

These are illustrated in Table 2.3, where it is shown that for the set of scores (1, 3, 6, 6), the *sum* of squared deviations is 18. Notice that in finding $\Sigma(X_i - M)^2$, M is first subtracted from X, then the difference is squared, and finally the squared differences are summed. Unlike the sum of the unsquared deviations, the sum of squared deviations does not generally equal zero. (It would if, and only if, all scores were identical). However, an important property of the mean is that the sum of squared deviations from the mean is smaller than the sum of squared deviations from any other value. This point is illustrated next by considering a variable called D. Let D take on values between 2 and 6. (Recall that $M = 4$ for these data). Table 2.4 and Fig. 2.7 show what happens to $\Sigma(X_i - M)^2$ as D varies.

For each value of D, $\Sigma(X_i - D)^2$ is calculated. These values are plotted in Fig. 2.7, which plots $\Sigma(X_i - D)^2$ "as a function of" D. (Table 2.4 shows $\Sigma(X_i - D)^2$ only for integer values of X, but the figure also

TABLE 2.4
Sum of squared deviations from D for data in Table 2.3 where $M = 4$.

X	D = 2		D = 3		D = 4		D = 5		D = 6	
	X − D	(X − D)²	X − D	(X − D)²	X − D	(X − D)²	X − D	(X − D)²	X − D	(X − D)²
1	−1	1	−2	4	−3	9	−4	16	−5	25
3	1	1	0	0	−1	1	−2	4	−3	9
6	4	16	3	9	2	4	1	1	0	0
6	4	16	3	9	2	4	1	1	0	0
	Sum	34	Sum	22	Sum	18	Sum	22	Sum	34

shows $\Sigma(X_i - D)^2$ for values between the integers. Stated in terms introduced in Chapter 1, Fig. 2.7 treats D as a *continuous* variable.)

FIGURE 2.7
Graph of the relation of $\Sigma(X_i - D)^2$ to D.

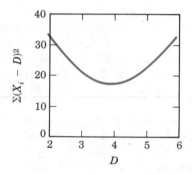

The point of Fig. 2.7 is that $\Sigma(X_i - D)^2$ is at its smallest value when $D = M$. We have illustrated this with a specific example, but it is also true in general. To recapitulate, M has an important property in relation to squared deviations. The sum of squared deviations of scores from their mean is less than the sum of squared deviations from any other value. The concept of squared deviations comes to play again in a measure of variability described in the next section.

2.4

MEASURES OF VARIABILITY

Measures of variability are measures of the degree to which scores in a distribution differ. Dispersion, spread, and heterogeneity are all synonyms of variability; they all refer to the degree to which scores are not alike. Measures of variability are important because they give information about how much certainty can be attached to inferences made from data. Suppose, for example, two experiments are performed, each involving one experimental and one control group. In both experiments the mean of the experimental group is 30 and the mean of the control group is 20. In Experiment A, however, the scores in the two groups cluster tightly around their means—there is no overlap in the two sample distributions. In Experiment B, however, the scores of the two groups overlap considerably—one-fourth of the scores in the control group are higher than the mean of the experimental group. (See Fig. 2.8).

FIGURE 2.8
Same means but different variability.

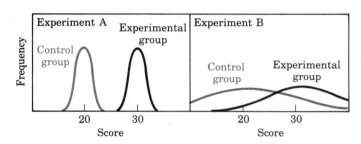

The mean of the experimental group is higher than the mean of the control group in both experiments. This suggests there may be a difference produced by the experimental treatment relative to the control. But which evidence is more convincing? From which set of results, A or B, would you be more confident in concluding that the two groups were sampled from *different* populations? Perhaps it is obvious that greater confidence could be attached to the A results than to the B results, even though the magnitude of the difference between the means is the same. Although spelled out in detail in later chapters, the point of this exercise is that inferences about sample means cannot be made in the absence of knowledge about variability. The greater the variability, the greater the uncertainty about sample means.

We have already seen the range as an example of a measure of variability. The range is easy to calculate, but it has not proved to be as useful as the *variance*. The **variance** of a set of scores is a measure of the degree to which the scores deviate from the mean. We have seen that in a set of scores, the sum of the signed deviations from the mean is equal to zero. We have also considered the sum of the squared deviation, $\Sigma(X_i - M)^2$. The variance is an *average* of these squared deviations, sometimes referred to as the *mean squared deviation*. Another measure of variability is the **standard deviation, s,** which is the positive square root of the variance.

2.4.1 FORMULAS FOR VARIANCE

The formula for the variance depends on whether the distribution is a population distribution or a sample distribution. For a *population* distribution, the variance is defined as

pop var

$$\sigma^2 = \frac{\Sigma(X_i - \mu)^2}{N} \qquad (2.3)$$

In Equation 2.3, σ is the Greek letter *sigma* (actually, lowercase sigma). Equation 2.3 should make clear why the variance is sometimes referred to as the mean squared deviation. Notice that the equation calls for the squared deviations (from the mean) to be summed and then for the sum to be divided by the total number of scores. The variance, σ^2, is therefore the average of the squared deviations from the mean.

For a *sample* distribution, the variance is defined as

$$s^2 = \frac{\Sigma(X_i - M)^2}{N - 1} \qquad (2.4)$$

Since population distributions are rarely known, experimenters typically use Equation 2.4 for sample variance rather than Equation 2.3 for population variance. You will notice that the right sides of the two equations differ in two ways. First, μ, the population mean, is used in the numerator of Equation 2.3, whereas M, the sample mean, is used in the numerator of Equation 2.4. Second, Equation 2.3 has N in the denominator, whereas Equation 2.4 has $N - 1$ in the denominator. The reason for this denominator difference is that s^2 is to be thought of as an *estimate* of σ^2, using M, and it has been found that dividing by $N - 1$ permits a better estimate of σ^2.

Equation 2.4 is referred to as the *definitional* formula for sample variance. It is not an efficient formula because it first requires M to be calculated, then subtracted from each score, each result squared, and so forth. Alternative formulas can be derived by expanding the square, $(X_i - M)^2$, and performing a few algebraic manipulations. These could result in a variety of *computational*, or *raw-score*, formulas. Several formulas are presented in Table 2.5.

Equations 2.5 and 2.6 are equivalent (always give the same result), as are equations 2.7 and 2.8. Equations 2.5 and 2.7 are the same except for square root, as are equations 2.6 and 2.8.

It is very important to distinguish similarly appearing terms in the computational formulas, ΣX^2 and $(\Sigma X)^2$. ΣX^2 refers to the sum of squared scores. Each score must first be squared and then summed. For the set of scores (1, 2, 3, 3)

$$\Sigma X^2 = 1^2 + 2^2 + 3^2 + 3^2 = 1 + 4 + 9 + 9 = 23$$

TABLE 2.5
Computational formulas for variance and standard deviation.

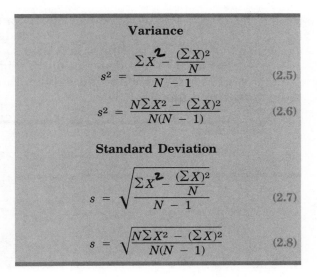

Variance

$$s^2 = \frac{\Sigma X^2 - \frac{(\Sigma X)^2}{N}}{N - 1} \qquad (2.5)$$

$$s^2 = \frac{N\Sigma X^2 - (\Sigma X)^2}{N(N - 1)} \qquad (2.6)$$

Standard Deviation

$$s = \sqrt{\frac{\Sigma X^2 - \frac{(\Sigma X)^2}{N}}{N - 1}} \qquad (2.7)$$

$$s = \sqrt{\frac{N\Sigma X^2 - (\Sigma X)^2}{N(N - 1)}} \qquad (2.8)$$

$(\Sigma X)^2$, however, requires the scores be summed first and then squared. For the same set

$$(\Sigma X)^2 = (1 + 2 + 3 + 3)^2 = 9^2 = 81$$

Let us consider these scores as a distribution, then, and demonstrate the use of Equation 2.6 in determining s^2.

$$s^2 = \frac{N\Sigma X^2 - (\Sigma X)^2}{N(N - 1)} = \frac{4(23) - 81}{4(3)} = \frac{92 - 81}{12} = \frac{11}{12} = .92$$

The standard deviation is

$$s = \sqrt{s^2} = \sqrt{.92} = .96$$

The reader needs to be sensitive to the differences between s^2 and σ^2, when a preprogrammed hand calculator or computer program is used to calculate variance. It is unsafe to assume s^2 rather than σ^2 is being calculated, even if the symbol s^2 is reported. Not all statisticians use the same symbols. How can you determine which formula is being used by your calculator? Often a booklet accompanying the calculator will describe the formula used. However, since computational formulas can be written a number of ways, you may not find an exact match with a formula in Table 2.5. If this is the case, examine the denominator. If $N - 1$ is found in the denominator of the reported formula, you can be sure that sample variance as defined in Equation 2.4 is being

used. If the formula is unavailable, the calculator can be given the data in Table 2.6 as a check. If you find σ^2 is calculated, the result may be multiplied by $N/(N - 1)$ to transform it to s^2.

See Chapter 14 for a description of widely available computer statistics packages and an explanation of how to match them with equations 2.4–2.8.

TABLE 2.6
Example calculations of variance and standard deviation.

X	$X_i - M$	$(X_i - M)^2$	X^2
2	−1	1	4
5	2	4	25
3	0	0	9
3	0	0	9
2	−1	1	4
3	0	0	9
$\Sigma X = 18$		$\Sigma(X_i - M)^2 = 6$	$60 = \Sigma X^2$

$$M = \frac{\Sigma X}{N} = \frac{18}{6} = 3.0$$

Definitional formula:

$$s^2 = \frac{\Sigma(X_i - M)^2}{N - 1} = \frac{6}{6 - 1} = \frac{6}{5} = 1.20$$

$$s = \sqrt{1.20} = 1.095$$

Computational formula:

$$s^2 = \frac{N\Sigma X^2 - (\Sigma X)^2}{N(N - 1)} = \frac{6(60) - (18)^2}{6(6 - 1)} = \frac{360 - 324}{6(5)} = \frac{36}{30} = 1.20$$

The advantage of using the computational formula is not clear with this example because M is an integer. If M were fractional, the advantage would be evident.

2.4.2 RELATIONSHIP BETWEEN THE RANGE AND STANDARD DEVIATION

It is possible to perform two rough checks on your calculation of the variance. First, we have seen that the variance is defined as a mean squared deviation. Since squared values must be positive, it follows that the variance must always be zero or positive. Any time a negative variance is obtained, therefore, there must be an error. Often errors stem from confusing ΣX^2 and $(\Sigma X)^2$. A second rough check on the calculation of the variance takes advantage of the fact that the standard deviation and the range are related. In the long run it will be found that the range is between 2.5 and 6 times the magnitude of the standard deviation. Stated another way, the ratio of range/s is usually between 2.5 and 6. However, the value of the ratio depends on sample size. If the sample size is less than 10, the ratio may be as low as 2.5. If the sample size is larger than 100, the ratio is usually 5 or more. If you find that the ratio of range/s departs markedly from these values, a recomputation of both range and s is in order.

2.4.3 WHICH MEASURE DO WE USE?

We have three measures of variability. Which do we compute? The range is by far the easiest to compute, since it only requires a simple difference between the largest and smallest scores. However, the source of the range's strength is also the source of its weakness. The range depends on the two extreme scores and therefore uses only a small amount of the information available in the data. Two distributions could have identical values for the range, yet one could be uniformly distributed between the two extremes and the other could have all but one of its scores located at the largest value. The range would not reflect the considerable difference in variability between these two distributions. In comparison with the other measures of variability, the range is a quick but rough index, and it is rarely reported.

The variance, on the other hand, is widely used in statistical inference. Its major drawback, however, is that it is expressed in squared units. For example, if the measure were pounds, the *mean* would be in pounds, and the frequency distribution could be graphed on an x-axis labelled Pounds, but the variance would be expressed in (pounds)^2, a unit with little intuitive value. However, the positive square root of the variance, which is the standard deviation, is expressed in the same units as the original measure. As such, the standard deviation, like the range, can be thought of as a *distance* on the x-axis. Thus, the standard deviation is an excellent descriptive statistic, but the variance is more prominent in inferential statistics.

2.5

STANDARD SCORES AND PERCENTILES

The meaning of one score depends on its relation to others. There are two ways to express the relationship of a score to other scores in the distribution. One, familiar to most readers, is the *percentile*. The **Xth percentile** is defined as that value at or below which X percent of the scores fall. To determine the Xth percentile, the scores must first be rank-ordered. We have already seen how this is done in calculating the median, which is the 50th percentile. The percentile is an excellent statistic for expressing a score in relation to a set of scores. However, it is not as useful in inferential statistics as is the standard score. For this reason, this text will devote primary attention to the standard score.

The **standard score**, *z*, is defined for a score *X* as the deviation of *X* from *M* divided by the standard deviation:

$$z = \frac{X - M}{s} \tag{2.9}$$

One way to think of *z* is as a distance: the distance *X* is from *M*, expressed in units of *s*. Thus, *z* serves to locate *X* on the *x*-axis, relative to *M*. A *z* score of -1 indicates that *X* is located one standard deviation below the mean. Scores above the mean will have a positive *z* score, scores below the mean will have a negative *z* score. We have seen that the range is usually 2.5 to 6 times the magnitude of the standard deviation. Expressed in terms of *z*, this means that scores will generally fall between a *z* of -3 and a *z* of $+3$ for symmetrical distributions.

Another way to think of *z* scores is to imagine all scores in a distribution as having been transformed to their corresponding value of *z*. This new distribution will have a mean of 0. This can be seen by substituting *M* for *z* in Equation 2.7. When all scores of a distribution have been transformed to *z*, the new distribution will have a standard deviation of 1.0. As seen from Equation 2.7, when the deviation from *M*, $X - M$, equals *s*, then *z* equals 1.

The *z* score is referred to as a "standard" score because it provides a meaningful way to compare scores from distributions that have different means and variances. For example, to say that a student received a score of 32 on an English test and 73 on math is virtually meaningless, since raw scores say nothing about relative position. But, to report both scores as having a *z* value of $+3.0$ is to say that both must be among the best in the class.

As a specific example of the calculation of *z*, suppose an individual received a score of 60 on a test where the class mean was 50 and the standard deviation was 10. Substituting appropriately,

$$z = \frac{X - M}{s} = \frac{60 - 50}{10} = \frac{10}{10} = 1.0$$

Similarly, a score of 45 on the same test would transform to

$$z = \frac{X - M}{s} = \frac{45 - 50}{10} = \frac{-5}{10} = -.5$$

These two examples are illustrated in Fig. 2.9.

FIGURE 2.9
Transforming X scores to z.

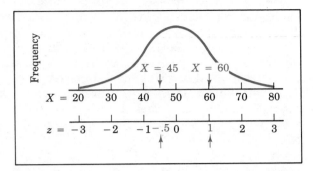

It is useful to be able to work in reverse; that is, to be able to determine what X value is associated with a particular z in a given distribution. For example, what score in the preceding distribution corresponds to a z of -1.5? To solve problems of this sort we would rewrite Equation 2.9 solving for X:

$$z = \frac{X - M}{s}$$
$$sz = X - M$$
$$X = sz + M \qquad (2.10)$$

Substituting as needed for the current problem:

$$X = sz + M = (10)(-1.5) + 50 = -15 + 50 = 35$$

Thus, a z score of -1.5 corresponds to a score of 35 in this distribution. Again, what score corresponds to a z of 3?

$$X = sz + M$$
$$X = (10)(3) + 50 = 80$$

2.6

SUMMARY

Data may be organized into frequency distributions which show the frequency that each category, or score value, occurs. In addition, distributions may be graphed as histograms or frequency polygons, which represent frequencies on the y-axis and score values on the x-axis.

Descriptive statistics, then, are numerical values calculated according to well-specified rules to provide measures of characteristics of distributions. Skewness is one measure of the shape of a distribution. A distribution with zero skew is symmetrical. A distribution with positive skew is a distribution which has a "tail" pointing along the x-axis toward increasingly positive scores. A negatively skewed distribution is one with a tail pointing towards increasingly negative scores.

Two other classes of measures of distributions are measures of central tendency and measures of variability. Measures of central tendency are measures of location on the x-axis. The mean, median, and mode are measures of central tendency. The mean, or arithmetic average, equals the sum of the scores divided by the total number of scores. The mode is the most frequently occurring score. The median is the value below which half of the scores fall. Statistical inference is most typically applied to the mean.

Measures of variability are measures of the degree to which scores in a distribution are scattered, that is, are dissimilar. The range is the difference between the largest and smallest score, and it is the least useful measure of variability. The variance is the mean squared deviation (from the mean), the sum of the squared deviations about the mean divided by the number of scores (population variance, σ^2) or divided by one less than the sample size (sample variance, s^2). The standard deviation, s, is the positive square root of the variance. The variance is typically used in statistical inference; the standard deviation is a useful descriptive statistic.

Percentiles and standard scores express relationships of one score to the other scores in the distribution. The Xth percentile is defined as the value below which X percent of the distribution falls. The standard score, z, is defined for score X as the deviation of X from M divided by the standard deviation.

KEY TERMS

Frequency distributions summarize a collection of data by showing the frequency with which each category, or score value, occurs.

In a **histogram**, the score values are represented along the x-axis and the y-axis represents the frequency or proportion of each score value.

Descriptive statistics are numerical values calculated according to well-specified rules to provide measures of characteristics of distributions.

The **range** is the difference between the largest and smallest scores. It is a measure of variability.

The **mode** is the score that has the greatest frequency. It is a measure of central tendency.

Measures of variability include the range and other statistics that reflect the degree to which scores are dissimilar, or vary.

Measures of central tendency are measures of the general location of the distribution on the x-axis. They are indexes of the "typical" score.

When a distribution is **positively skewed**, the scores extend farther from the mode in the direction of large values, rather than in the direction of small values. When a distribution is **negatively skewed**, the scores extend from the mode farther in the direction of small rather than large values.

The **mean** of a set of scores is defined as the sum of all scores in the set divided by the total number of scores. It is a measure of central tendency.

The **median** is the middle score value of the group of scores, i.e., the 50th percentile. It is a measure of central tendency.

The **variance** is an average of the squared deviations from the mean. It is a measure of variability.

The **standard deviation**, s, is the positive square root of the variance. It is a measure of variability.

The **Xth percentile** is defined as that value below which X percent of the scores fall.

The **standard score**, z, is defined for a score X as the deviation of X from M divided by the standard deviation.

QUESTIONS

1. Construct a histogram for each of the following. Be sure to label the x and y axes.

   (A)   Number of Shoppers Choosing Soap Powder

Brand X	37
Brand Y	82
Brand Z	44

   (B)   Average Length of Men's Hair (in inches)

Year	
1964	1.5
1966	2.0
1968	4.0
1970	4.5
1972	4.5
1974	3.5
1976	3.0
1978	2.5
1980	2.0

1, 2, 2, 3, 3, 5, 5, 5, 7, 7, 9

2. Consider the following data to be the number of aggressive acts of one sibling against another: 3, 9, 5, 7, 2, 1, 5, 3, 5, 2, 7.
(A) Construct a frequency distribution of the data.
(B) Construct a frequency histogram of the data.
(C) Construct a frequency polygon of the data.
(D) What is the range?
(E) What is the mode? Median? Mean?

3. Identify which distribution is uniform; bell-shaped; bimodal; unimodal.

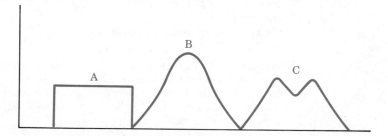

4. (A) Which distribution has the smallest variance?
(B) Which mean is twice as large as A's?
(C) Which distribution is negatively skewed?
(D) If a distribution is positively skewed, are most of the scores high or low?

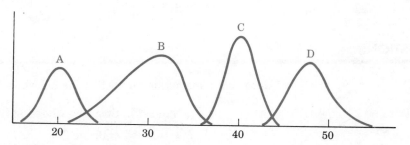

5. Identify which distributions have positive, negative, and zero skewness.

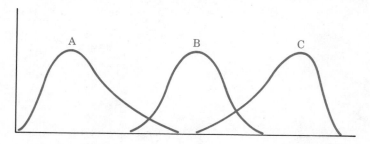

6. What is meant by the term *measures of variability*?

7. Why is it that the mean can be influenced by changing one value of one of the observations and this is not true of the mode or median? Give an example.

8. Which is the best measure of central tendency when drawing statistical inference? Why?

9. Suppose two experiments are conducted and there are two groups in each experiment. The average score in Group A, Experiment 1 is the same as the average of Group A, Experiment 2. Likewise the average score for Group B, Experiment 1 equals the average score of Group B, Experiment 2. Can we say the experiments had the same effect? Why?

10. Why are measures of variability important?

11. (A) Construct a frequency distribution from the following data for the number of baseball games won in the American and National Leagues as of August 2 (combine all 26 teams).
    (B) What is the median number of games won?
    (C) What is the mode?

American League		National League	
Eastern Division		**Eastern Division**	
Milwaukee	59	Philadelphia	58
Boston	58	St. Louis	58
Baltimore	54	Pittsburgh	54
Detroit	51	Montreal	54
New York	50	New York	45
Cleveland	50	Chicago	41
Toronto	48		
Western Division		**Western Division**	
California	59	Atlanta	61
Kansas City	57	San Diego	56
Chicago	52	Los Angeles	56
Seattle	52	San Francisco	50
Oakland	44	Houston	46
Texas	40	Cincinnati	38
Minnesota	35		

Divide by *
N & N-1

12. Using the data from Question 11,
    (A) Calculate and fill in:

	American League			National League	
	East	West		East	West
Mean # of games won			Mean # of games won		
Standard deviation			Standard deviation		
Range			Range		

(B) In which league's division is there the greatest variability in the number of games won? In which is there least variability?

13. Using the data from questions 11 and 12,
    (A) Of all clubs, which team has the highest z score, relative to its own division?
    (B) Which team has the lowest z score, relative to its own division?

14. Compare and contrast range, variance, and standard deviation. Which is preferred as a descriptive statistic? Why?

15. Does the 50th percentile equal the mode? Median? Mean?

16. Consider the following data as scores on the first statistics test: 25, 98, 97, 96, 33, 55, 76, 62, 88, 43, 97, 95, 97, 21, 29, 35, 86, 92, 97, 100, 65.
    (A) Determine the range, mode, median, and mean.
    (B) What is the variance and standard deviation?
    (C) What is the "standard" score for the person who made 62 on the test? 100? 25? 33? 55?
    (D) What would I make on the test if I had a $z = -1.00$?

17. The following test scores are for an Introductory Psychology midterm and final:

Student	Midterm	Final
A	48	76
B	37	82
C	26	62
D	44	93
E	43	89
F	39	78
G	22	80
H	45	84

(A) Find the mean and standard deviation for each test.
(B) Which student had the highest z score on the midterm?
(C) Which student had the lowest z score on the final?
(D) Which student performed closest to the averages, when both tests are considered?
(E) Which student had the highest z score for both tests summed?

21 25 29 33 35 43 55 62 65 76 86 88 92 95 96 97 97 97 97 98 100

1487

$$Var = s^2 = \frac{N\Sigma X^2 - (\Sigma X)^2}{N(N-1)}$$

$=_{441}$

$\frac{21(112,421) - 2,211,169}{21 \cdot 20}$

$\frac{199672}{420}$

3 Probability

P robability is a familiar concept. Such terms as *certain* and *impossible* describe the extremes of its range. In fact, different terms imply different degrees of probability. The reader will probably agree that the order of the following terms reflects decreasing probability: *certain, likely, probable, chancy, improbable,* and *unlikely.* A more precise ordering of probability statements is afforded by numbers; for example, "30% chance of showers today," or "the odds of winning are 2 to 1."

The concept of probability is developed three ways in this chapter. In addition to describing the concept with words, we use mathematics and graphs. For instance, probability is treated mathematically as a ratio, the numerator of which is some value between zero and the value of the denominator. Graphically, probability can be illustrated in two dimensions as an *area*. It is useful to understand probability all three ways, verbally, mathematically, and graphically, and to be able to translate probability ideas from one mode of expression to another. In the sections that follow, verbal definitions are given first, followed by mathematical formulas and, finally, pictorial representations of the definitions and formulas.

3.1

DEFINITIONS

The basic unit of probability theory is the *event,* which is the outcome of an experiment. The event is said to have some probability p that is a number between 0 and 1, inclusive ($0 \leq p \leq 1$). To say that event A has probability 1.0 of occurring is to say it is certain to occur. To say the event has probability 0 of occurring is to say it is certain not to occur. For our purposes, a particular **elementary event** is said to have occurred when a member of the population has been sampled and a label or number has been assigned to it. For example, in an experiment with dice, we may toss one die and count the number of dots on the top side. The outcome of the toss is the elementary event. We might ask the probability that a particular elementary event would occur. In the dice example, we could ask the probability that one tossed die will show a 5.

More typically, however, probability statements are applied to event classes. An **event class** is a *collection* of elementary events. Event classes are defined when a rule is stated that clearly specifies how every elementary event is to be categorized or numbered. For example, "odd number" is an event class. Its definition requires that all odd numbers, 1, 3, 5, . . . , be treated as members of the same class.

Most event classes that are defined on numerical data for statistical inference are of the form "larger than *n*," or "smaller than *n*." For instance, "What is the probability that the sample mean will be larger than 0?" This question defines an event class that includes all positive numbers.

The **probability** of an event class depends on the probability of each elementary event that is a member of the class. In all applications of probability theory considered in this text, it is assumed that no elementary event is more likely than any other. Under these conditions, the probability of an event class given a single draw from the population is simply the proportion of elementary events that are members of that class. If all elementary events belong to a given class, its probability is 1.0. For example, if a deck of cards is composed entirely of red cards, the probability of drawing a red card is 1.0. If half of the elementary events belong to the event class (e.g., half the cards are red), its probability is .50.

Another way to consider this is to treat probability as a *relative frequency*. The probability of an event class depends on the number of elementary events that are members of the class *relative* to the total number of elementary events. For example, the probability of getting a 5 with one toss of a die is $1/6 = .17$ because there is one 5 and six total possible events. Similarly, the probability of an odd number given the toss of a die is $3/6 = .50$ because three of the possible events (1, 3, and 5) are odd numbers.

3.1.1 MUTUALLY EXCLUSIVE EVENTS

In many applications the probability of two or more events will be considered at the same time. What is, for example, the probability that a person drawn from the population of all people will be both male and color-blind? That is, what is the probability of the *joint occurrence* of male and color-blind? Calculations of this sort require that each member of the population be classified two ways; in this case, according to gender and color-blindness. When two or more event classes have been defined, it is important to determine whether they are mutually exclusive. Two event classes are said to be **mutually exclusive** if no elementary event is a member of both. Maleness and color-blindness are *not* mutually exclusive because color-blind males do exist. However, "color-blind" and "not color-blind" are mutually exclusive. Other examples of pairs of classes that are not mutually exclusive are: female and schizophrenic, third-grader and highly motivated, introverted and intelligent.

If two event classes are mutually exclusive, the probability of their joint occurrence is zero. The probability of drawing an individual that is both color-blind and not color-blind is zero. Such an individual cannot exist. If two event classes are not mutually exclusive, the probability of their joint occurrence is equal to the proportion of elementary events that are members of both classes. Since about 5% of the U.S. population is both male and color-blind, the probability that one individual, randomly drawn, will be a color-blind male is about .05.

3.1.2 INDEPENDENT EVENTS

Two events are **independent** if knowledge that one has occurred provides no information about whether the other has occurred (or will occur). Classic examples of independent events are coin and die tosses. If a coin and a die are tossed together, knowledge that the coin is heads provides no basis for predicting the number of dots on the top of the die. An example of psychological events that appear to be independent are intelligence and pitch discrimination ability. Knowing a person's intelligence test score is not helpful in predicting his ability to discriminate among auditory tones.

 ✳   If two events are independent, the probability of their joint occurrence is equal to the product of their two separate probabilities. For example, the probability that the coin toss will be heads is 1/2 and the probability that the die will be 5 is 1/6. Since the two events are independent, the probability that tossing them together will result in both heads and 5 is (1/2)(1/6) = 1/12.

However, if two events are *not* independent, the probability of their joint occurrence does not equal the product of their two probabilities. For example, gender and color-blindness are not independent. Males constitute about half the population and about 6% of the total population have some form of color-blindness. If gender and color-blindness were independent, the proportion of people that are both male and color-blind would be (.50)(.06) = .03. But, in fact, the proportion of people that are both color-blind and male is greater than .03 and closer to .05. Another way to express the dependence between gender and color-blindness is to point out that whereas about 10% of males are color-blind, only about 1% of females are. Clearly, knowing a person's gender is of value when predicting whether they will be color-blind.

If the probability of the joint occurrence of two dependent events is not equal to the product of their two separate probabilities, what does it equal? In order to answer this question we need to become familiar with *conditional probabilities*.

3.1.3 CONDITIONAL PROBABILITY

To express the probability of one event conditional on another (**conditional probability**) is to state the probability one will occur *given* that the other is known to have occurred. That is, the probability that A has occurred given that B has occurred is the conditional probability of A on B. For instance, what is the probability that a U.S. male is unemployed *given* that he is a college graduate? This conditional probability, A given B, may be thought of as redefining the population, paring it down to that proportion in which B is known to have occurred. Stated another way, B defines a subset of the original population and the conditional probability question becomes, "What proportion of B is also A?" The conditional probability, unemployed given male college graduate, asks the proportion of male college graduates that are unemployed, that is, of all male college graduates what proportion are unemployed?

Another way to calculate the same conditional probability, A given B, is by dividing the proportion of elementary events that are *both A* and B by the proportion of elementary events that are B. For example, the probability that a male is unemployed given that he is a college graduate is obtained by dividing the proportion of all males that are both college graduates and unemployed by the proportion of all males that are college graduates. Stated still another way, probability is a relative frequency and the conditional probability of A given B expresses the frequency of A relative to B, as opposed to the frequency of A relative to both B and not B. As another example of conditional probability, we may consider the probability that an individual is color-blind given that he is male. In the previous section we learned that this conditional probability is .10. On the other hand, the probability that an individual is color-blind given that she is female is .01.

We will return to the question of how to calculate the probability of the joint occurrence of two dependent events in Section 3.2.

3.1.4 PROBABILITY DISTRIBUTIONS

How are these concepts used in statistical inference? How are they used to draw conclusions about parameters from statistics? As we shall see, evaluations from statistics take the form of evaluating their likelihood when certain conditions are true. For example, we might evaluate the likelihood that 53% of a *sample* would prefer one cola if only 50% of the *population* had that preference. Stated another way, we might ask the probability of obtaining the sample we did, if the result of each taste-test were like the flip of a coin—each preference being equally likely. What we do in such an evaluation is assume certain conditions and from these assumptions derive probability statements. These, then, are the traditional steps:

(A) Make certain assumptions about the population;
(B) Derive the likelihood of a set of hypothetical outcomes; and

(C) Compare the outcome we obtain with the probability computed for it in (B).

A critical link between probability theory and statistical inference is the probability distribution. <u>A **probability distribution** *on a discrete variable* is defined when a rule is stated that assigns a probability to each possible event class.</u> The rule can be specified simply with a list that associates a probability for each event. Consider the outcome of tossing two dice and summing their separate dots. Table 3.1 lists the probability of each possible sum.

TABLE 3.1
Example of a probability distribution on a discrete variable: sum of the toss of two dice.

Event (Sum of Two Dice)	Probability
2	1/36 = .028
3	2/36 = .056
4	3/36 = .083
5	4/36 = .111
6	5/36 = .139
7	6/36 = .167
8	5/36 = .139
9	4/36 = .111
10	3/36 = .083
11	2/36 = .056
12	1/36 = .028
Total	36/36 = 1.000

Table 3.1 is the probability distribution for the sum of two dice. Notice that every possible event is listed along with its probability. (Also, note that the sum of the probabilities is 1.0.) Given this probability distribution, we can consider some of its characteristics. For example, the mode of the distribution, the most likely event, is 7. That is, the highest probability (.167) is associated with the total 7. Stated still another way, the probability that two dice will total 7 is .167 and no other sum is more likely.

How were the probabilities in Table 3.1 calculated? The probabilities were generated from the assumptions that for each die each outcome (1, 2, . . . , 6) is equally likely and that the outcomes of the two dice are independent. From these assumptions the probability of each event can be determined by computing its relative frequency. When two six-sided dice are tossed, there are (6)(6) = 36 possible, equally likely, elementary events. We can designate these elementary events with a pair of numbers, e.g., (3, 6), the first number of which is the outcome for die #1, the second the outcome for die #2. Of these 36 elementary events, only one (1, 1) results in a sum of 2. Therefore, the probability of throwing a 2 with two dice is $1/36 = .028$. There are two ways to get a 3, (1, 2) and (2, 1), so the probability of a 3 is $2/36 = .056$. Similarly, the probability of a 4 is $3/36 = .083$, because there are three ways to obtain a sum of 4: (1, 3), (3, 1), and (2, 2). A second way a probability distribution can be specified is with a mathematical formula. An example of this form will be given in Chapter 4.

So far we have illustrated probability distributions by using an example on a discrete variable. Probability statements from probability distributions on a continuous variable could also be treated mathematically, but this would require knowledge of calculus, which is beyond the background of many of the users of this book. For our purposes, it is sufficient to note that some probability distributions in classic statistical inference are on a continuous variable. When we use these classic distributions in our applications, we take advantage of the fact that most pertinent information has already been computed and entered into tables (see Appendix A for examples), or is easily computed as part of computer statistics packages (discussed in Chapter 14).

3.2

PROBABILITY FORMULAS

In this section probability concepts are given more precision by stating them mathematically. We begin by defining the probability of an event and then give formulas related to the concepts of mutually exclusive events, independent events, and conditional probability.

The probability of event A, $P(A)$, depends on the number of elementary events belonging to A and the total number of elementary events. Here we use $n(A)$ to stand for the *number* of elementary events that belong to A, and we use E to stand for the *total number* of elementary events. In these terms, then, when all elementary events are equally likely,

$$P(A) = \frac{n(A)}{E} \tag{3.1}$$

Equation 3.1 simply states that $P(A)$ is equal to the proportion of all available elementary events that belong to A. Thus, if event A is tossing an odd number with the roll of one die, then $n(A) = 3$ because there are three elementary events that are odd (1, 3, and 5), and $E = 6$ because there are six possible elementary events (1, 2, 3, 4, 5, 6). Therefore, the probability of an odd number, event A, is $P(A) = 3/6 = .50$.

3.2.1 MUTUALLY EXCLUSIVE EVENTS

When two events, A and B, are considered simultaneously, we can ask the probability of their joint occurrence, $P(A$ and $B)$. Readers familiar with formal treatments of set theory may have seen $P(A$ and $B)$ written as $P(A \cap B)$. If the two events are mutually exclusive, the number of elementary events that are simultaneously members of both is equal to zero. Therefore, the probability of their joint occurrence is equal to zero. We use $n(A$ and $B)$ to stand for the *number* of elementary events that are simultaneously members of both A and B. In these terms,

$$P(A \text{ and } B) = \frac{n(A \text{ and } B)}{E} = \frac{0}{E} = 0$$

when A and B are *mutually exclusive.*

3.2.2 INDEPENDENCE AND CONDITIONAL PROBABILITY

When A and B are *not* mutually exclusive, the probability of their joint occurrence depends on whether they are independent. When A and B are independent

$$P(A \text{ and } B) = P(A)P(B) \tag{3.2}$$

The probability of their joint occurrence equals the product of their separate probabilities. This was illustrated earlier with die and coin. Let A be the event that the coin toss is heads: $P(A) = 1/2$. Let B be the event that the die is 5: $P(B) = 1/6$. What, then, is the probability that in one toss, both the coin will be heads *and* the die will be 5? Since A and B are independent,

$$P(A \text{ and } B) = P(A)P(B) = \left(\frac{1}{2}\right)\left(\frac{1}{6}\right) = \frac{1}{12}$$

The *conditional probability* that A occurs given that B occurs, symbolized $P(A|B)$, "the probability of A given B," is

$$P(A|B) = \frac{n(A \text{ and } B)}{n(B)} \tag{3.3}$$

Equation 3.3 states that the probability of A given B equals the number of elementary events that are both A and B divided by the number of elementary events that are B. Stated another way, $P(A|B)$ is equal to the proportion of elements in B that are also in A.

We seek now to express the right side of Equation 3.3 in terms of probabilities. We do this because some important relations are revealed and we gain insight into the concept of independence. First, we can define $P(A$ and $B)$ using Equation 3.1:

$$P(A \text{ and } B) = \frac{n(A \text{ and } B)}{E}$$

We can then solve for $n(A$ and $B)$

$$n(A \text{ and } B) = P(A \text{ and } B)E$$

Second, we can define $P(B)$ using Equation 3.1.

$$P(B) = \frac{n(B)}{E}$$

Solving this for $n(B)$

$$n(B) = P(B)E$$

Substituting these results into Equation 3.3 allows us to write $P(A|B)$

$$P(A|B) = \frac{P(A \text{ and } B)E}{P(B)E}$$

The Es cancel, yielding

$$P(A|B) = \frac{P(A \text{ and } B)}{P(B)} \tag{3.4}$$

Equation 3.4 is a classic relation known as Bayes' theorem. It states $P(A|B)$ as a ratio of probabilities. More specifically, Bayes' theorem states that the probability of A given B equals the probability of their joint occurrence divided by the probability of B. Suppose, for example, that $P(B) = .50$ and that $P(A$ and $B) = .20$. Then, according to Bayes' theorem, $P(A|B) = (.20)/(.50) = .40$.

Another important result occurs when we solve Equation 3.4 for $P(A$ and $B)$

$$P(A \text{ and } B) = P(A|B)P(B) \tag{3.5}$$

The importance of Equation 3.5 can be appreciated when it is compared with Equation 3.2, which states $P(A$ and $B)$ when A and B are independent. Notice that Equation 3.5 is a general expression that becomes Equation 3.2 in the special case when A and B are independent. To show this, we consider independence in more detail.

When A and B are *independent*

$$P(A|B) = P(A) \tag{3.6}$$

Equation 3.6 may be thought of as *defining* independence. A and B are said to be independent when $P(A|B) = P(A)$. Another way to say this is that if $P(A)$ is unchanged by knowledge of B, A must be independent of B. For example, if A and B are independent and $P(A) = .50$, then $P(A|B)$ must also be equal to .50. Knowledge about the occurrence of B does not alter the probability of A. If we toss a coin and a die, the probability of heads on the coin is unaffected by the outcome of the die:

$$P(\text{Heads}) = P(\text{Heads}|5 \text{ on the die}) = \frac{1}{2}$$

Finally, consider Equation 3.5 when A and B are independent. Since $P(A|B) = P(A)$,

$$P(A \text{ and } B) = P(A|B)P(B)$$

becomes

$$P(A \text{ and } B) = P(A)P(B)$$

which is the same as Equation 3.2 for the probability of the joint occurrence of two independent events.

3.2.3   PROBABILITY DISTRIBUTIONS

We have said that a probability distribution is defined for an experiment when a rule is stated that permits the determination of the probability of possible events. Usually the rule is stated mathematically, but we delay offering mathematical examples until the next chapter.

3.3

GRAPHIC REPRESENTATION OF PROBABILITY

Thus far concepts in probability have been treated verbally and mathe-matically. In this section these same concepts are treated graphically. We begin with elementary events and move through other topics in the same order as before. *Elementary events* are sometimes thought of as points in space.

FIGURE 3.1

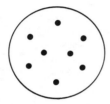

A *population,* then, is a collection of such points. Usually the collection is drawn in the shape of a circle, but the shape of the collection is irrelevant.

FIGURE 3.2

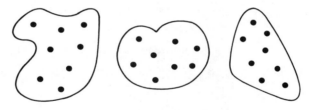

Here we find it helpful to conceive of the collection as a square (Figure 3.3),

FIGURE 3.3

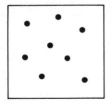

and to think of the elementary events as *squares* rather than points (Figure 3.4).

FIGURE 3.4

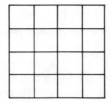

Populations can have a small number of elementary events,

FIGURE 3.5

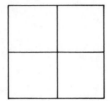

or a larger number of elementary events.

FIGURE 3.6

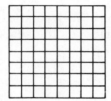

If the size of the population becomes very large,

FIGURE 3.7

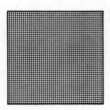

we are back to a collection of dots too small to show.

FIGURE 3.8

The advantage of drawing populations this way is that we can think of populations and events as having *extent* over an area. This permits us to conceive visually as well as mathematically.

Now, event classes (sets or collections of elementary events) can be thought of as *rearrangements* of the elementary events in space. Elementary events that are members of event class A, represented here as ▨, initially distributed without regard to order,

FIGURE 3.9

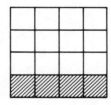

can be rearranged

FIGURE 3.10

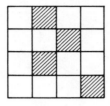

Another way to state this is to say that defining an event class amounts to specifying where elementary events are to be located in space.

FIGURE 3.11

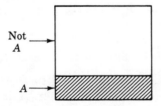

Not
A

A

Probability of an event, which has been treated as a relative frequency, now becomes proportion of a space. In the preceding graph, $P(A) = 4/16 = 1/4$ and $P(\text{not }A) = 3/4$. In the next example, $P(A) = 7/8$ appears as a population that is $7/8$ A.

FIGURE 3.12

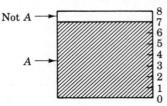

Not A

A

3.3.1 MUTUALLY EXCLUSIVE EVENTS

Events that are mutually exclusive are events that *do not overlap*. Events A and not A, graphed in Fig. 3.12 on the previous page, do not overlap. Neither do events B and not B, graphed as follows,

FIGURE 3.13

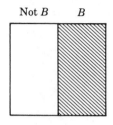

Now consider A and B simultaneously:

FIGURE 3.14

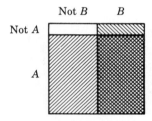

In this case, A and B do overlap; A and B are not mutually exclusive. In fact, this graph illustrates a situation in which any of four joint events may occur. These are: event(A and B), graphed as ⊠; event(A and not B), graphed as ▧; event(not A and B), graphed as ◨; and event(not A and not B), graphed as ☐.

3.3.2 CONDITIONAL PROBABILITY AND INDEPENDENCE

Now it is easy to see conditional probability. $P(A|B)$ is the proportion of B that is also A. It is not the proportion A is of the entire square (population). It is the proportion of ◨ that is also ▧.

FIGURE 3.15

picture doesn't fit description

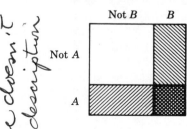

In this example, $P(A|B) = 1/3$ because the part that is both A and B, ⊠, covers 1/3 of the area of B, ◨. Mathematically,

$$P(A|B) = \frac{P(A \text{ and } B)}{P(B)}$$

In this case $P(B) = 1/2$, $P(A \text{ and } B) = 1/6$, so

$$P(A|B) = \frac{\frac{1}{6}}{\frac{1}{2}} = \frac{1}{3}$$

Now, let us change $P(B)$ but keep A and B independent. We could let $P(B) = 7/8$:

FIGURE 3.16

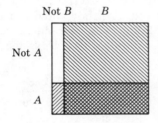

Notice that despite the change in $P(B)$, $P(A|B)$ remains 1/3 because the part of B that is both A and B remains 1/3 of the area of B.

In the preceding graphs, A and B are *independent*. We know this because $P(A|B) = P(A)$ regardless of the size of B. Graphically, this means that A covers the same proportion of the whole square as it does of B, regardless of the area of B.

Now consider what one graph would look like if two events, A and B, are *not* independent. This is illustrated next.

FIGURE 3.17

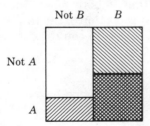

As before, A is shaded as ▨ and B is shaded as ▧. Notice that in this case, B covers half the total area: $P(B) = 1/2$. However, A covers half of B plus 1/4 of not B, or 3/8 of the total: $P(A) = 3/8$. What is $P(A|B)$? $P(A|B) = 1/2$ because the proportion of B that is also A is 1/2. Since $P(A|B) = 1/2$ but $P(A) = 3/8$, we conclude that A and B are not independent. Another way to see this is to note that $P(A \text{ and } B) = 1/4$

(covers one fourth of the total square). But if A and B were independent, $P(A$ and $B)$ should equal the product of $P(A)$ times $P(B)$, $P(A)P(B)$ = $(3/8)(1/2)$ = $3/16$, which is not $1/4$. Therefore, in this case, A and B cannot be independent. It is interesting to notice that when A and B are independent, as illustrated in Fig. 3.16, the four possible joint events (A and B, A and not B, etc.), can be partitioned with just two perpendicular lines. However, when A and B are not independent, as illustrated in Fig. 3.17, the partitioning into all possible joint events requires more than two lines. A way to think of independence graphically, then, is in terms of whether the graphic depiction of all possible joint events can be achieved with but two perpendicular lines.

An example of data showing a lack of independence is found in U.S. voting behavior. According to the Bureau of the Census, 20% of 20-year-old blacks reported that they voted in the 1976 election. However, about 40% of the 20-year-old nonblacks reported voting in that election. These data could appear in a table as:

	Nonvoting	Voting	Sum
Black	.152	.038	.190
Nonblack	.486	.324	.810
Sum	.638	.362	1.000

The four entries in the cells of this table are the proportions of the four joint events. The proportion of the total population that is both black and nonvoting is shown to be .152. The proportion of the total population that is both black and voting is shown to be .038, etc. The sums of the rows and the columns are the proportions for the single events. The proportion of the total population that is black is .190. (Thus, the proportion of the total that is both black and voting is .038 because .038 is 20% of .190.) The proportion of the total population that voted is .362, etc. Graphically, these data could be shown as:

FIGURE 3.18

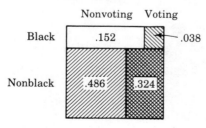

in which is nonblack and ▨ is voting. Are race and voting independent in these data? No. How do we know this? Because the probability of voting given black is not the same as the probability of voting overall. The probability of voting is .362, but the probability of voting given black is .038/.190 = .20.

Now consider a situation in which A and B are independent.

FIGURE 3.19

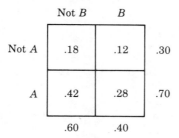

In this figure $P(A)$ = .70 and $P(B)$ = .40. These probabilities are shown in the margins. The entries inside the figure are the proportions of the *joint* events of A and B, A and not B, and so forth. In this example, the probability that A and B occur together is .28. We could show this graphically, so that probabilities correspond to areas as:

FIGURE 3.20

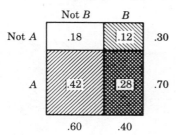

Again, the proportions inside the square refer to probabilities of *joint* events. In this example, A and B are independent. Again, we know this because $P(A|B) = P(A)$. $P(A)$ is shown in the margins as .70. In order to calculate $P(A|B)$ from these figures, we need to find the proportion of B, $\diagdown$, that is both A and B, $\boxtimes$. Consulting the figure we see that $P(A \text{ and } B)$ = .28, so that

$$P(A|B) = \frac{P(A \text{ and } B)}{P(B)} = \frac{.28}{.40} = .70$$

We have verified that $P(A|B) = P(A)$, as it should, since A and B are independent.

3.3.3 PROBABILITY DISTRIBUTIONS

Now consider populations in which each elementary event has a number assigned to it (i.e., consider numerical data). A small population with 4 elementary events might be represented as in Figure 3.21:

FIGURE 3.21

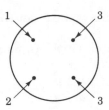

Again, using squares instead of points,

FIGURE 3.22

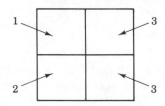

we could arrange this population according to number and frequency:

FIGURE 3.23

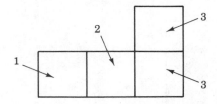

A larger population

FIGURE 3.24

1	5	6	3
7	4	2	4
3	6	3	4
8	5	5	5

could be rearranged

FIGURE 3.25

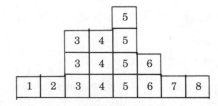

If the axes are labelled, as in Figure 3.26,

FIGURE 3.26

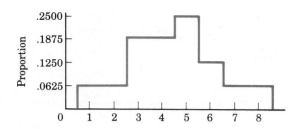

a histogram representing a *probability distribution* is produced. The point is that it is useful to think of probability distributions as populations in space with elementary events ordered as to size. Questions of event probability then become questions of area covered by the event. For example, to ask

$$P(X < 3) = ?$$

is to ask what proportion of the probability distribution is located to the left of 3. More generally, for any probability distribution, to ask

$$P(X < a) = ?$$

is to ask what proportion of the distribution is to the left of a. We illustrate this with a continuous variable.

FIGURE 3.27

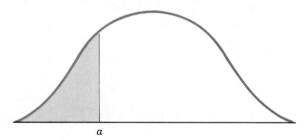

The proportion of area below a (shaded) is the proportion of values of X that are greater than b but less than c. Similarly $P(b < X < c)$ may be viewed as:

FIGURE 3.28

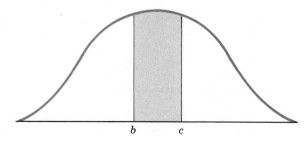

This is the proportion of values of X that are greater than b but less than c. We will see many examples of this type of problem in the following chapters.

KEY TERMS

The basic unit of probability is the **elementary event**, which is the outcome of an experiment. The elementary event is represented graphically as an area or point in space.

An **event class** is a collection of elementary events and is represented graphically as a portion of a population area.

The **probability** of an event class is defined in terms of relative frequency, conveyed mathematically as a ratio of the number of elementary events comprising the event class divided by the total number of elementary events. Graphically, probability is portrayed as a proportion of an area.

Two event classes are **mutually exclusive** if they share no elementary event. Graphically, two mutually exclusive events are events that do not overlap. Mathematically, we would say that two possible events are mutually exclusive if the probability of their joint occurrence is zero.

Two events are **independent** if knowledge that one has occurred provides no information that the other has occurred. Mathematically, if two events A and B are independent, the probability of their joint occurrence, $P(A$ and $B)$, is equal to the product of their separate probabilities, $P(A$ and $B) = P(A)P(B)$. Graphically, if two events are independent they can be accurately depicted using two perpendicular straight lines in a square-shaped population.

The **conditional probability** of A given B, $P(A|B)$, is the probability that A has occurred given that B has occurred. Mathematically, $P(A|B) = P(A$ and $B)/P(B)$. Graphically, the probability of A given B is the proportion of the area of B that is also A.

A critical link between probability theory and statistical inference is the **probability distribution**. A probability distribution is defined when assumptions are made that permit the determination of probabilities of experimental outcomes.

QUESTIONS

1. When we roll a die, what are the probabilities of getting:
   (A) A 1?

(B) An odd number?

(C) A 3, 4, 5, or 6?

2. If H represents "heads" in a coin flip and T stands for "tails," the possible outcomes for 2 successive coin flips are HH, HT, TH, and TT. Provided that the coin is "fair," we can assume that each of the four possibilities is equally likely. What is the probability of getting 0 heads? 1 head? 2 heads?

3. Suppose you had the job of picking delegates for a multi-party political convention. The group of people from which you must choose is comprised of 17 Democrats, 10 Socialists, 20 Republicans, and 3 Independents. If you choose a delegate by pulling a piece of paper from a hat containing all potential delegates' names, what is the probability that the delegate you chose will be:

(A) A Democrat?

(B) A Socialist or Republican?

(C) An Independent?

(D) Neither a Socialist nor an Independent?

4. You are the foreman for a shipping-receiving dock. Suppose you are expecting a shipment of 400 bottles of cola. You have been told that 124 of the bottles have defective caps. What is the probability that when you randomly choose 1 bottle:

(A) It will have a defective cap?

(B) It will not have a defective cap?

5. In a random sample of 1,000,000 words taken from psychology textbooks, it was found that 318,000 were nouns, 204,000 were adjectives, and 119,000 were verbs. Find the probability that a word randomly selected from this sample would be:

(A) An adjective.

(B) An adjective or a noun.

(C) Neither a noun, verb, nor adjective.

6. Johnny Ray of the Pittsburgh Pirates has been at bat 403 times and has gotten 123 hits. What is the probability that he will get a hit the next time he bats?

7. Suppose we conduct a study of college dropout rates and SES (socioeconomic status). A represents a college dropout and B represents the event that the person's family is considered poor. Describe (in words) the probabilities symbolically represented below:

(A) $P(\text{not } A)$

(B) $P(\text{not } B)$

(C) $P(A \text{ and } B)$

(D) $P(\text{not } A \text{ and not } B)$

8. The events Y and Z are mutually exclusive. Given $P(Y) = .21$ and $P(Z) = .33$, calculate:

(A) $P(\text{not } Y)$

(B) $P(\text{not } Z)$

(C) $P(Y \text{ or } Z)$

9. If two event classes are mutually exclusive, the probability of their joint occurrence is ____.

10. Three gamblers are talking about the status of their money following a night of gambling. The first claims that the probabilities for having more money, less money, or the same amount of money as before the evening started is .11, .31, and .40, respectively. The second gambler claims that his probabilities are .19, .33, and .48. The third claims that the probabilities are .19, .42, and .51. Can all three gamblers be correct?

11. If Q is the event that a person has a high IQ score and R is the event that a person can easily discriminate tones, state in *words* what probability is expressed by each of the following:
    (A) $P(Q|R)$
    (B) $P(R|Q)$
    (C) $P(Q|\text{not } R)$
    (D) $P(\text{not } Q|R)$
    (E) $P(\text{not } Q|\text{not } R)$
    (F) $P(\text{not } R|\text{not } Q)$

12. Sixty college students are asked whether they would marry while still in school. The results are:

	Yes	No
Males	32	11
Females	8	9

You are to examine the questionnaires by choosing one at random, and let Y = the answer is Yes. Let M indicate that the response was made by a male student. Determine the following probabilities:
    (A) $P(M)$
    (B) $P(\text{not } M)$
    (C) $P(Y)$
    (D) $P(\text{not } Y)$
    (E) $P(M|Y)$
    (F) $P(Y|M)$
    (G) $P(\text{not } M|\text{not } Y)$
    (H) $P(\text{not } Y|\text{not } M)$

13. One thousand families were polled as to how many children each had. The results are as follows:

Number of Children	Frequency
0	150
1	150
2	300
3	200
4	100
5	50
6 or more	50

(A) Construct a probability distribution.
(B) What is the probability that a family has 3 or more children?
(C) What is the probability of a family having 2 or less children?
(D) What is the probability of a family having either 1 or 4 children?

14. Billy the Kid has a passion for red M&Ms. He does not care if they are plain or peanut, as long as they are red. Based on a random sample of 500 plain and 500 peanut M&Ms:

Color	Plain	Peanut
Brown	270	230
Tan	58	71
Red	89	102
Yellow	46	45
Green	37	52

(A) What is the probability that a randomly chosen M&M would be red? What is the probability that a randomly chosen peanut M&M would be red?
(B) If both samples (plain and peanut) are combined, what is the probability of a green M&M? What is the probability that it would be a plain green M&M?
(C) Assuming our sample represents the population of all M&Ms, does color appear to be independent of type (plain or peanut)? Justify your answer.

15. (A) Complete the following table:

Sex

	Male	Female	
Beer		.20	.60
Wine			
		.50	

Beverage Preference

(B) What % of men prefer beer?
(C) What % of women prefer wine?
(D) Is beverage preference independent of sex? Prove your answer.

16. A psychologist wanted to find out if watching TV had an effect on children's sex-role development. She conducted an experiment where children were placed in a room full of toys to play with (containing half "typically male" and half "typically female" toys). She noted which toys each child chose. Two groups of children participated: one group was composed of children who watched an average of less than 1 hour of TV per day; the other group watched an average of 4 hours of TV per day.

Her results are:

	Number Choosing	
	Same-sex Toys	Opposite-sex Toys
Less than 1 hour	32	8
4 hours	48	12

(TV Watching — left side label)

(A) Convert the data into a probability table.
(B) Is amount of television viewing related to choice of toys? Why or why not?

17. A researcher wondered if people's favorite colors were related in any way to their hair color. A sample of 50 redheads, 50 blondes, and 50 brunettes was asked to name their favorite color. The results are as follows:

Favorite Color

Hair Color	Blue	Yellow	Green	Red
Red	8	18	21	3
Blonde	22	7	16	5
Brunette	10	13	15	12

(A) What is the probability of a redhead choosing blue?
(B) What is the probability of green being the favorite color of a brunette?
(C) Given that a person chose red, what is the probability he or she is blonde? What is the probability he or she is brunette?
(D) What is the probability that either blue *or* yellow were chosen, given that the person was *not* a redhead?

18. A recent poll asked voters whether or not they favored an increase in U.S. aid to foreign countries. The results were:

Favored Increased Aid

	Yes	No
Democrat	2870	1560
Republican	998	1052

(A) What is the probability of a person favoring increased aid, given he or she is a Democrat?
(B) What is the probability of a Republican favoring increased aid?
(C) If people's opinions on this issue had been independent of their po-

litical party, how many Republicans would you have expected to favor increased foreign aid?

(D) Given the results of the poll, what conclusions would you draw?

19. Assuming A and B are independent, complete the following table:

	B	Not B	
A			$P(A) = .30$
Not A			$P(\text{not } B) = .40$

20. Complete the following table:

	B	Not B	
			$P(A) = .65$
A			$P(A \text{ and } B) = .20$
Not A			$P(A|B) = .80$

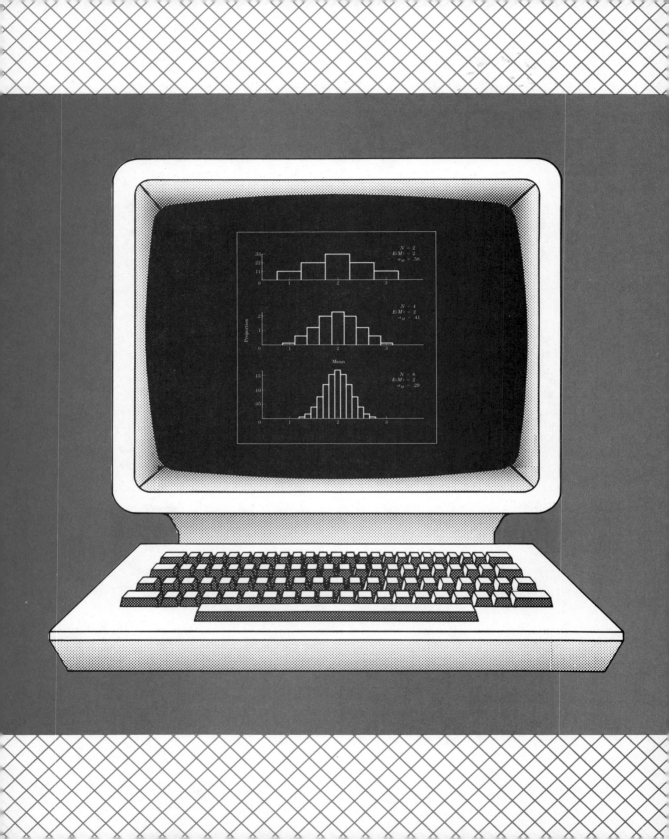

4 Sampling Distributions

Whon an experiment is performed, some statistic, let us call it G, is calculated. The statistic G describes a certain property of a sample drawn from a population. However, since populations typically have many different members, it is likely that another sample would yield a different value of G. Stated another way, a repetition of the experiment would not likely produce the identical value of G. The statistic G typically has variance. Suppose, for example, an experiment was performed in which 20 eighth graders made an average of 4.55 errors on a reading task. It is unlikely that another set of 20 eighth graders would produce an average error score exactly equal to 4.55, even if the conditions were identical. For this reason, we are interested not so much in the observed value of G as we are in what it would be in the long run and in how variable G is.

To make probability statements about G, then, we consider the distribution of G. More precisely, we consider the sampling distribution of statistic G. The concept of sampling distribution lies at the heart of statistical inference. *Sampling distributions* are probability distributions, which, as we saw in Chapter 3, are used to make probability statements about experimental outcomes. When inferences are made about populations from samples, that is, about parameters from statistics, the inferences involve sampling distributions. In this chapter we develop the concept of sampling distribution with the help of a specific population. We consider the sampling distributions of a number of statistics, including M, s^2, and differences between means. We then compare characteristics of sampling distributions, sample distributions, and population distributions. We show, for instance, how the means of population distributions and sampling distributions are related. This chapter then illustrates how specific probability statements about statistics can be made from knowledge about sampling distributions.

4.1

SPECIFIC EXAMPLE WITH A SMALL KNOWN POPULATION

4.1.1 SAMPLING DISTRIBUTION OF THE MEAN

Populations are usually large and parameters are usually unknown, but for the moment consider a small known population. We invent this simple population so that we may more easily develop the basic principles of sampling distributions. The population we invent is composed entirely of three scores: 1, 2, and 3. Even though the population is small, the principles derived can be extended to populations and samples of any size. Suppose an experiment is performed that involves drawing random samples from this population. A _random sample_ is defined as one which results when each member of the population has an equal chance of being sampled. Suppose our samples have two observations each.

Now consider the possible outcomes of this experiment. Each sample consists of exactly two observations (events). The first observation can be either 1, 2, or 3. After the first observation is made the second observation can be either 1, 2, or 3. Notice that the second observation can be the same as the first. This can occur because we are sampling with "replacement," i.e., all scores are available for each sample, even if sampled earlier. Thus, there are $3 \times 3 = 9$ possible outcomes. These are presented in Table 4.1.

TABLE 4.1
All possible samples of size 2 from population (1, 2, 3).

First Event	Second Event	Sample	Label	Sum	Mean
1	1	1,1	A	2	1.0
1	2	1,2	B	3	1.5
1	3	1,3	C	4	2.0
2	1	2,1	D	3	1.5
2	2	2,2	E	4	2.0
2	3	2,3	F	5	2.5
3	1	3,1	G	4	2.0
3	2	3,2	H	5	2.5
3	3	3,3	I	6	3.0

For each sample a mean, M, has been calculated. These means are shown in the right-hand column of Table 4.1. For example, for the first sample, labelled A for convenience, M equals $(1 + 1)/2 = 1$. For B, $M = (1 + 2)/2 = 1.5$, and so forth.

The mean of the population, μ, is 2.0 ($\mu = (1 + 2 + 3)/3 = 2$) but the means, M, of all possible samples vary from 1 to 3. One of the sample means (1/9 of the total) equals 1.0. Two of the sample means (2/9 of the total) equal 1.5, etc. The distribution of these sample means is shown in Table 4.2 and in the top right side of Fig. 4.1.

TABLE 4.2
Distribution of sample means.

Sample M	Frequency	Labels	Proportion
1.0	1	A	.11
1.5	2	B,D	.22
2.0	3	C,E,G	.33
2.5	2	F,H	.22
3.0	1	I	.11
Sum	9	Sum	1.00

FIGURE 4.1
Population, sampling, and sample distributions.

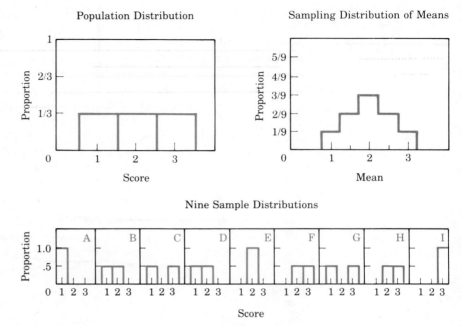

This distribution of sample means is called the sampling distribution of the mean. In general, the **sampling distribution of the mean** obtained from samples of size N is defined as the distribution of means of all possible samples of size N drawn from the population.

The sampling distribution of the mean is illustrated in Fig. 4.1 after the population distribution and before the sample distributions to emphasize the sampling distribution as an intermediary between the population and the samples. It is the sampling distribution that permits inferences about parameters from statistics.

4.1.2 SAMPLING DISTRIBUTION OF THE VARIANCE

Consider next the variances of the nine samples. These are calculated in Table 4.3, using Equation 2.4:

$$s^2 = \frac{\Sigma(X_i - M)^2}{N - 1}$$

Since sample size is 2, the denominator of the formula for sample variance becomes 1: ($N - 1 = 2 - 1 = 1$) and our calculations are simplified, since only the numerators need be calculated. As can be seen, there are three values of the variance in these samples. Three

TABLE 4.3
Variances of all possible samples of size 2 from population (1, 2, 3).

Sample Label	Sample	M	s^2
A	1,1	1.0	$[(1 - 1)^2 + (1 - 1)^2]/1 = 0$
B	1,2	1.5	$[(1 - 1.5)^2 + (2 - 1.5)^2]/1 = .5$
C	1,3	2.0	$[(1 - 2)^2 + (3 - 2)^2]/1 = 2$
D	2,1	1.5	$[(2 - 1.5)^2 + (1 - 1.5)^2]/1 = .5$
E	2,2	2.0	$[(2 - 2)^2 + (2 - 2)^2]/1 = 0$
F	2,3	2.5	$[(2 - 2.5)^2 + (3 - 2.5)^2]/1 = .5$
G	3,1	2.0	$[(3 - 2)^2 + (1 - 2)^2]/1 = 2$
H	3,2	2.5	$[(3 - 2.5)^2 + (2 - 2.5)^2]/1 = .5$
I	3,3	3.0	$[(3 - 3)^2 + (3 - 3)^2]/1 = 0$

Sampling Distribution of the Variance

s^2	Frequency	Samples	Proportion
0	3	A,E,I	.33
.5	4	B,D,F,H	.44
2	2	C,G	.22
Sum	9	Sum	1.00

samples (3/9 of the total) have no variance (both scores in the sample are identical). Four samples have variance equal to .5 and the rest have variance equal to 2. The distribution of these variances, shown in Fig. 4.2, is the sampling distribution of variances for samples of size 2 from this population. <u>Notice that none of the sample variances equals the population variance:</u>

$$\sigma^2 = \frac{[(1 - 2)^2 + (2 - 2)^2 + (3 - 2)^2]}{3} = \frac{(1 + 0 + 1)}{3} = \frac{2}{3} = .67$$

FIGURE 4.2
Sampling distribution of the variance.

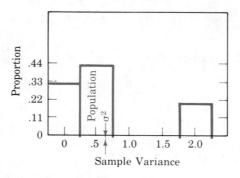

<u>Notice also that, unlike the sampling distribution of the mean, the sampling distribution of the variance is skewed.</u>

4.1.3   GENERAL DEFINITION OF SAMPLING DISTRIBUTION

By now it may be obvious that a sampling distribution may be defined for any statistic. In general, the **sampling distribution** of statistic G obtained from samples of size N is defined as the distribution of G obtained from all possible samples of size N drawn from the population.

4.2

DESCRIPTIVE STATISTICS OF THE THREE DISTRIBUTIONS

4.2.1   SHAPE

Often it is instructive to compare characteristics of the population distribution with those of the sample and sampling distributions. Consider shape first. The population we began with was rectangular in form. The nine sample distributions, in contrast, take a variety of forms, and the sampling distribution of the mean is symmetrical (see Fig. 4.1). On the other hand, the sampling distribution of the variance appears bimodal and positively skewed (see Fig. 4.2). Keep these differences in mind, because regardless of the shapes of the population distribution and the sample distribution, it is the sampling distribution that is used to make statistical inferences.

4.2.2   MEANS

The mean of the population we are considering, μ, equals 2.0 and the means of the sample distributions vary from 1 to 3. What is the mean of the sampling distribution of the mean? This would be calculated for our (1, 2, 3) population from the means in Table 4.1 as follows:

$$\text{mean of sample means} = \frac{(1 + 1.5 + 2 + 1.5 + 2 + 2.5 + 2 + 2.5 + 3)}{9}$$

$$= \frac{18}{9} = 2.0$$

We find that the mean of the sampling distribution of means is identical to the population mean, μ.

What is the mean of the sampling distribution of the variance? For our (1, 2, 3) population, the mean would be calculated from the variance in Table 4.3 as follows:

$$\text{mean of sample variances} = \frac{(0 + .5 + 2 + .5 + 0 + .5 + 2 + .5 + 0)}{9}$$

$$= \frac{6}{9} = .67$$

Here we find that the mean of the sampling distribution of the variance is identical to the population variance, σ^2.

We have now encountered two very important facts that are true of sampling distributions in general. First, we found in our example that the mean of the sampling distribution of the mean is equal to the population mean. This is always true. That is, the mean of the sampling distribution of the mean will equal μ. Second, it is always true that the mean of the sampling distribution of the variance equals the population variance. We refer to the mean of the sampling distribution of statistic G as the **expected value** of G, labelled $E(G)$. The expected value of M is $E(M)$ and the expected value of s^2 is $E(s^2)$.

In summary, then, we have seen that

$$E(M) = \mu \tag{4.1}$$

and that

$$E(s^2) = \sigma^2 \tag{4.2}$$

4.2.3 VARIANCES OF SAMPLING DISTRIBUTIONS

Thus far we have been concerned with means of sample means and means of sample variances. Now we consider variances of sample means. To obtain a variance of sample means is to calculate a measure of the extent to which the means scatter. The more sample means vary, the less certain we are about the information any single mean gives. That is, the more they vary, the less confidence we have that any particular sample mean is near the population mean. The more sample means vary, the less confidence we have that a particular sample mean represents the population as a whole. The concept of variability of a sample statistic is so important that it is given a special name: standard error. In general, the **standard error** of statistic G is defined as the standard deviation of the sampling distribution of G. The standard error of G is labelled σ_G. In this chapter, only the standard error of the mean, σ_M, is considered, but the concept extends significantly to other statistics as well. In Chapter 2 we argued that the sample mean is more "stable" than the median or mode, by which we meant that sample means tend to vary less than sample medians and modes. We can now see that that was really a claim about the relative sizes of the standard errors of the three statistics. <u>Typically, the standard error of the mean is smaller than the standard error of the median and the standard error of the mode</u>.

So what *is* the standard error of the mean for our hypothetical population (1, 2, 3)? Looking at the means in Table 4.1, we would calculate their variance. To do this we treat each mean as a separate score that is entered into Equation 2.3 for variance of scores. That equation calls for the *mean* of the scores, which in this case is the mean of the means:

$$\sigma_M^2 = \frac{(1 - 2)^2 + (1.5 - 2)^2 + \ldots + (2.5 - 2)^2 + (3 - 2)^2}{9}$$

$$= \frac{(1 + .25 + 0 + .25 + 0 + .25 + 0 + .25 + 1)}{9}$$

$$= \frac{3}{9} = .33$$

$$\sigma_M = \sqrt{.33} = .577$$

(Notice that Equation 2.3 for population variance is used, not Equation 2.4 for sample variance, because in this case we are considering all possible samples from a known population—not just a single sample of such outcomes.)

Now compare σ^2 and σ_M^2. For our example population, we have found that $\sigma^2 = 2/3$ and $\sigma_M^2 = 1/3$; that is, the value of the square of the standard error of the mean is half the population variance. This is no accident. In general,

$$\sigma_M^2 = \frac{\sigma^2}{N}$$

where N = sample size. Thus, in our example population

$$\sigma_M^2 = \frac{\sigma^2}{N} = \frac{\frac{2}{3}}{2} = \frac{1}{3} = .33$$

Now, since the standard error is the standard deviation of the sampling distribution, not its variance, the standard error of the mean is

$$\sigma_M = \sqrt{\frac{\sigma^2}{N}}$$

which can be rewritten as

$$\sigma_M = \frac{\sigma}{\sqrt{N}} \tag{4.3}$$

(To be precise, Equation 4.3 can be shown only when σ^2 is finite. But populations dealt with in psychology and education may be considered to have finite variance, so this qualification has no practical consequence.)

When σ_M is unknown, it must be estimated from the data. Since $E(s^2) = \sigma^2$, an estimate of σ_M (which is s_M) is obtained by substituting s for σ in Equation 4.3, yielding

$$s_M = \frac{s}{\sqrt{N}} \tag{4.4}$$

We have explained that σ_M is important because it is a measure of uncertainty about M. If σ_M is large, uncertainty is great. But Equation 4.3 states that the size of the standard error of the mean depends only on σ, the population standard deviation and the size of the sample, N. As σ increases, the numerator of Equation 4.3 increases and so does σ_M. However, as N, the sample size, increases, the denominator of Equation 4.3 increases and σ_M decreases. Thus, we have relatively large σ_M and greater uncertainty concerning M when σ is large and N is small. On the other hand, we have relatively small σ_M and greater certainty about M, when σ is small and N is large.

The concepts and formulas discussed so far in this chapter are summarized in Table 4.4.

TABLE 4.4
Notation for the three distributions.

Statistic	Sample	Sampling Distribution	Population
Mean	M	$E(M)$	μ
Standard deviation	s	σ_M	σ

Important findings about M and s

$$E(M) = \mu \tag{4.1}$$

$$E(s^2) = \sigma^2 \tag{4.2}$$

$$\sigma_M = \frac{\sigma}{\sqrt{N}} \tag{4.3}$$

$$s_M = \frac{s}{\sqrt{N}} \tag{4.4}$$

4.2.4 EXAMPLE

Let us illustrate these concepts more concretely. In Chapter 1, we discussed an experiment in which 200 observations produced a mean reaction time of 425 milliseconds. We do not know μ, the mean of the population from which this sample is drawn, but our best estimate of μ is 425 milliseconds. The value of σ^2 is also unknown, but again our best estimate is s^2. Let us suppose that $s^2 = 3600$, $s = \sqrt{3600} = 60$. Assume for the moment that $\mu = 425$ and $\sigma = 60$. What can be stated about means of samples drawn from this population? We know something about how much they will vary. We know that the standard deviation of the sample means should be

$$\sigma_M = \frac{\sigma}{\sqrt{N}}$$

$$= \frac{60}{\sqrt{200}} = \frac{60}{14.14} = 4.24$$

We have found, then, that the standard deviation of sample means under these conditions should be 4.24. But we can go further. Recall from Chapter 2 that the range of a distribution is usually about 6 times the standard deviation. If the distribution is symmetrical, the distribution should extend from about 3 standard deviations below its mean to about 3 standard deviations above its mean. For our current example, we can expect virtually all of the distribution of our sample means to extend from 3 standard deviations below 425 to 3 standard deviations above 425. Since in this case the standard deviation is 4.24, we expect virtually all of our means to fall between

$$425 - (3)(4.24) = 425 - 12.72 = 412.28$$

and

$$425 + (3)(4.24) = 425 + 12.72 = 437.72$$

In other words, our sample means should tend to fall between approximately 412 and 438, if all of the preceding were true. In general terms, we would expect a sample mean *outside* these values to be unlikely. In the next chapter we describe methods that provide more precise probability statements about sample means. That development depends on ideas discussed next.

4.3

OTHER IMPORTANT SAMPLING DISTRIBUTIONS

4.3.1 SAMPLING DISTRIBUTION OF $(M - \mu)/\sigma_M$

In Chapter 2 we saw that a score could be expressed as a standard score z, by dividing its deviation from the mean by the standard deviation. That is, from Equation 2.9:

$$z = \frac{X - M}{s}$$

The concept of standard score may be applied to *any* score, including statistic G. Thus, we might standardize G as

$$z_G = \frac{G - E(G)}{\sigma_G}$$

Notice that all we have done to obtain z_G is to substitute G for X, replace M by $E(G)$, the mean of the sampling distribution of G, and replace s with σ_G, the standard error of G.

In particular, then, we might standardize M by substituting M for G in the last equation. This would yield

$$z_M = \frac{M - \mu}{\sigma_M}$$

When sample means are standardized this way, an important sampling distribution emerges. As will be seen in Chapter 5, this sampling distribution plays a role in making inferences about sample means. Here we illustrate the sampling distribution of $(M - \mu)/\sigma_M$ with the (1, 2, 3) population. The sampling distribution of $(M - \mu)/\sigma_M$ for population (1, 2, 3) is presented in Table 4.5 on page 76.

The sampling distribution of this statistic, graphed in Fig. 4.3, is symmetrical around a mode of zero.

Inferences about means make use of the sampling distribution illustrated in Fig. 4.3. The distribution will be treated in detail in the next chapter.

TABLE 4.5
Sampling distribution
of $(M - \mu)/\sigma_M$.

Sample Label	Sample	M	$M - \mu$	$(M - \mu)/\sigma_M$
A	1,1	1.0	−1.0	−1.73
B	1,2	1.5	−.5	−.87
C	1,3	2.0	0	0
D	2,1	1.5	−.5	−.87
E	2,2	2.0	0	0
F	2,3	2.5	.5	.87
G	3,1	2.0	0	0
H	3,2	2.5	.5	.87
I	3,3	3.0	1.0	1.73

$(M - \mu)/\sigma_M$	Frequency	Sample	Proportion
−1.73	1	A	.11
−.87	2	B,D	.22
0	3	C,E,G	.33
.87	2	F,H	.22
1.73	1	I	.11
Sum	9	Sum	1.00

FIGURE 4.3
Sampling distribution
of $(M - \mu)/\sigma_M$.

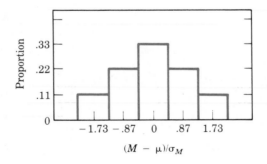

$(M - \mu)/\sigma_M$

4.3.2   SAMPLING DISTRIBUTION OF $M_1 - M_2$

Inferences about *pairs* of means, on the other hand, depend on the sampling distribution of another statistic. Inferences about pairs of means are made in experiments involving two or more conditions. The traditional two-condition experiment involves a comparison between one experimental and one control group. An experimenter interested in comparing the performance of the two groups may wish to examine the size of the difference between the average score of each group. This difference between means would then be considered the statistic of interest and its sampling distribution would be used to make statistical inferences.

An example of the sampling distribution of the difference between means is shown in Table 4.6.

TABLE 4.6
Sampling distribution of the difference between means.

					Second Sample					
		A	**B**	**C**	**D**	**E**	**F**	**G**	**H**	**I**
	M	1.0	1.5	2.0	1.5	2.0	2.5	2.0	2.5	3.0
A	1.0	0	−.5	−1	−.5	−1	−1.5	−1	−1.5	−2
B	1.5	.5	0	−.5	0	−.5	−1	−.5	−1	−1.5
C	2.0	1	.5	0	.5	0	−.5	0	−.5	−1
D	1.5	.5	0	−.5	0	−.5	−1	−.5	−1	−1.5
E	2.0	1	.5	0	.5	0	−.5	0	−.5	−1
F	2.5	1.5	1	.5	1	.5	0	.5	0	−.5
G	2.0	1	.5	0	.5	0	−.5	0	−.5	−1
H	2.5	1.5	1	.5	1	.5	0	.5	0	−.5
I	3.0	2	1.5	1	1.5	1	−1.5	1	.5	0

First Sample (row label, left side)

Summary

$(M_1 - M_2)$	Frequency	Proportion
2.0	1	.012
1.5	4	.049
1.0	10	.123
.5	16	.198
0	19	.235
−.5	16	.198
−1.0	10	.123
−1.5	4	.049
−2.0	1	.012
Sum	81	1.000

Note: Entry in Table 4.6 is the difference between the mean of sample 1 and the mean of sample 2, $(M_1 - M_2)$.

Table 4.6 shows how the sampling distribution could be constructed for an experiment when two samples each of size 2 are drawn from population (1, 2, 3). Since there are nine possible samples for each draw, there are $9 \times 9 = 81$ possible sample pairs. The entries in the top part of Table 4.6 are the differences when the mean of the second sample is subtracted from the mean of the first. Even when the two means are calculated on samples drawn from the same population, their differences can be as large as 2. A difference this large is unlikely (of 81 pairs only one difference is $+2$, another is -2). On the other hand, 19 of the differences (about 1/4 of the total) are equal to zero, the mode of this distribution.

The sampling distribution is summarized in the bottom half of Table 4.6 and graphed in Fig. 4.4.

FIGURE 4.4
Sampling distribution
of $M_1 - M_2$.

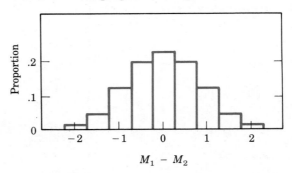

The mean of this distribution is equal to zero. Again, this is true of our example and, more importantly, true in general. <u>That is, when two samples are drawn from the same population and the difference between their means is calculated:</u>

$$E(M_1 - M_2) = 0$$

<u>Notice, however, that if two samples are taken from two populations with *different* means, the expected value of their difference is *not* equal to zero.</u> This fact is used when inferences are made about mean differences and the sampling distribution of mean differences is treated in detail in Chapter 7.

4.4

EXAMPLES OF PROBABILITY STATEMENTS FROM SAMPLING DISTRIBUTIONS

In Chapter 3 we saw that the probability of an event, $P(A)$, could be defined as a ratio in which the number of elementary events comprising A is divided by the total number of elementary events. We can now combine this with our knowledge of sampling distributions to make probability statements about statistics. So far we have seen four useful sampling distributions: of M, s^2, $(M - \mu)/\sigma_M$, and $(M_1 - M_2)$. We are now in a position to make probability statements about these four. For example, it makes sense to ask the probability that the sample mean equals the population mean; that is,

$$P(M = \mu) = ?$$

We could ask the probability that the sample variance is greater than 1,

$$P(s^2 > 1) = ?$$

We could ask

$$P(M_1 - M_2) > 1 = ?$$

However, in each case the probability could be calculated only after the sample size is specified, because the sampling distribution is always defined in terms of a specific N. Thus, for example, it is more correct to state

$$P(M = \mu | N = 2) = ?$$

That is, what is the probability M equals μ, given that $N = 2$?

For population (1, 2, 3) when $N = 2$, the sampling distribution of M is reported in Table 4.2. There we saw nine equally likely outcomes, three of which (samples C, E, and G) produced a mean of 2. Thus,

$$P(M = \mu | N = 2) = \frac{3}{9} = .33$$

We can ask of the same population, what is the probability that $s^2 = 2$, given that $N = 2$? Consulting Table 4.3, we see two outcomes for which s^2 is equal to 2 (samples C and G). Therefore,

$$P(s^2 = 2 | N = 2) = \frac{2}{9} = .22$$

It is also meaningful to determine the probability that a statistic will fall within a specified interval. For example, for our situation the probability that a sample mean will be greater than 1 and smaller than 3:

$$P(1 < M < 3 | N = 2) = \frac{7}{9} = .78$$

because there are seven sample means (B, C, D, E, F, G, H; see Table 4.2) that fall in that interval.

Determining the probability that a statistic will fall within a specified interval is equivalent to determining the proportion of the sampling distribution that falls in that interval. We could turn this around and seek to determine the upper and lower values that would be needed in order to obtain a given proportion of the distributions. Suppose, for example, the lower value were 1.25. What would the upper value have to be in order to incorporate exactly 5/9 of the sampling distribution of the mean? We will see many examples of this kind of probability problem in the next chapter.

4.5

EXAMPLES OF INFLUENCES OF SAMPLE SIZE ON SAMPLING DISTRIBUTIONS

In Section 4.2 it was shown that the size of the standard error of the mean depends on N, and throughout this chapter we have been careful to define sampling distributions of statistics in terms of N. Now is the time to illustrate the influence of sample size on sampling distributions. We do this by returning to population (1, 2, 3) and considering several sample sizes. To do so we will consider sample sizes that are

TABLE 4.7
Influence of sample size on sampling distribution of the mean.

| | N = 2 | | | N = 4 | | | N = 8 | |
M	Frequency	Proportion	M	Frequency	Proportion	M	Frequency	Proportion
1.0	1	.11	1.00	1	.012	1.000	1	.0002
						1.125	8	.0012
			1.25	4	.059	1.250	36	.0055
						1.375	112	.0171
1.5	2	.22	1.50	10	.123	1.500	266	.0405
						1.625	504	.0768
			1.75	16	.198	1.750	784	.1195
						1.875	1016	.1549
2.0	3	.33	2.00	19	.234	2.000	1107	.1687
						2.125	1016	.1549
			2.25	16	.198	2.250	784	.1195
						2.375	504	.0768
2.5	2	.22	2.50	10	.123	2.500	266	.0405
						2.625	112	.0171
			2.75	4	.059	2.750	36	.0055
						2.875	8	.0012
3.0	1	.11	3.00	1	.012	3.000	1	.0002
Sum	9	1.00	Sum	81	1.000	Sum	6561	1.0000

actually larger than the population itself. In normal applications of statistics, of course, the sample size may be considered to be much smaller than the population. Nevertheless, the principles that emerge from the examples given here are true of sampling distributions from populations large and small.

Table 4.7 compares the sampling distribution of means of samples from population (1, 2, 3), when sample sizes are 2, 4, or 8.

As can be seen here, when the sample size is 4 there are 81 possible samples (1, 1, 1, 1 is one; 1, 1, 1, 2 is another; and so forth). Of these, only one yields a mean of 1, four yield a mean of 1.25, etc. As with $N = 2$, when $N = 4$ the mode (and mean, as we know from Equation 4.1), is 2.0. When sample size is 8, there are 6561 possible samples. Again the mode and mean are 2.0. These three distributions are plotted in Fig. 4.5

FIGURE 4.5

Influence of sample size on sampling distribution of M.

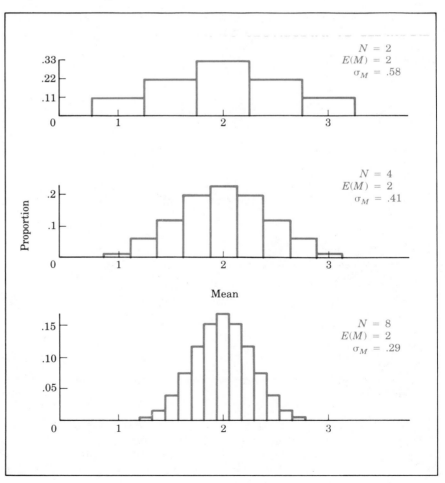

Notice that all distributions are symmetrical and that as N increases, the graph becomes more and more bell-shaped. Again, although demonstrated here with a specific example, it can be stated that the sampling distribution of means approaches this bell-shaped form as N increases *regardless of the shape of the original population.*

Another feature to notice in Fig. 4.5 is that the sample means become more and more clustered around μ as N increases. Another way to show this is to compare the probability that a sample mean will fall between 1.5 and 2.5 as N increases.

When $N = 2$, 3 of the 9 sample means fall between 1.5 and 2.5. Therefore,

$$P(1.5 < M < 2.5 | N = 2) = \frac{3}{9} = .33$$

When $N = 4$, 51 of the 81 sample means fall between 1.5 and 2.5. Therefore,

$$P(1.5 < M < 2.5 | N = 4) = \frac{51}{81} = .63$$

But when $N = 8$, 5715 of the 6561 sample means fall between 1.5 and 2.5. Therefore,

$$P(1.5 < M < 2.5 | N = 8) = \frac{5715}{6561} = .87$$

This decrease in variability of sample means as sample size increases is expected from our knowledge of the standard error of the mean. As we have seen

$$\sigma_M = \frac{\sigma}{\sqrt{N}}$$

Since $\sigma = .67$ for population (1, 2, 3), σ_M equals .58, .41, and .29 for Ns of 2, 4, and 8, respectively. Fig. 4.5 graphically demonstrates the decrease in uncertainty about M as N increases.

4.6

SUMMARY

In this chapter we have defined an important new distribution—the **sampling distribution**, a theoretical probability distribution defined for statistic G as the distribution of G obtained for all possible samples of size N. We have seen that the sampling distribution resides conceptually between the population distribution and a sample distribution, because it is used to make inferences about populations from samples.

The mean and standard deviation of the sampling distribution have special roles in statistical inference. The mean of the sampling distribution of statistic G is called the **expected value** of G, $E(G)$. The standard deviation of the sampling distribution of statistic G is called the **standard error** of G, σ_G. The standard error is important because its magnitude is a measure of uncertainty about G. The larger its value, the greater the uncertainty about the corresponding population parameter.

Four sampling distributions were illustrated: the sampling distribution of the mean, the variance, $(M - \mu)/\sigma_M$, and $(M_1 - M_2)$. It was shown, in general, that $E(M) = \mu$, that $E(s^2) = \sigma^2$, and that when M_1 and M_2 are from the same population, $E(M_1 - M_2) = 0$. It was also shown that $\sigma_M = \sigma/\sqrt{N}$. This means that σ_M increases with increased σ but decreases as N increases. Specific examples of the influence of N on the sampling distribution of the mean were also given. In general, the **sampling distribution of the mean** obtained from samples of size N is defined as the distribution of means of all possible samples of size N drawn from the population. It was shown graphically how the sampling distribution of the mean becomes less variable and more bell-shaped, even for a population distribution that is rectangular in shape.

QUESTIONS

1. The sampling distribution of a statistic obtained from samples of size N can be defined as ____.

2. Which is used to make statistical inference:
   (A) Sample distribution?
   (B) Sampling distribution?
   (C) Population distribution?
   (D) Proportion distribution?

3. Generally, the mean of the sampling distribution of the mean equals ____. Also, the mean of the sampling distribution of the variance generally equals ____.

4. Describe the relationship among σ_μ, σ, and N.

5. Define the standard error of a statistic.

6. Typically, which is smaller:
   (A) Standard error of mean?
   (B) Standard error of median?
   (C) Standard error of mode?

7. If the standard error of the mean is squared, the result is ____.

8. Suppose a researcher conducts a memory experiment and finds that people can remember an average of 6.8 nonsense syllables. Now suppose the exact experiment is repeated, but different subjects participate. This time the average number of remembered syllables is 6.4. Is this unusual? Which average is the "true" population mean?

9. Just as there are several ways of calculating measures of central tendency and variability, it is possible to measure the skewness of a distribution: This is called the *skewness coefficient*. How would you define the sampling distribution of the skewness coefficient?

10. The standard deviation of the sampling distribution of the mean is called ____. What kind of information does it give?

11. Given a rectangular population, what would be the shape of its sampling distribution of the mean?

12. As sample size increases, the standard error of the mean ____. Therefore, if every possible member of the population were sampled, the standard error of the mean would be equal to ____.

13. If random samples of size $N = 120$ are drawn from an infinite population with $\sigma^2 = 3080.25$, compute the standard error of the mean.

14. For each of the following, fill in the missing information:
    (A) $s^2 = 144$ (B) $s = 25$ (C) $s^2 = $ (D) $s = $ (E) $s = 7.5$
    $s_M = $ $s_M = $ $s_M = 2.18$ $s_M = 9.2$ $s_M = .25$
    $N = 10$ $N = 20$ $N = 50$ $N = 100$ $N = $

15. Based on a sample of 100 rats, the average number of bar presses in an experiment was found to be 20.5 with a standard deviation of 3.8. If a repetition of the experiment yields a new sample mean of 19.8, should the experimenter be surprised? What if the new mean was found to be 21.8?

16. Random samples of size 2 are chosen from the small population 4, 5, 6, 7, 8, 9. What is the mean and standard deviation of this population?

17. A sample of size 2 is taken from the preceding population *with replacement*—that is, we draw one value, replace it, and draw again.
    (A) List the 36 possible samples of size 2 that can be drawn using sampling with replacement, and calculate the means.
    (B) Construct the sampling distribution of the mean for the samples of size 2.
    (C) Calculate the mean and the standard deviation of the probability distribution formulated in (B).

18. A survey of undergraduates at ten large universities was conducted to find out about student dating behavior. One question asked how many dates each student had per month. Each of the 10 samples questioned 100 students. The means for each university are given as follows:

School	Average Number of Dates	Standard Deviation
1	6.2	1.5
2	6.8	2.4
3	5.9	2.1
4	6.1	1.9
5	6.2	3.0
6	6.4	1.8
7	6.0	1.5
8	6.5	2.0
9	6.3	2.2
10	6.0	1.7

Estimate μ (the population mean) for each school.

19. For each probability statement, write a sentence that expresses the question:
(A) $P(M_1 - M_2) = 0 = ?$
(B) $P(s^2 < 5) = ?$
(C) $P(M = \mu) = ?$
(D) $P(6 < M < 8 | N = 100) = ?$

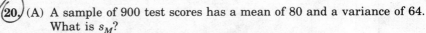

20. (A) A sample of 900 test scores has a mean of 80 and a variance of 64. What is s_M?
(B) If the sample had been only 50, what could s_M be?
(C) Another sample of scores from the same test yields a mean of 82.5. Would this mean be considered likely if the sample size was 900? If it was 50? What does this say about the relationship between sample size and variability of sample means?

5 Inferences about a Single Mean

The chapter on descriptive statistics covered a number of ways that data may be described by calculating measures of central tendency and measures of variability. Drawing inferences from such statistics requires probability theory, covered in Chapter 3, and an understanding of sampling distributions, covered in Chapter 4. We now begin to consider the problem of drawing inferences from sample means. That is, we begin to consider the problems involved when an investigator has calculated a mean from a set of data in the hopes of drawing some general conclusions about the phenomenon being studied. Experiments typically involve several conditions, and experimenters compare several means at the same time. However, the major ideas of statistical inference are more clearly developed in the simple, if less typical, setting involving a single mean.

These ideas are developed here with the help of a specific experiment. In the process of trying to reach a decision about the results of this experiment, we will introduce the concept of the _null hypothesis_. As we shall see, the null hypothesis specifies an hypothesized value of a population parameter, in this case the value of μ. Inferences about M, then, involve determining the likelihood of the observed value of M, assuming that the null hypothesis is true. In order to consider such probability statements, we introduce two theoretical distributions: the normal distribution and the t distribution.

5.1

PRETEST-POSTTEST EXPERIMENTS

A classic type of experiment in which an inference is made about a single mean is an experiment using the _pretest-posttest_ design. In a pretest-posttest experiment, subjects are first given a test or measurement of some kind (the pretest). Then, the experimental treatment is given followed by another measurement (the posttest). For example, an experimenter interested in evaluating a particular weight-management program would weigh the subjects before the program starts and again after the program has ended. Another experimenter may seek to determine whether alcohol consumption is influenced by attendance at a workshop in which stress-coping strategies are explained. In this case, the experimenter would obtain a measure of alcohol consumption both before and after the workshop is conducted. Investigators using the pretest-posttest design seek to determine whether a significant change occurred between the pre- and posttest measurements. To do so, a _difference_ score could be calculated for each subject by subtracting that subject's posttest score from that subject's pretest score. A mean of the difference scores would then be calculated and an inference about treatment effects would involve an inference about the mean difference score.

5.1.1 AN EXAMPLE

Katahn examined the effectiveness of a weight-management program in which physical fitness training was combined with training designed to alter eating behavior (Katahn, Pleas, Thackrey, and Wallston, 1982.) The program lasted for six months and involved numerous subjects, but for our purposes only the first week of the first eight subjects (7 female and 1 male) is considered. Their results are shown in Table 5.1.

TABLE 5.1
Katahn's weight-management data (weight in pounds).

Subject	Pretest	Posttest	D
1	261.50	256.50	+5.00
2	212.75	214.00	−1.25
3	218.25	215.00	+3.25
4	185.75	184.75	+1.00
5	204.25	203.75	+.50
6	152.00	151.75	+.25
7	207.00	204.50	+2.50
8	293.75	290.50	+3.25
		Sum	14.50

The weights of the subjects before the program began are reported (in pounds) in the column labelled Pretest. The weights after one week are given in the column labelled Posttest, and the difference scores are given in the D column. (D = pretest − posttest weights. Notice that decreases in weight are shown by positive scores and increases in weight are shown by negative scores).

The mean of a set of scores is

$$M = \frac{\Sigma X}{N}$$

In this case, we will designate the mean as M_D to indicate that the mean is from a set of D scores:

$$M_D = \frac{\Sigma D}{N} = \frac{5 + (-1.25) + \ldots + 3.25}{8} = \frac{14.5}{8} = 1.8125$$

(Remember that in the determination of a sum, adding a negative number is the same as subtracting the same quantity). On the average, then, 1.8125 pounds were lost during the first week of Katahn's program.

5.1.2 THE EXPERIMENTAL AND NULL HYPOTHESES

Do we conclude from these data that the program is effective during the first week? It turns out to be more useful to frame this question in the negative: Do we conclude that the program is *ineffective*? Stated another way, pretest-posttest experiments are said to involve *two* hypotheses. One hypothesis, called the **experimental hypothesis**, is that the treatment is effective. The other hypothesis, called the **null hypothesis** and referred to as H_0, is that the treatment is ineffective. According to the null hypothesis, M_D departs from zero only by chance. More precisely, according to H_0 the set of D scores was sampled from a population with μ_D equal to zero. Thus, the principal difference between the two hypotheses is in their assumed value of μ_D. According to H_0, $\mu_D = 0$. But according to the experimental hypothesis, μ_D does not equal 0. We saw in the previous chapter how sample means vary around μ, depending on σ^2 and N. Stated in terms of Katahn's experiment, then, we may say that even if the weight program were ineffective ($\mu_D = 0$), it is unlikely that the obtained sample mean weight change will equal zero exactly. Even the fact that the observed mean shows an average weight *loss*, rather than an average weight gain, is not very informative. If H_0 were true, a positive mean would be expected about 50% of the time; a positive mean is as likely as a negative mean (at least if the population distribution is symmetrical).

How, then, do we proceed? We proceed by *testing* the null hypothesis. We test the null hypothesis rather than the experimental hypothesis because H_0 specifies a single value for μ_D. And we test the null hypothesis by assessing the likelihood of certain results if H_0 were true. The question becomes: If the null hypothesis were true, is the observed mean likely or unlikely? If we can conclude that the observed mean is *unlikely* under the null hypothesis, we are then encouraged to conclude that the null hypothesis is untrue and that the experimental hypothesis is supported. On the other hand, if we conclude that the observed mean is *likely* under H_0, we do not reject H_0.

In the Katahn example, our evaluation of treatment effectiveness is an evaluation of the hypothesis that μ_D is zero. In order to evaluate this hypothesis, we seek to determine the likelihood of obtaining M_D equal to or greater than 1.81, if μ_D were zero. We learned in Chapter 4 that this is a question about the sampling distribution of the mean.

However, in Chapter 4 we were able to make probability state-ments about means because the population distribution was given. In all our applications of probability theory, the population distribution is unknown. At this point, then, we have two options. One option is to assume that the populations we sample from are _normal_ distributions. Normal distributions are described in the next section of this chapter. When the population distribution is assumed to be normal, precise statements about the likelihood of sample means are permitted. Statis-tical inferences based on the normal distribution are covered in chap-ters 5–11. Our second option is to make no assumption about the form of the population distribution. Statistical inferences that do not depend on the assumption of a normal population distribution are covered in chapters 12 and 13.

We will return to the Katahn experiment after becoming ac-quainted with the normal distribution.

5.2

NORMAL DISTRIBUTION

5.2.1 CHARACTERISTICS OF THE NORMAL DISTRIBUTION

The normal distribution is a hypothetical bell-shaped probability dis-tribution:

$$Y = \frac{1}{\sqrt{2\pi\sigma^2}} \, e^{-(X - \mu)^2/2\sigma^2} \tag{5.1}$$

A simpler formula is obtained when the distribution is standardized. You will recall that this occurs when each value of X is transformed to a standard score. (In this case, $z = (X - \mu)/\sigma$). The standardized form of the normal distribution is

$$Y = \frac{1}{\sqrt{2\pi}} \, e^{-z^2/2} \tag{5.2}$$

Equation 5.2 is expressed in terms of z and the mathematical constants e ($e = 2.7182 \ldots$) and π ($\pi = 3.1459 \ldots$). These equations are rarely used directly because virtually all pertinent computations have been performed and are tabled (see Table A-1 in Appendix A and Pearson & Hartley, 1966) or are readily obtained on computers. Both formulas are illustrated in Fig. 5.1.

FIGURE 5.1

Graphs of normal distributions. Top: A normal distribution in terms of X. Bottom: A standardized normal distribution.

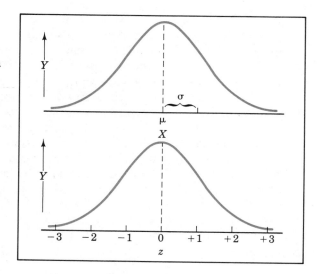

The top graph in Fig. 5.1 illustrates Equation 5.1, a normal distribution plotted on an x-axis. This graph also shows an interval equal to one σ. This distance, of course, is equal to one z on the standardized normal distribution, illustrated in the bottom graph of Fig. 5.1. As can be seen there, the bulk of the distribution extends from $-3z$ to $+3z$. Suppose, for example, that $\mu = 100$ and $\sigma = 15$. The bulk of the distribution would extend from

$$\mu - 3(z) = 100 - 3(15) = 100 - 45 = 55$$

to

$$\mu + 3(z) = 100 + 3(15) = 100 + 45 = 145$$

Fig. 5.1 shows the normal as a unimodal, symmetrical distribution. Its mean, median, and mode coincide. How the area under the normal is distributed is shown in Fig. 5.2.

FIGURE 5.2

Percentage of normal distribution between units of z.

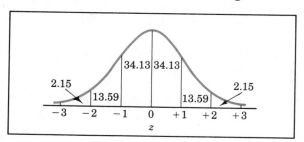

The graph shows the area under the normal distribution between several selected points. The values come from tabled values for the standardized normal distribution. The graph shows that 34.13% of the area is located between μ ($z = 0$) and $z = +1$. Since the distribution is symmetrical, 34.13% of the area is also located between $z = 0$ and $z = -1$. This means that over two-thirds (68.26%) of the area of the normal distribution falls between $-1z$ and $+1z$. The graph also shows that another 13.59% falls between $+1z$ and $+2z$. This means that 95.44% falls between $-2z$ and $+2z$. Finally, virtually all the area (99.74%) lies between $-3z$ and $+3z$.

Why all this concern about area? Because, as we saw in Chapter 3, probability statements depend on knowledge of the distribution of the area under probability distributions. With Fig. 5.2, for example, we have established that if X is sampled from a population that is normally distributed, the probability that X will fall between μ and $z = +1$ is .3413. That is, mathematically

$$P(\mu < X < 1z) = .3413$$

Graphically:

FIGURE 5.3

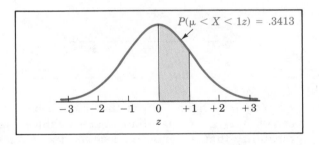

Figs. 5.2 and 5.4 also show that

$$P(-1z < X < +1z) = .6826$$

FIGURE 5.4

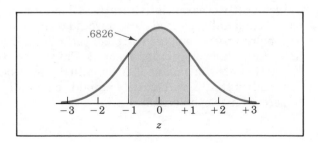

We have also shown that

$$P(+1z < X < +2z) = .1359$$

FIGURE 5.5

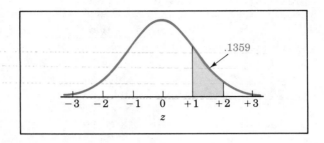

and that

$$P(-3z < X < +3z) = .9974$$

FIGURE 5.6

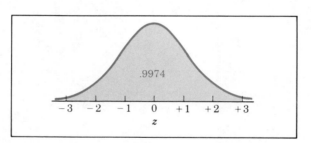

You may have noticed that our probability examples have all been on *intervals* rather than on exact values. We have considered the probability that X would be between two values, but not the probability that X equals a particular value. The reason for this is that when X is normally distributed, it is a continuous variable. When X is continuous, it is virtually meaningless to ask the probability that X will take on a particular value. The probability that X will take on *any* value exactly is essentially zero. Consider, for example, the probability that a randomly selected person would weigh exactly 155.3271 pounds. The more precise our measurement, the less likely we will find someone who weighs a specified amount. This is also true of the mean of X. Even though the mean of X is called the *expected value* of X, the probability that X will exactly equal the mean is approximately equal to zero, if X is continuous.

$$P(X = \mu) \cong 0$$

(On any continuous measure, *nobody* is perfectly average!) For this reason, statistical probability statements on continuous variables do not take the form:

$$P(X = a)$$

Rather, probabilities are calculated on intervals. Instead of asking the probability that X is equal to μ, we seek the probability that X is *near* μ, by which is meant the probability that X falls between some value b just below the mean and some value c just above:

$$P(b < X < c)$$

Likely events, therefore, are those in the vicinity of μ, unlikely events are those far away. For the normal distribution we have found

$$P(-3z < X < +3z)$$

This is a highly likely event, for it is nearly 1.0

$$P(-3z < X < +3z) = .9974$$

The probability that X falls outside this interval is correspondingly small

$$P(X < -3z \quad \text{or} \quad X > +3z) = 1 - .9974 = .0026$$

5.2.2 HOW TO USE A TABLE OF THE NORMAL DISTRIBUTION

The percentages reported in figures 5.2–5.6 were derived from tables of the standardized normal distribution. One such table is given in Table A-1 in Appendix A and repeated in part in Table 5.2.

TABLE 5.2
Selected rows and columns from Table A-1. Standardized normal distribution.

(A) z	(B) Area between Mean and z	(C) Area beyond z
.50	.1915	.3085
1.00	.3413	.1587
1.50	.4332	.0668
2.00	.4772	.0228
2.52	.4941	.0059
3.00	.4987	.0013

Tables A-1 and 5.2 have three columns. Values of z are in column A. In columns B and C are proportions of the area under the normal distribution corresponding to z. In column B is the proportion between the mean and z ($z = 0$). This is illustrated in Fig. 5.7.

FIGURE 5.7
Column B of
Table 5.2.

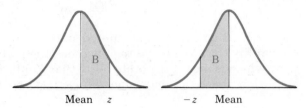

Two distributions are shown. Since the normal distribution is symmetrical, the proportion of the area between the mean and a positive value of z is the same as the proportion of the area between the mean and the negative value of z. Suppose, for example, that $z = 1.0$. Consulting Table 5.2, we first find $z = 1$ in column A. On that row we find the value .3413 in column B. This represents the proportion of the area between the mean and a z of 1.0. It also represents the proportion between the mean and a z of -1.0. We have already seen this fact in figures 5.2 and 5.3.

Column C in Table 5.2 reports the proportion of the area under the normal distribution beyond z. This is illustrated in Fig. 5.8.

FIGURE 5.8
Column C of
Table 5.2.

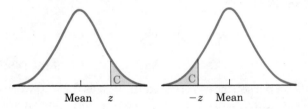

For example, for $z = 1.0$ we see in column C the value .1587. That is, .1587 of the area lies beyond a z of 1.0. Again, since the distribution is symmetrical this holds for both positive and negative values of z. The proportion of the area of the normal distribution that lies above 1.0 is .1587 and the proportion that lies below -1.0 is .1587. Notice that since half of the normal distribution lies on either side of the mean (recall that the mean and median have identical values in the normal distribution), the values in column C are obtained by subtracting the value in column B from .500.

As another example of the use of Table 5.2, consider how the table might be used to find the area between two different values of z. For instance, what is the proportion between $z = 1$ and $z = 2$? We saw in figures 5.2 and 5.5 that .1359 of the area of the normal distribution falls between $z = 1$ and $z = 2$. How might we calculate this from the information in Table 5.2? By subtraction. Notice in Table 5.2 that .4772 falls between $z = 2$ and the mean (row $z = 2$, column B). Since .4772 falls between $z = 2$ and the mean, whereas only .3413 falls between $z = 1$ and the mean, it follows that $.4772 - .3413 = .1359$ must fall between $z = 1$ and $z = 2$, as illustrated earlier.

A more general application for Table A-1 is considered next. Since the table is for the *standardized* normal distribution, it can be used for a normal distribution with any μ and σ^2. However, the scores would first have to be transformed to z. Let us assume, for example, that scores on a Graduate Record Examination (GRE) are normally distributed with $\mu = 100$ and $\sigma = 16$. With such an assumption we can make any number of probability statements about sampling from the distribution; that is, about the likelihood of sampling individuals with particular scores. What, for example, is the probability of sampling a score whose value is between 92 and 116? In order to use Table A-1, we must first transform these scores into z scores. First, the score of 92:

$$z = \frac{X - \mu}{\sigma} = \frac{92 - 100}{16} = \frac{-8}{16} = -.5$$

Next, the score of 116:

$$z = \frac{X - \mu}{\sigma} = \frac{116 - 100}{16} = \frac{16}{16} = 1.0$$

Thus, we rephrase our original question to ask: What is the probability of drawing a score between a $z = -.5$ and a $z = +1.0$? This is the same as assessing the proportion of the standardized normal distribution between these two z scores. But our table only reports proportions between z scores and the mean, not between two zs. In order to solve such problems, it is usually very helpful to graph the desired area. Fig. 5.9 illustrates the problem at hand.

FIGURE 5.9
The area between $z = -.5$ and $z = +1.0$.

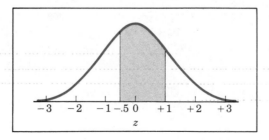

We can see in Fig. 5.9 that the desired area can be expressed as the sum of two areas: the area between −.5 and the mean plus the area between +1.0 and the mean. Consulting Table 5.2 we see the first proportion to be .1915. We have already calculated the second area; it is the proportion of the total that is between $z = 0$ and $z = 1$; it is .3413. Therefore, the proportion between $z = -.5$ and $z = +1.0$ is

$$.1915 + .3413 = .5328$$

In terms of the original problem, then, we can say that the probability that a randomly sampled individual will have a GRE score between 92 and 116 is .5328.

5.2.3   CENTRAL LIMIT THEOREM

In Chapter 4, the sampling distribution of the mean was shown to be influenced by N, the size of the sample. In Fig. 4.5 it was shown that as N increased, the sampling distribution became more and more bell-shaped. In fact, the sampling distribution of the mean was approaching a normal distribution. To show this, Fig. 4.5, with $N = 8$, is graphed again in Fig. 5.10 along with a normal distribution with the same mean and variance.

FIGURE 5.10
The sampling distribution of the mean from a rectangular distribution plotted with a normal distribution.

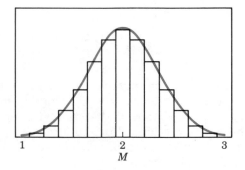

The quality of the fit is obvious. It will be remembered that the population from which these means were sampled was a rectangular distribution. This illustrates a remarkable principle about the mean that is known as the **central limit theorem**. Basically, the theorem states that the sampling distribution of the mean will approach a normal distribution as N increases, *regardless of the form of the population distribution.** This fact *could* be used to make statistical inferences about sample means.

---

*If the population distribution is normal, the sampling distribution of the mean is normal regardless of the value of N.

To illustrate this point, reconsider the sample mean in Katahn's experiment (M_D = 1.8125). Let us assume in this case that the sampling distribution of the mean is normally distributed. Let us assume further that the null hypothesis is true (μ_D = 0). Given these assumptions, we are able to calculate the probability that a given sample mean will equal 1.8125 or larger. However, we need one further statistic—the standard deviation of the sampling distribution of the mean. How might we obtain this value? Recall that

$$\sigma_M^2 = \frac{\sigma^2}{N}$$

and that we have an estimate of σ^2, which is s_D^2, the variance of the observed scores in the Katahn data. Using Equation 2.6 to calculate the variance of these eight scores, we find (shown in detail after Table 5.4) that their variance equals 4.138. For purposes of this exercise, we will assume that the variance of the population from which these scores were sampled equals 4.138. If this is true

$$\sigma_{M_D}^2 = \frac{4.138}{8} = .51725$$

and the standard deviation of the sampling distribution, the standard error of the mean, is

$$\sigma_{M_D} = \sqrt{\sigma_{M_D}^2} = \sqrt{.51725} = .7192$$

Let us review what has been assumed so far. Katahn's mean came from a sampling distribution that is normally distributed with mean equal to zero and standard deviation equal to .7192. If all these assumptions were true, we could calculate the probability of obtaining a mean of 1.8125 or larger. How? By using the table of the standardized normal distribution, just as we did in the previous section. We first transform 1.8125 into a z score. Recall that a score in a set of scores is transformed into a z score by dividing its deviation from the mean of the scores by the standard deviation of the set of scores. In this application the sample mean (1.8125) is the score we wish to transform to z. The mean of these scores is μ, and the standard deviation is the standard error of the mean:

$$z = \frac{X - \mu}{\sigma} = \frac{M_D - \mu}{\sigma_{M_D}} = \frac{1.8125 - 0}{.7192} = 2.52$$

Now, our question becomes: What is the probability of getting a z of

2.52 or greater? Consulting Table 5.2 we first find 2.52 in column A, and then examine the value in that row under column C. There we see that .0059 of the area extends beyond a z of 2.52. Stated another way, if all the preceding conditions were true, the probability of getting a mean as large or larger than 1.8125 would be .0052.

Although the preceding exercise is useful for demonstrating the use of the normal distribution in making probability statements about means, there are two major problems with the analysis. First, it was assumed that the sampling distribution of the mean is normally distributed. We were encouraged to make this assumption because the central limit theorem states that the sampling distribution of the mean approaches the normal distribution, even if the population distribution is not normal, as N increases. Thus, *provided N is large*, the normal distribution could be used to test hypotheses about treatment effects. However, there is considerable uncertainty about how "large" N has to be before the sampling distribution is approximately normal. The rate with which a sampling distribution approaches normality depends upon the shape of the population distribution. It is obvious from Fig. 5.10 that if the population distribution is rectangular, the rate is very fast. With N as small as 8, the approximation is excellent. However, if the population distribution were severely skewed, N would have to be considerably larger before the normal distribution could be used to approximate the sampling distribution.

A second problem with the preceding analysis is that it requires estimating the population variance. That is, in finding z, $z = (M_D - \mu)/\sigma_{M_D}$ we used an estimate of σ^2 from our sample. It turns out that when σ^2 is estimated, the ratio of $(M - \mu)/\text{est. } \sigma_M$ is not distributed normally when N is small. On the other hand, provided that the population distribution is normal, the distribution of this statistic is known. It is called the t distribution. For these reasons, it is the t distribution that is to be used in testing inferences about a mean, when it is assumed that the population distribution is normal.

5.3

THE *t* DISTRIBUTION

In Chapter 4, the sampling distribution of the statistic $(M - \mu)/\sigma_M$ was considered with the (1, 2, 3) population, and the distribution of this statistic was found to be approximately bell-shaped. (See Fig. 4.3.) It is known that if the population distribution is normal, this statistic is also distributed normally. However, we have just seen that when σ_M is estimated by $s_M = s/\sqrt{N}$ we have the new statistic

$$\frac{M - \mu}{s/\sqrt{N}}$$

which is not normally distributed. However, its distribution is known, and it is called the *t* distribution.

5.3.1 THE *t* DISTRIBUTION FAMILY

If a set of N scores is drawn from a normal population distribution with mean μ and variance σ^2, then the statistic

$$\frac{M - \mu}{s/\sqrt{N}}$$

is distributed as t. This can be stated as

$$t = \frac{M - \mu}{s/\sqrt{N}} \tag{5.3}$$

The *t* distribution is graphed in Fig. 5.11.

FIGURE 5.11
The *t* distributions for three values of N.

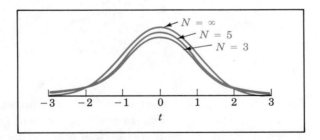

There are two items to note about Fig. 5.11. First, notice that t is not a single distribution, rather it is a *family* of distributions. There is a different distribution of t for each sample size. Fig. 5.11 shows the distributions of t for sample sizes of 3, 5, and ∞. Secondly, the t distributions are all somewhat bell-shaped and have a mean (expected value) of zero. This is similar to the standardized normal. In fact, as N approaches ∞, the t distribution approaches the standardized normal. The distribution labelled $N = \infty$ in Fig. 5.11 is identical to the standardized normal plotted in Fig. 5.1.

The differences between the normal and t with small N do not appear very substantial in Fig. 5.11, but Fig. 5.11 cannot show very well the difference that is critical: the difference in the "tails" of the distribution, when t is less than -2 or greater than $+2$. These regions are important because observations that fall there are judged unlikely and reflect on the null hypothesis that $t = 0$. Notice that the t has a higher proportion of its area in the tails than does the normal. For this reason, a large value of t is more likely than the corresponding value of normal z. This point will become more apparent as the use of the t table is described.

5.3.2　USE OF THE *t* TABLES

Table A-2 in Appendix A can be used to make probability statements from the t distribution. The values of t associated with different regions under the t distribution are reported for several sample sizes. The columns of Table A-2 are labelled α, by which is meant the proportion of the area in one tail of the distribution. This is similar to column C in Table A-1. The part designated α is shown in Fig. 5.12.

FIGURE 5.12

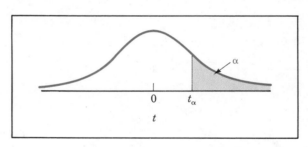

The rows of the columns are labelled *df*, which stands for *degrees of freedom*. The concept of degrees of freedom will be discussed in detail in Section 5.4. For now it is sufficient to note that for the t distribution derived from

$$t = \frac{M - \mu}{s/\sqrt{N}}$$

The number of degrees of freedom is equal to the sample size less one:

$$df = N - 1$$

Thus, to use Table A-2 to find a t associated with a sample of size N, we examine the row labelled $N - 1$. For example, if $N = 2$ we would consult the row for which $df = 1$ (since $df = N - 1 = 2 - 1 = 1$).

A portion of Table A-2 is repeated as Table 5.3.

TABLE 5.3
Selected rows and columns from Table A-2.

df	.10	.05	.01
1	3.078	6.314	31.821
7	1.415	1.895	2.998
30	1.310	1.697	2.457
∞	1.282	1.645	2.326

Consider the entry for 1 *df* and $\alpha = .10$. This value is 3.078. That is, for 1 *df*, .10 of the area under the *t* distribution is to the right of $t = 3.078$. This is shown in Fig. 5.13.

FIGURE 5.13

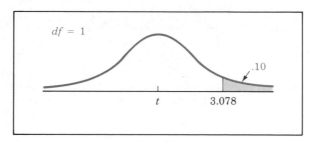

Let us state Fig. 5.13 in terms of sampling: If we draw a sample of size 2 ($N = 2$, $df = N - 1 = 2 - 1 = 1$) from a normal population and insert the observed M and s into Equation 5.3, the probability of obtaining a value of *t* of 3.078 or greater is .10. And conversely, ninety percent of the sample *t*s would be less than 3.078.

Holding *df* constant at 1 ($N = 2$), consider next $\alpha = .01$. Table 5.3 tells us that the *t* associated with $\alpha = .01$ is 31.821. That is, for 1 *df*, .01 of the area under *t* is to the right of $t = 31.821$. This is shown in Fig. 5.14.

FIGURE 5.14

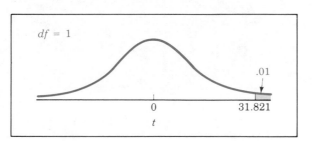

In sampling terms, if we were to draw a sample of size 2 from a normal distribution and insert M and s into Equation 5.3, the probability of getting a t as large as or larger than 31.821 is one in a hundred.

Another example from Table 5.3 is $df = 30$, $\alpha = .05$. Here we find $t = 1.697$.

FIGURE 5.15

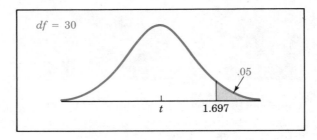

Notice in Table 5.3 how t decreases for any value of α as df gets larger. The last row of Table 5.3, and of Table A-2 in Appendix A, has $df = \infty$. Recall that with $df = \infty$, the t is the same as the normal distribution. Thus, the normal distribution appears in the last row of the t table. Also note that the values of t associated with 30 df are already close to the corresponding values for the normal. Stated another way, the t distribution closely approximates the normal, when N is 31 or greater.

5.3.3 t SCORES FROM KATAHN'S DATA

Table 5.4 repeats Katahn's D scores and includes D^2, so that s_D may be calculated from a computational formula.

TABLE 5.4
Katahn's D scores.

Subject		D	D^2
1		+5.00	25.0000
2		−1.25	1.5625
3		+3.25	10.5625
4		+1.00	1.0000
5		+.50	.2500
6		+.25	.0625
7		+2.50	6.2500
8		+3.25	10.5625
	Sum	14.50	55.2500

Using the sums shown in Table 5.4, we calculate the variance of D by substituting our Ds for the Xs in Equation 2.6.

$$s_D^2 = \frac{N\Sigma X^2 - (\Sigma X)^2}{N(N-1)} = \frac{N\Sigma D^2 - (\Sigma D)^2}{N(N-1)} = \frac{8(55.25) - (14.5)^2}{8(7)} = 4.138$$

$$s_D = \sqrt{s_D^2} = \sqrt{4.138} = 2.034$$

and the standard error of the mean is

$$s_{M_D} = \frac{s_D}{\sqrt{N}} = \frac{2.034}{\sqrt{8}} = .719$$

How are we to calculate *t* for these data? We have established that

$$t = \frac{M_D - \mu_D}{s_{M_D}}$$

but μ_D is unknown. We proceed by calculating *t* as if the null hypothesis is true. If the null hypothesis is true, $\mu_D = 0$. (H_0 = the treatment is ineffective. On the average there is no weight change.) We therefore calculate *t* with $\mu_D = 0$ to see how *t* is distributed according to the null hypothesis:

$$t = \frac{M_D - 0}{s_{M_D}} = \frac{1.8125 - 0}{.719} = 2.52$$

The relevant question is: If the null hypothesis is true, how likely is a *t* of this value? We consult Table 5.3. There are eight scores in Katahn's data, so

$$df = N - 1 = 8 - 1 = 7$$

We see in Table 5.3 that the *t* associated with $\alpha = .05$ is 1.895.

FIGURE 5.16

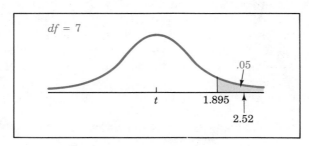

This tells us that the probability of getting a *t* of 1.895 or larger is .05. But the *t* from Katahn's data is 2.52, a value greater than 1.895.

Let us repeat what has been established so far: if (1) Katahn's *D* scores were sampled from a normal distribution and if (2) the null hypothesis were true, then (3) the probability of getting a *t* as large or larger than 1.895 is .05. We have further established that the value of *t* for Katahn's data is larger than 1.895. We are encouraged to conclude that Katahn's mean *D* score is *unlikely*, if the null hypothesis were true. You will recall that the null hypothesis ($\mu_0 = 0$) is the hypothesis that Katahn's weight-management program is ineffective. The alternative, experimental hypothesis is that there is a weight loss (a positive mean *D* score). Given the choice between the two hypotheses, we are encouraged to decide in favor of the experimental hypothesis; that is, to *reject* the null hypothesis.*

The use of computer statistics packages to calculate a *t* test on a single mean is discussed in sections 14.1.2, 14.2.2 and 14.3.2.

5.3.4   REVIEW

The statistic

$$t = \frac{M - \mu}{s/\sqrt{N}}$$

can be used to test hypotheses about μ. From a set of *N* scores, *M* and *s* can be calculated and inserted into the formula for *t* along with the hypothesized value of μ. If the hypothesis is true, the obtained value of *t* is likely to be in the vicinity of 0. However, the more *t* departs from 0, the less credible is the hypothesis. The *t* distribution is dependent on sample size. Large values of *t* are more likely for small samples than for large samples. In order to evaluate the obtained value, it can be compared with table values.

We have seen that we may reject the null hypothesis when the obtained mean would be very unlikely if the null hypothesis were true. But more typically the outcome is unclear. Suppose, for instance, in Katahn's experiment the *t* had been less than 1.5. What would we conclude? Statistical decisions, made in uncertainty, are subject to error. What can be done to reduce the chance of making erroneous statistical decisions? These problems are considered in the next chapter.

---

*Although the decision to reject the null hypothesis is the appropriate *statistical* decision to make about Katahn's data, in the next chapter we discuss the distinction between statistical decisions about data and *explanations* of that data. There we will argue that although rejecting the null hypothesis is the appropriate statistical decision, it might be incorrect to conclude that the weight-management program *caused* the mean weight change. As will be discussed in Section 6.2.3, the absence of a control group in the Katahn experiment permits the data to be interpreted a number of ways.

5.4

THE CONCEPT OF DEGREES OF FREEDOM

Virtually every statistic has associated with it a degree of freedom that influences inferences made about it. Here we consider this important concept generally and then return to explain the number of degrees of freedom in a *t* test about a single mean. In general, **degrees of freedom** may be thought of as a measure assigned to a set of observations. It is a measure of the extent to which the values of the observations are free to vary and still satisfy some constraint. For example, suppose we have a set of four numbers that must sum to 30. The first three numbers can be *any* value, but once they are fixed, the fourth is also determined. If, for example, the first three numbers are 13, 10, and 3, the fourth must be 4 in order for the total to be 30. We may say that three of the four are free to vary; that is, there are three degrees of freedom. If the first three are 40, -13, and 12, the fourth must be -9, and so forth. In our applications, *df* has much the same meaning.

The *t* test on a single mean involves an estimate of σ^2. This fact is the constraint on the set of possible observations and determines the magnitude of *df*. Recall that the calculation of s^2 requires the determination of $(X - M)$ for all X and the $\Sigma(X_i - M) = 0$. We may think of this property as placing a constraint on the set of observations. Once M is fixed, all deviations from it, by definition, sum to zero. Therefore, the *t* test on a single mean has $df = N - 1$, because all scores except one are free to vary and still meet the constraint.

5.5

SUMMARY

A sampling distribution that can be used to make inferences about the mean of a set of scores is the sampling distribution of

$$\frac{M - \mu}{s/\sqrt{N}}$$

This statistic is distributed as t with $N - 1$ degrees of freedom provided two assumptions are true: (1) the scores are sampled from a normal distribution, and (2) the scores are independently sampled. The t distribution is symmetrically shaped with mean, median, and mode all equal to zero. As *df* increases, the *t* distribution approaches the normal distribution.

Statistical inferences about a sample mean involve a test between a **null hypothesis**, which specifies a value of μ and an **experimental hypothesis**. Basically, the test requires an evaluation of the likelihood of the observed *M*, if the null hypothesis (the specified value of μ) is true.

This chapter also introduced the **central limit theorem**, a theorem about sampling that states the sampling distribution of the mean will approach the normal distribution as sample size increases, regardless of the form of the population distribution. Finally, the concept of **degrees of freedom** is defined as a measure of the extent to which the values of observations are free to vary.

REFERENCES

Katahn, M.; Pleas, J.; Thackrey, M.; and Wallston, K. A. 1982. "Relationship of eating and activity self-reports to follow-up weight maintenance in the massively obese." In *Behavior Therapy* 13:521–28.

Pearson, E. S., and Hartley, H. D. 1966. *Biometrika tables for statisticians.* Cambridge, England: Cambridge University Press.

QUESTIONS

1. Describe the central limit theorem.

2. Suppose you wanted to test the mean of a sample of scores, but you have reason to believe that the population distribution of these scores is positively skewed. Does this present a problem in trying to make statistical inference? Why or why not?

3. Typically, there are two hypotheses associated with experimental questions. Name and describe both.

4. On a standard normal distribution, what percent of the area under the curve falls between −1 and 1? −2 and 2? −3 and 3?

5. Assume that height is a normally distributed variable.
   (A) Is it possible to draw a graph of the area under the curve for the probability of someone being *exactly* 62″ (i.e., 62.0000 . . . ″) tall?
   (B) What is the probability of finding a person exactly 62″ tall?

6. Under what sampling condition would values of t and z be almost identical?

7. A researcher studying the fetal alcohol syndrome has reason to believe that alcohol consumed during pregnancy leads to lower birth weights. Pregnant female rats are injected with alcohol and the weight of their pups is measured. Extensive prior research with this strain of rats indicates that on the average, the rat pups weigh 8.40 grams. The researcher's sample of 100 pups weighs an average of 7.98 grams, with a standard deviation of 1.3.
   (A) Calculate t for the experimenter's average sample weight.
   (B) Is it likely that this sample of pups is the same weight as a normal population of pups? That is, is the null hypothesis of *no difference* likely?
   (C) What can be concluded from this experiment?

8. Suppose that in Question 7 the standard deviation of the experimenter's pups had been 4.2.
   (A) Calculate the new t.
   (B) Is the null hypothesis likely under this condition?

9. The statistic $\dfrac{M - \mu}{s/\sqrt{N}}$ is distributed as t with $(N - 1)$ degrees of freedom provided two assumptions are true. What are these two assumptions?

10. Using Table A-2 (the t distribution) in Appendix A, complete the following table for a one-tailed test:

(A)

df	t	p
1	$t \geqslant 12.706$	
2	$1.886 \leqslant t \leqslant 2.920$	
10		.01
20		.005
20	$1.325 \leqslant t \leqslant 2.086$	
29	$0 \leqslant t \leqslant 1.699$	

(B) For each problem in part A, sketch diagrams like figures 5.10, 5.11, and 5.12, and shade in the appropriate areas.

11. (A) In what ways are the curves for $t(df = 1)$ and $t(df = 30)$ similar? Dissimilar?
    (B) By comparing these two curves, tell why it is more likely to find large values of t, when the sample size is small.

12. Describe what is meant by the concept of degrees of freedom.

13. Suppose the "Do Nothing, It's All in Your Head" Society decided to start a weight-management program. The Society took pretest measures on 10 subjects, asked them to do nothing for 8 weeks, then took posttest weight measurements. Given the individual weights that follow, state the null hypothesis, the experimental hypothesis, and calculate t. On

the basis of the t result and the t table, what can you conclude about the effect of treatment?

Subject	Pretest	Posttest
1	291.25	291.00
2	200.00	200.25
3	193.50	193.00
4	207.00	206.00
5	251.25	249.75
6	243.75	244.00
7	261.00	261.75
8	275.50	275.00
9	195.00	195.25
10	300.25	300.00

14. Given the following sample data, calculate t. If the null hypothesis is that the sample is drawn from a population where $\mu_D = 0$, can we conclude that the mean of this data is consistent with the null hypothesis? Why or why not?

$$6.3, -1.83, 4.2, .83, -5.00, 3.99, 2.56, 3.45$$

15. The Knock-Down Sleep Aid Company wanted to test the effects of its latest product. A reaction time task is given to a group of 12 subjects before and after taking the drug. Using the t statistic, what can we state about the hypothesis that the drug had no effect?

Reaction Time Scores

Subject	Before	After
1	.7897	1.5643
2	.4111	2.9974
3	.9762	2.0134
4	.8314	1.9976
5	.5555	1.5762
6	1.0013	.9978
7	.9700	3.0127
8	.3112	.9984
9	.4497	1.2146
10	.5000	2.0010
11	.7294	1.8793
12	.8432	.9984

16. Students are usually required to take the Scholastic Aptitude Test (SAT) before they are admitted to a college or university. Suppose you and several of your friends decided to re-take the SAT after 2 years of college. Based on the following scores, what can you conclude about the effectiveness of college with respect to SAT scores? Calculate t as a basis for your conclusion.

Before College SAT (Total)	After College SAT (Total)
800	805
750	747
900	910
1095	1100
1106	1100
925	936
1000	995
1150	1153

17. For a given sample, the mean is 4.87, the standard deviation is 1.96, and $N = 35$. Calculate t. What proportion of the t distribution lies to the right of the calculated t? What proportion is to the left? What proportion of the area under the curve lies between 0 and the calculated t?

18. A fourth grade teacher has a hunch that his students are brighter than average. He obtains the IQ scores of the 21 students in his class and finds their average IQ to be 102.5 with a standard deviation of 18.7. The average of this IQ test among a large normative sample is 100. Evaluate the teacher's hunch.

19. It is commonly believed that the grades of first semester freshmen are lower than their grades in later semesters. Based on the following sample of nine students, does this seem to be true?

Student	First Semester GPA	Fourth Semester GPA
A	2.0	2.5
B	2.8	2.9
C	3.2	2.8
D	3.5	3.6
E	1.6	2.5
F	2.5	2.8
G	3.7	3.6
H	2.9	3.3
I	3.1	3.2

20. A gardening supply company wanted to test a new fertilizer that was claimed to increase the yield of tomato plants. Ten home gardeners agreed to use the new product, fertilizer A, on half of their plants, and use an established product, fertilizer B, on the other half of their plants. The company recorded how many pounds of tomatoes were produced by each plant. The average number of pounds of tomatoes is reported as

Gardener	Fertilizer A	Fertilizer B
1	20.2	18.2
2	19.4	21.4
3	12.8	17.9

Gardener	Fertilizer A	Fertilizer B
4	25.3	19.8
5	18.6	20.3
6	22.3	21.8
7	20.8	20.1
8	14.9	15.6
9	16.7	15.2
10	18.5	22.4

(A) Assuming the null hypothesis (that $\mu_A - \mu_B = 0$), calculate a t value for the difference between fertilizer A and B.

(B) Is the probability of getting a t value of this size or larger less than .05?

(C) What would you conclude about the claims made for fertilizer A?

21. Proponents of the Speed-Eye reading course claim their students read significantly faster after completing the program. They offer the following data to support their claim. What do you conclude?

Reading Speed (Words per Minute)

Student	Before	After
1	225	230
2	350	360
3	168	197
4	241	253
5	337	350
6	426	432
7	508	515
8	197	220

22. Dr. Gregor Zilstein (famous yet fictitious psychologist) decided to test his ability to pour 12 ounces of his favorite beverage (which is bottled in 32 ounce containers) into a 16 ounce cup. He tried 8 times and found that the mean pour was 10.9 ounces with a .16 ounce standard deviation. Test at a .05 level of significance whether the mean underfill is significant.

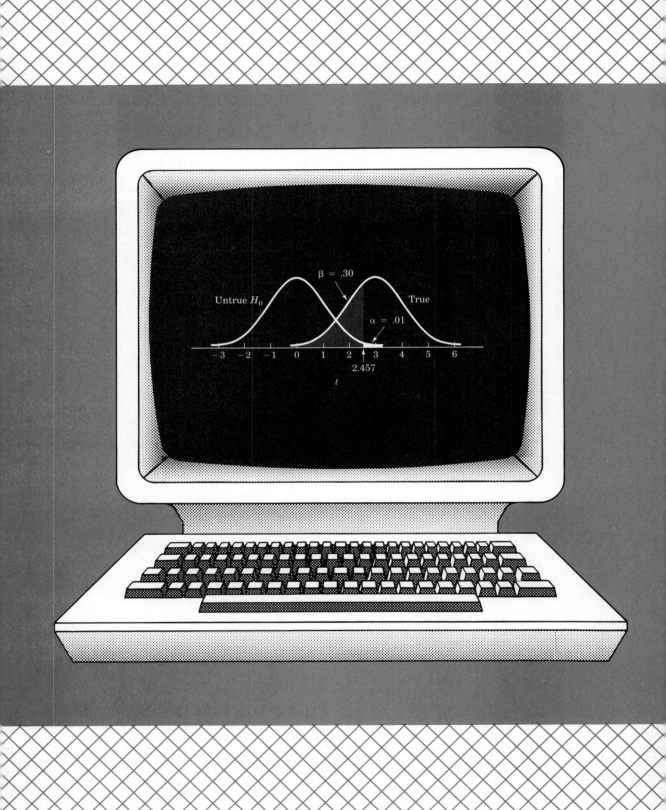

6 Hypothesis Testing and Decision Making

T he previous chapters describe four steps used in statistical inference. The first assumes that the observed data are observations
drawn from some population, that sampling has been random,
and an assumption is (or is not) made about the form of the population
distribution. Second, two well-defined hypotheses are stated: an experimental hypothesis countered by a null hypothesis. The sampling
assumptions and the null hypothesis together permit the third step—
computation of outcome probabilities. Fourth, the observed statistic is
compared to the computed probability values. These four steps are
outlined in Fig. 6.1.

     Fig. 6.1 also shows a fifth step, applying decision rules, which is
the subject of this chapter. We will discuss rules experimenters use to
decide whether to reject the null hypothesis, and we discuss how this
statistical decision is translated into a decision about the phenomenon
under study. We saw in Chapter 5 that such decisions are risky and
made in uncertainty. Incorrect conclusions are, of course, possible. The
experimenter, for example, could reject a null hypothesis that is true.
In this chapter we analyze the types of errors that can be made in
statistical inference and discuss several factors that influence the
likelihood of such errors.

6.1

DECISION MAKING

6.1.1 DECISION RULES

By convention, statistical choices are binary; that is, if an observed
event is improbable under an hypothesis, that hypothesis is rejected. If
an observed event is not improbable, the hypothesis is not rejected.
Convention often defines *improbable* as a probability of .05 or less. In
Katahn's study, for instance, we had calculated t to be equal to 2.52.
We had also established that under the null hypothesis, the probability
of getting a t equal to or greater than 1.895, with 7 df, was .05. Since
the observed t was greater than 1.895, the conventional decision would
be to reject the null hypothesis. We conclude that the null hypothesis is
untrue and that the alternative hypothesis, that there would be weight
loss, is supported. We had also considered our decision had t been 1.50
instead of 2.52. Now we can state that if t had been *any* value less than
1.895, we would not reject the null hypothesis. Therefore, had t been
1.50, we could not reject the null hypothesis at the conventional level
of confidence and we could not conclude that the program was effective
in the first week. Stated another way, if the null hypothesis were true,
a t (7 df) of less than 1.895 would not be unlikely and any value of t less

FIGURE 6.1
Five steps of statisti-
cal inference.

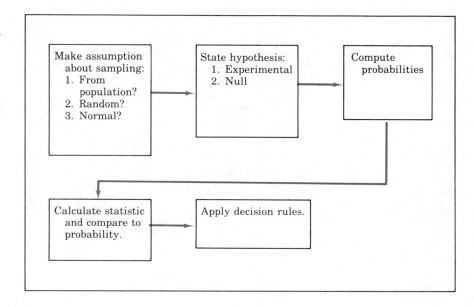

than 1.895 would lead to a decision to retain the null hypothesis
(retain H_0).

Results that permit rejecting H_0 at the conventional level of
confidence are said to be **statistically significant**. We have stated
that the conventional level of confidence is .05. This is sometimes
referred to as the **significance level**, or **alpha (α) level**. The value of a
statistic associated with the α level is called the **critical value** of that
statistic. In Katahn's study, the critical value of $t(7\ df)$ at $\alpha = .05$ was
1.895. Occasionally, the significance level is set at .01. An experimenter might elect to set α at the .01 level in order to reduce the
chance of *incorrectly* rejecting H_0. If the null hypothesis is in fact true,
the probability of incorrectly rejecting H_0 is equal to α. To guard
against this kind of error, an experimenter might decide to set α at .01.
It is understood that this decision (setting the value of α) is made
before the statistics are calculated, that is, before step 3 in Fig. 6.1.

6.1.2   TWO KINDS OF EXPERIMENTAL HYPOTHESES

In Katahn's study, the experimental hypothesis was that between the
pretest and the posttest there would be a loss in body weight. This is an
example of a **directional experimental hypothesis**, because it specifies the direction the effect would take relative to the null hypothesis.
Katahn expected a weight loss, not a weight gain. A **nondirectional
experimental hypothesis** would be one in which the direction is
unspecified. The experimenter predicts *some* effect, but does not pre-

dict whether it will be positive or negative. For example, an investigator may seek to determine whether a new drug influences learning without predicting whether the influence will be beneficial or harmful. In this case, the experimental hypothesis could be that the drug has *some* effect on a measure of learning, and the null hypothesis would be that it has no effect.

The distinction between directional and nondirectional hypotheses is important because the two kinds of hypotheses yield different critical values for the same level of α. To see this, let us compare Katahn's directional hypothesis with a nondirectional hypothesis for the same study. Using the directional hypothesis, we determine the critical value of t with $\alpha = .05$ as in Fig. 6.2.

FIGURE 6.2

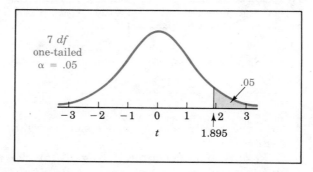

Since Katahn predicted a *positive* value of t, only the positive tail of the t distribution is considered and the critical value of t is that value above which 5% of the t distribution falls. Thus, in this case, the critical value of t is 1.895, because a t distribution with 7 *df* has 5% of its area above 1.895. The decision rule here, then, is to reject the null hypothesis, if t falls at or above 1.895. Now contrast this critical value with critical values obtained for a nondirectional hypothesis with the same level of α. For a nondirectional hypothesis, the critical value of t would be determined as in Fig. 6.3.

FIGURE 6.3

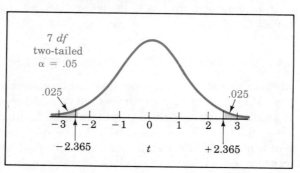

The critical values for a nondirectional hypothesis are set to allow either a positive or a negative outcome. To do so, the critical values are set so that half of α is above the positive value of t and half is below the negative value. For our selected α of 5%, this means 2.5% of the t distribution will be above the positive critical value of t and 2.5% will be below the negative critical value. Fig. 6.3 shows the critical values of t (from Table A-2) to be $+2.365$ and -2.365, when $\alpha = .05$ and $df = 7$. In this case, the decision rule is to reject H_0, if t is larger than $+2.365$ or smaller than -2.365. Comparing figures 6.3 and 6.2, it can be seen that if t is positive, a smaller absolute value of t is needed to reject H_0 when a positive directional hypothesis is being tested rather than when a nondirectional hypothesis is being tested.

As another example of different critical values produced by directional as opposed to nondirectional hypotheses, suppose $\alpha = .01$ and $df = 29$. With a positive directional hypothesis, the critical value of t is 2.462.

FIGURE 6.4

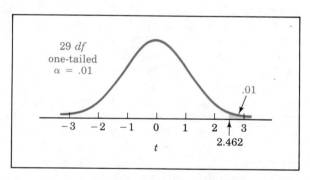

Here the null hypothesis is rejected only if t is equal to or greater than $+2.462$. With the nondirectional hypothesis, the critical values of t are $+2.756$ and -2.756.

FIGURE 6.5

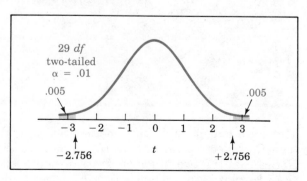

Here H_0 is rejected, if t is larger than $+2.756$ or smaller than -2.756.

The critical values of t (29 df) shown in figures 6.4 and 6.5 were obtained from Table A-2 in Appendix A.

Tests of directional hypotheses are said to be **one-tailed tests**. Tests of nondirectional hypotheses are said to be **two-tailed tests**. Table A-2 is set up so that αs associated with critical values of t for one-tailed tests are given at the top of the table and αs for two-tailed tests are given at the bottom of the table. Note in the table, for instance, that 2.583 is the critical value both for $df = 16$, $\alpha = .01$, one-tailed, and $df = 16$, $\alpha = .02$, two-tailed.

6.1.3   TWO KINDS OF ERRORS

It has been pointed out that α is set at .05 or .01 (or some other value) as a way of protecting against the error of incorrectly rejecting a true null hypothesis (this amounts to incorrectly concluding that there is a treatment effect when there is none). If α is .05, the probability of this error when H_0 is true is .05. If α is .01, the probability of incorrectly rejecting a true H_0 is .01. Considering only the null hypothesis, we might conclude that the best way to avoid errors of statistical inference would be to set α at a very small value. However, this approach ignores another kind of error. This error could occur if the null hypothesis were untrue, that is, if the correct decision is to reject H_0 and conclude that there is an effect. In a pretest-posttest experiment, for instance, the error would occur when the intervening treatment *does* produce a change in the measures (when $\mu_D \neq 0$), but H_0 is not rejected.

In other words, there are two possible conditions: either the null hypothesis is true or it is false. For each possible state of affairs, there are two possible decisions: reject H_0 or not reject H_0. Table 6.1 shows these two possible conditions and the two kinds of decisions.

TABLE 6.1
Possible outcomes of statistical decisions.

		Statistical Decision	
		Reject H_0	Do not reject H_0
State of the Population	H_0 true	Type I error	Correct decision
	H_0 untrue	Correct decision	Type II error

There are four possible outcomes: rejecting a true H_0, not rejecting a true H_0, rejecting an untrue H_0, or failing to reject an untrue H_0. Notice that the two kinds of errors are labelled type I and type II. A **type I error** occurs when a true H_0 is rejected, and a **type II error** occurs when an untrue H_0 is not rejected. Suppose that an experimenter is testing the effectiveness of a program designed to reduce cigarette smoking. A pretest-posttest design is used and a difference score is obtained for each subject by subtracting a posttest measure of cigarette consumption from a pretest measure. The experimental hypothesis would be that the program reduces smoking and the null hypothesis, then, would be that the program produces no change (that the difference scores are sampled from a population with μ_D equal to zero). Now, suppose the null hypothesis is *true*, that is, that the program is ineffective. If the experimenter rejects the null hypothesis (for example, concludes the program is effective), a type I error has been made. If, however, the experimenter does not reject H_0, a correct decision has been made. (This is row 1 column 2 of Table 6.1). Now suppose the null hypothesis is *not* true, that is, that the treatment is effective. In this case, μ_D is not equal to zero and an error would be made if H_0 were not rejected. This failure to reject an untrue null hypothesis is a type II error.

Table 6.2 shows the symbols used to refer to the probability of each of these errors.

TABLE 6.2
Probabilities of type I and type II errors.

State of the Population	Statistical Decision	
	Reject H_0	Do not reject H_0
H_0 true	α	$1 - \alpha$
H_0 untrue	$1 - \beta$	β

As we have already seen, the probability of incorrectly rejecting a true H_0 (type I error) is equal to α. The new symbol, β **(beta)**, refers to the probability of a type II error (failing to reject an untrue H_0).

A preferred decision strategy, of course, would be one which minimized both kinds of errors. Unfortunately, however, an attempt to reduce the probability of a type I error will result in an increase in the probability of a type II error. That is, if the experimenter sets α at .01 instead of .05 in order to reduce the chance of incorrectly rejecting a true H_0, the result will be an increase in the probability of failing to reject an untrue H_0.

6.1.4   THE RELATIONSHIP BETWEEN ALPHA AND BETA

To see how giving a value to α can influence the probability of making a type II error (β), let us consider a specific situation in which H_0 is false. In this example, $df = 30$ and a one-tailed test is used. To calculate β for this situation, we have to consider two distributions. One is the distribution of t assuming H_0 to be true, the other is the true distribution of t. We use the distribution assuming H_0 is true to determine the critical value of t that is to be used in reaching a statistical decision. We then turn to the other (true) distribution to determine the proportion of times t will not exceed this critical value. This proportion is the probability that a type II error will be made. Fig. 6.6 displays the two distributions.

FIGURE 6.6
Computation of power of one-tailed t test, when α = .05, df = 30, and expected value of t = 3.0.

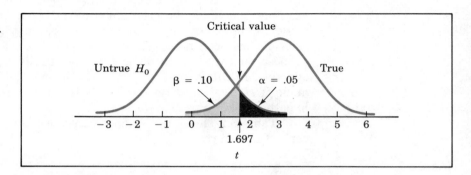

The distribution on the left is the distribution of t, *assuming H_0 is true* (which it is not in this example). Notice that its mean (expected value) is at t = 0. The distribution on the right shows the true situation; that is, the true distribution of t in this example. (Its mean is 3.0.) We first consider the situation in which α is set at .05. From Table A-2 we see that the t associated with α = .05, df = 30, using the one-tailed test, is 1.697. Our decision, then, is to reject H_0 when t is larger than 1.697. What, then, is the probability of making a type II error—the probability of failing to reject H_0? To calculate this, we must examine the true distribution (the distribution on the right in Fig. 6.6). We must determine the proportion of the true distribution that is below 1.697, because this is the proportion of times t will fail to exceed the critical value. We see that about 10% of this true distribution lies below 1.697. This part is shown as ☐. If 1.697 were used as the critical value for rejecting H_0, therefore, under these circumstances H_0 would not be rejected 10% of the time. Since failure to reject an untrue H_0 is a type II error, the probability of making a type II error is .10. The probabilities are summarized for this situation in Table 6.3.

Now, suppose α had been .01 instead of .05. Table A-2 shows that the critical value of t (30 df, α = .01, one-tailed test) is 2.457. The altered critical value along with the consequences are shown in Fig. 6.7.

TABLE 6.3
Decision probabilities when α = .05.

		Statistical Decision	
		Reject H_0	Do not reject H_0
State of the Population	H_0 true	.05	.95
	H_0 untrue	.90	.10

FIGURE 6.7
Computation of power of one-tailed t test, when α = .01, df = 30, and expected value of t = 3.0.

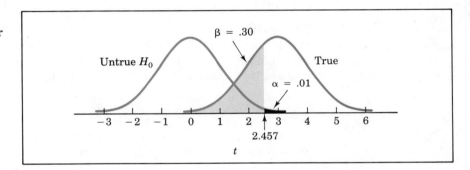

By increasing the critical value, we now see that a higher proportion of the true distribution fails to exceed the critical value. The probability of failing to reject an untrue H_0 is thus increased. In fact, we see that about 30% of the true distribution falls below the critical value. In other words, β = .30. The decision probabilities are summarized in Table 6.4, which should be compared with Table 6.3.

TABLE 6.4
Decision probabilities when α = .01.

		Statistical Decision	
		Reject H_0	Do not reject H_0
State of the Population	H_0 true	.01	.99
	H_0 untrue	.70	.30

In summary, we have found that steps taken to decrease the probability of a type I error, namely decreasing α from .05 to .01, have had the effect of increasing the probability of making a type II error. When α was .05, the probability of a type II error was .10. When α was .01, the probability of a type II error was .30.

Although demonstrated with a specific example, this finding is true in general. Decreasing the probability of a type I error is accomplished by decreasing α. But a decrease in α is accomplished by increasing the magnitude of the critical value. The larger the critical value, the smaller the chance that a true difference can be detected. Thus, an attempt to decrease the probability of a type I error by reducing α will invariably increase the probability of a type II error, if H_0 is false.

6.1.5 POWER OF A TEST

The probability of failing to reject an untrue H_0 is β. Since the decision rule is either to reject or not reject, the probability of correctly rejecting an untrue H_0 is $1 - \beta$. The **power** of a statistical test is defined as the probability of correctly rejecting an untrue H_0:

$$\text{power} = 1 - \beta$$

We see that as β for a test increases, the power of that test decreases. We have seen that β depends on α. In our specific example, when $\alpha = .05$ and $\beta = .10$, power $= 1 - .10 = .90$. However, when $\alpha = .01$ and $\beta = .30$, power $= 1 - .30 = .70$. Therefore, as α increases, the power of the t test also increases. The power of a statistical test is important because support for the experimental hypothesis is not obtained unless H_0 can be rejected. Experiments are usually designed as tests of an experimental hypothesis (a test of a treatment effect, a test of a difference between conditions, etc.) and if H_0 is false, the investigator desires to know it.

We have seen that the power of a test is influenced by the value of α set by the experimenter. As α increases, the power of the t test increases. The power of the t test is also influenced by N, σ^2, directionality of the experimental hypothesis, and the magnitude of the difference between μ_D under the null hypothesis and the true value of μ_D. The relationships between power and these factors are described in the next three sections.

6.1.6 POWER, SAMPLE SIZE, AND VARIANCE

In Chapter 4, we saw that the uncertainty about a mean is influenced by N and σ^2. Uncertainty is decreased either by an increase in N or a decrease in σ^2. We can now show how uncertainty about the mean influences the power of a t test on that mean. We consider N first. The computation of power for the t distribution graphed in Fig. 6.6 was based on an N of 31 (30 df). Suppose $N = 8$ instead of 31. Consulting Table A-2 we see that for 7 df, $\alpha = .05$, and a one-tailed test, the critical value of t is 1.895. Fig. 6.8 compares the t distribution under H_0 with an alternative true t distribution. In this particular example, as in Fig. 6.6, the expected value of t is not 0, as stated by H_0. Instead, the expected value of t is 3. The distributions in figures 6.6 and 6.8 differ in shape, because of the influence of the size of N on t.

FIGURE 6.8
Computation of power of one-tailed t test, when $\alpha = .05$, $df = 7$, and the expected value of $t = 3.0$.

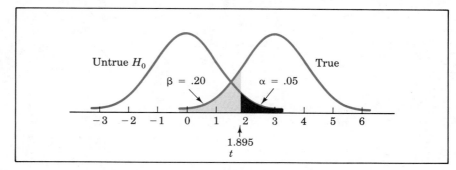

The power of the t test is the probability that it will correctly reject the untrue H_0, or $1 - \beta$. We see in Fig. 6.8 that 80% of the true t distribution is above 1.895. That is, under these circumstances 80% of the time H_0 would be correctly rejected. The power of the test is therefore .80. Overall, then, when $\alpha = .05$, the test is one-tailed, and the magnitude of the true μ is held constant, we find that power increases as N increases. This is summarized for our examples in Table 6.5.

TABLE 6.5
Comparison of power with two different dfs.

df	$1 - \beta$
30	.90
7	.80

Next we illustrate the relationship between power and σ^2. Let us assume first that $s^2 = \sigma^2 = 100$ and that N remains at 8. If s^2 is 100, what must the value of M be in order for t to be significant at $\alpha = .05$? Recall again that the equation for t is

$$t = \frac{M - \mu}{s/\sqrt{N}}$$

In order for M to be significant with $N = 8$ and $\alpha = .05$, it must exceed the critical value of 1.895. We can substitute what we know (the values for t, s, and N) and what is assumed under H_0 ($\mu_D = 0$) into the equation for t and then solve for M.

$$1.895 = \frac{M - 0}{\frac{10}{\sqrt{8}}}$$

$$M = 1.895 \, \frac{10}{\sqrt{8}} = 6.70$$

We find that in order to reject H_0 that μ_D is zero, under these conditions, M must be 6.70 or larger. Now consider a larger σ^2. Suppose, we *did* obtain a mean of 6.70, but $s^2 = \sigma^2 = 200$ instead of 100. In this case,

$$t = \frac{6.70 - 0}{\frac{\sqrt{200}}{\sqrt{8}}} = 1.34$$

We obtain a t smaller than the critical value, and H_0 cannot be rejected. In other words, a mean sufficiently large to permit rejection of H_0 when σ^2 is one value does not permit the rejection of H_0 when σ^2 is doubled. Power is lost when σ^2 is increased.

Although demonstrated with specific examples, we have illustrated very important general relations among power, N, and variance. Power increases as N increases and decreases as variance increases.

6.1.7 POWER AND THE MAGNITUDE OF THE MEAN

Perhaps it is reasonable to expect that the probability of rejecting an untrue H_0 depends on just how untrue it is. This can be demonstrated in the present circumstances by contrasting power when the true distribution is in two positions. In Fig. 6.6, the expected value of the true distribution was $t = 3.0$. Holding α, directionality, and df constant, we can compare Fig. 6.6 with the situation in which the expected value of the true t distribution is $t = 1.0$. This is shown in Fig. 6.9.

FIGURE 6.9

Computation of power on one-tailed t test, when $\alpha = .05$, $df = 30$, and expected value of $t = 1.0$.

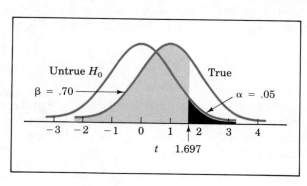

We see that 30% of the true distribution is larger than the critical value. Thus, under these conditions the untrue H_0 will be correctly rejected 30% of the time—the power is .30. Power is considerably reduced when the two distributions are close together. Comparison of power under these two conditions is summarized in Table 6.6.

TABLE 6.6

Comparison of power with two values of the difference between H_0 and the true expected value of t.

Difference	Power
3.0	.90
1.0	.30

Again, a general point is intended. Power decreases with decreased distance between the true μ_D and the value of μ_D under H_0. This point may be appreciated by considering the Katahn data again. In that study, the subjects lost an average of 1.8 pounds in the first week. Suppose the subjects had lost .5 pound. That is, suppose the sample distribution of D scores had been sampled from a population with $\mu_D = .5$ but with the same variance of D scores. Suppose further that the obtained M_D equals μ_D. Under these conditions, testing the H_0 that $\mu_D = 0$ with the obtained $M_D = .5$,

$$t = \frac{.5 - 0}{.719} = .70$$

t does not exceed the critical value of 1.895 and H_0, *although incorrect*, cannot be rejected. A type II error would be made even though $M_D = \mu_D$ in this case. Suppose, on the other hand, that μ_D were equal to 3.0; that is, on the average subjects lose 3 pounds in the first week of the program. In this case, the likelihood of rejecting H_0 would be high. In general, then, the greater the obtained mean departs from the mean under H_0, all else being equal, the greater the power of the t test.

In fact, this suggests another way to conceive of power. *Power is the ability to discriminate differences.* If the difference between the true value of μ_D and the value of μ_D under H_0 is small, a more powerful t test would be needed to detect the difference than if the difference were enormous. Powerful magnifying lenses are needed to perceive small distances; so too are powerful t tests needed to detect small mean differences. Anything that increases uncertainty about the mean (small N, large variance) diminishes the resolving power of the t test.

6.1.8 POWER AND DIRECTIONALITY

Earlier the critical value of t was shown to depend on whether the experimental hypothesis is directional or nondirectional. The positive critical value of t when the test was one-tailed was smaller than when the test was two-tailed. This suggests that the directional hypothesis yields a more powerful t than does a nondirectional hypothesis. However, this is true only when the experimenter has selected the correct directional hypothesis. When the wrong directional hypothesis has been selected, the probability of correctly rejecting an untrue H_0 is substantially less than when it is in the correct direction. To see this, return to the situation depicted in Fig. 6.9 and imagine the true distribution to have an expected value of -1.0 instead of the $+1.0$ shown there. When the true distribution is in the predicted direction, the probability of correctly rejecting the untrue H_0 is .30. However, when the true distribution is in the direction opposite to that predicted, the probability would be less than .02.

6.1.9 SUMMARY OF FACTORS INFLUENCING DECISION ERRORS

When the null hypothesis is false, we have seen that the power of the t test on a single mean is decreased by (A) decreased α, (B) decreased N, and (C) increased σ^2. In addition, it was shown that a one-tailed test is more powerful than a two-tailed test, provided the directional hypothesis is in the correct direction. Finally, we saw that the power of the t test depends on the magnitude of the difference between the true value of μ and the value of μ under H_0.

On the other hand, if the null hypothesis is true, and the assumption of normality is valid, the only factor to influence the probability of a type I error is the value of α set by the investigator. It is important to stress that N does not influence the probability of a type I error. Having gained some familiarity with the effect of N on power, students will occasionally confuse its effect on the type I error. As N increases, so does power, typically, but the probability of rejecting a true null hypothesis is unaffected.

6.2

DECISION MAKING IN THE RESEARCH ENVIRONMENT

To review the decision process as described so far: the experimenter states an hypothesis, counters it with the null hypothesis, conducts an experiment with α already set, or assumed, computes the appropriate statistic and its probability, and finally makes a decision about the experimental hypothesis. The decision has been portrayed as binary: H_0 is either rejected or retained. The decision process is purposely described without reference to the particular setting in which the research is conducted. In this sense, the procedure is intended to be *universal*—suitable for or characteristic of any research setting. And the decision is described as if it is made without benefit of any other information. In this section, the perspective is altered somewhat. We place the decision in a research environment where nonstatistical information influences the conclusion drawn about the research. From this vantage point we consider three matters: (A) assessing the reliability of the result, (B) nonstatistical influences on power, and (C) the distinction between statistical decisions and explanations.

6.2.1 ALPHA AND RELIABILITY

Another way to express the need for statistical inference is to say that the investigator is interested in knowing how *reliable* the results are. The investigator wishes to know whether the same conclusion would be drawn (or would be likely), if the experiment were repeated. In a sense, statistical inference is a substitute for repeating the experiment. For this reason, decision rules are conventionally stated as if no other information were available, as if the experimental hypothesis had never been tested, or as if the experimenter were planning a single test of the hypothesis. And in some cases expenses or circumstances preclude more than one experiment. An experiment may be too costly, or require a particular kind of subject difficult to obtain, or a condition difficult or hazardous to achieve. In such cases, a decision about the experimental hypothesis may have to depend on a single statistic.

More typically, however, the experimenter has the option of repeating the experiment. Often, the experimental hypothesis has already been tested before in the investigator's own lab or in someone else's lab. Scientists often replicate the research of others. In the scientific community, the reliability of a result is also assessed by repetition. Thus, when a single statistical result is obtained, it is often evaluated in a complex milieu that contains a past history of similar research. For example, it is well established that a partially reinforced conditioned response is more resistant to extinction than is a response reinforced 100% of the time. If a particular experiment fails to obtain a significant difference between a group receiving 100% reinforcement and one receiving partial reinforcement, the experimenter would probably be reluctant to reach a decision. Instead, the experimenter would likely withhold judgment until the experiment is repeated in such a way that power of the statistical test is increased. One way to increase power, we have learned, is to increase N. Therefore, the experimenter may elect to replicate the experiment with a larger number of subjects. Another way to influence power is discussed in the next section.

6.2.2 NONSTATISTICAL INFLUENCES ON POWER

The concept of variance, as treated so far, seems to imply that for the experimenter, variance is somewhere *out there*, out in the population, beyond control. Yet, it is valuable to recognize that variance is to some degree under the experimenter's control. An individual score, X, may be thought of as having two components, a true score and some error. (The true score is what X would be if there were no error). Formally:

$$X = \text{true score} + \text{error} \qquad (6.1)$$

This is a classic conception of psychological measurement held by the first experimental psychologists in the 1870s. According to them, error may be considered random, and an *estimate* of the true score may be obtained by averaging over a large number of separate measures. For example, Hermann Ebbinghaus studied memory and developed a relearning score as a measure of memory strength. His idea was that a particular relearning score was composed of a true measure of memory strength plus some error. One source of error, for example, might be a momentary lapse in attention. Other sources of error might erroneously inflate the relearning score. In order that these positive and negative errors be cancelled out, Ebbinghaus repeated the same memory conditions dozens of times and calculated an average memory strength score for that condition. Error was assumed to be variable (in fact, to be normally distributed with $\mu = 0$). Ebbinghaus sought to "control" this error variance by having a large number of observations. A similar point has been made earlier: uncertainty about a mean can be reduced with large N.

Another way to "control" variance is to control the sources of error. For example, Ebbinghaus sought to control error by stabilizing the learning environment. The observations were made at the same time of day in a quiet room. He chose learning material that was relatively homogeneous. He rigidly controlled the timing of the presentation of the material, and so forth. In a contemporary conditioning laboratory, animals are trained in soundproof chambers that have automatic control of stimulus events and other factors. All these are efforts to reduce extraneous contributions to the measurements, and thereby reduce the variability attributable to error.

An important link has been put into our inferential chain. The quality of our statistical inference depends on power; power depends on variance; and variance depends on control over extraneous contributions to our measures. These considerations participate in the experimenter's and others' evaluation of experimental results. If H_0 cannot be rejected, uncontrolled sources of variance may be partially responsible. Returning to our partial reinforcement study, for example, the experimenter may choose to replicate the study using conditions designed to reduce error variance. For example, sources of extraneous noise may be reduced. A more homogeneous group of subjects may be selected. Measures of behavior may be automated to reduce reading errors, etc. Each of these efforts would be aimed at reducing variance and increasing confidence about the resulting M.

6.2.3   DISTINCTION BETWEEN STATISTICAL DECISIONS AND EXPLANATIONS

Katahn found that the observed mean weight loss is unlikely under the hypothesis that the difference scores were sampled from a population with $\mu_D = 0$ (the null hypothesis). Consider again what is to be concluded. We can reject H_0 with some confidence. But can we conclude that the weight-management program produced the weight loss? It is important to distinguish between the decisions that can be made about H_0 and explanations that are offered for the results. It is possible to make a correct statistical decision and draw an unwarranted conclusion. Consider, for example, an alternative account of Katahn's results. In Katahn's study, all subjects were volunteers. All were at least 50 pounds overweight and had responded to notices requesting participants in a weight-management program. It is *conceivable* that the act of volunteering was part of a number of activities the subjects were beginning to initiate that would have resulted in weight loss. In other words, the subjects might have lost weight during the same time period even if they had not entered the program. To control for this possibility, a second group would be treated the same way as the experimental group, except that no program would be given. (They too would be overweight volunteers, would be weighed daily, and so forth.) This group would be called the *control* group. With this experimental design, instead of comparing M_D against 0, the experimenter would compare the experimental mean with the control mean. If a significant difference is observed, the experimenter has greater confidence about the cause of the result. The point here is not to argue for an alternative explanation for Katahn's data. The point simply is that the *statistical* decision has to be distinguished from conclusions about cause and effect. A result can be statistically significant but ambiguous as to cause.

6.3

SUMMARY

Results that permit rejecting H_0 at the conventional level of confidence are said to be **statistically significant**. The conventional level of confidence at which H_0 is rejected is referred to as the **significance level**, or **alpha (α) level**. The value of a statistic associated with the α level is called the **critical value** of that statistic. A **directional**

experimental hypothesis specifies the direction the results are hypothesized relative to the null hypothesis. A **nondirectional experimental hypothesis** would be one in which the direction is unspecified.

A **type I error** occurs when a true H_0 is rejected, and a **type II error** occurs when an untrue H_0 is not rejected. **Beta (β)** is the probability of a type II error.

Tests of directional hypotheses are **one-tailed tests**. Tests of nondirectional hypotheses are **two-tailed tests**.

The **power** of a statistical test is defined as the probability of correctly rejecting an untrue H_0: power $= 1 - β$. The power of a test is the ability to discriminate differences. In general, the more powerful the test, the better its ability to detect differences between the value of the parameter and the value hypothesized by the null hypothesis.

If the null hypothesis is false, the power of the t test on a single mean is decreased by (A) decreased $α$, (B) decreased N, and (C) increased $σ^2$. A one-tailed test is more powerful than a two-tailed test, provided the directional hypothesis is in the predicted direction. The power also increases with increased difference between the true value of $μ$ and the value of $μ$ under H_0.

If the null hypothesis is true, and the assumption of normality is valid, the only factor influencing the probability of a type I error is the value of $α$ set by the experimenter.

Finally, statistical decision making was placed in the context of the research environment, where additional, nonstatistical information influences decisions about experimental results. It was pointed out that the experimenter often has the option of repeating the experiment as a procedure for evaluating its reliability. In addition, the reliability of experimental results are often evaluated against a background of similar research.

Power is influenced by variance, and variance is often somewhat under the control of the experimenter. Variance can sometimes be reduced by systematic control of extraneous contributions to measurement.

Finally, an important distinction was made between statistical decisions and claims about cause and effect. It is possible to make a correct statistical decision and draw an erroneous conclusion as to cause.

QUESTIONS

1. Outline the five steps of statistical inference. In which stage is a type I error possible?

2. Suppose a calculated $t = 2.798$ and the table $t = 2.365$. Which t is the "critical value"? What does a critical value represent?

3. (A) Define what is meant by the critical value of a statistic.
   (B) For the t statistic, what is the relationship between the critical value and sample size, at a given α level?
   (C) What is the relationship between the critical t value and the α level, at any given sample size?

4. Differentiate between a directional and a nondirectional hypothesis. What is a "one-tailed" test?; a "two-tailed" test?

5. What is the benefit of making a directional prediction about the outcome of an experiment? What is a possible disadvantage?

6. If a test statistic is statistically significant, is it still possible to make a type I error? Justify your answer.

7. Indicate (using symbols) the probability that each of the following would occur, and identify the type of statistical error committed by the experimenter in each case where an error was made. The notation H_0 : $\mu = 20$ is read, "the null hypothesis is that μ equals 20."
   (A) μ (the true mean) $= 20$, and $H_0 : \mu = 20$ was rejected.
   (B) μ (the true mean) > 20, and $H_0 : \mu = 20$ was retained.
   (C) μ (the true mean) < 20, and $H_0 = \mu = 20$ was rejected.

8. For each of the following, indicate whether the statement is a correct decision or an error. If an error, indicate whether it is type I or type II.

True State of Population	Decision
H_0 is true	Do not reject H_0
H_0 is untrue	Reject H_0
μ is 0	Calculated $t = 7.93$, reject H_0 that $\mu = 0$
μ is 0	Calculated $t = .023$, do not reject H_0
H_0 is untrue	Do not reject H_0

9. In a particular experiment, a researcher wants to be very sure not to make the mistake of rejecting H_0, if H_0 is really true. If the level is set at .00001, what kind of risk is being taken?

10. All other factors being equal:
    (A) As power increases, β ____.
    (B) As α decreases, power ____.
    (C) As α increases, β ____.
    (D) As power increases, α ____.

11. Power increases as sample size increases. Explain why this is true, by considering the effect of sample size on the standard error of the mean and on critical values.

12. Suppose a team of psychologists is examining applicants to determine whether they are "mentally stable" to become executives in a high pres-

sure job. What type of error would they commit, if they erroneously accepted the hypothesis that an applicant is stable enough to become an executive? What type of error would they commit, if they erroneously rejected the hypothesis that an applicant is mentally stable enough to become an executive?

13. Home pregnancy test kits have been on the market for a few years. Assume that most women would prefer to make the error of thinking they are pregnant when in fact they are not, as opposed to thinking they are not pregnant when in fact they are.

(A) Assuming the null hypothesis is "pregnant," should manufacturers of these kits try to minimize type I or type II errors?

(B) Below is some hypothetical product testing data: (each cell entry is a frequency. For example, 5 pregnant women showed negative test results.)

	Test Result	
	Negative	Positive
Actual State Pregnant	5	388
Not pregnant	432	12

(I) What is the probability of the test yielding a correct decision for a pregnant woman?

(II) What is the probability of a type I error?

(III) What is the probability of a type II error?

14. A psychologist is testing the effectiveness of computer-assisted instruction (CAI).

(A) What hypothesis is he testing, if he is committing a type II error when he erroneously concludes that the CAI is effective?

(B) What hypothesis is he testing, if he is committing a type I error when he erroneously concludes that the CAI is effective?

15. Suppose you conducted an experiment and had a strong reason to believe the effects should be statistically significant (you might have prior research evidence and/or compelling theoretical arguments for finding an effect). After all the data are collected, you are disappointed to find you cannot reject H_0 at the .05 level: your t is only 1.702, while the critical value for t was 1.812. What changes might you consider, if you planned to repeat the experiment?

16. Name three ways to evaluate the reliability of an experiment.

17. If the results of your experiment are statistically significant, you are justified in concluding that your experimental treatment *caused* the difference. True or false? Justify your answer.

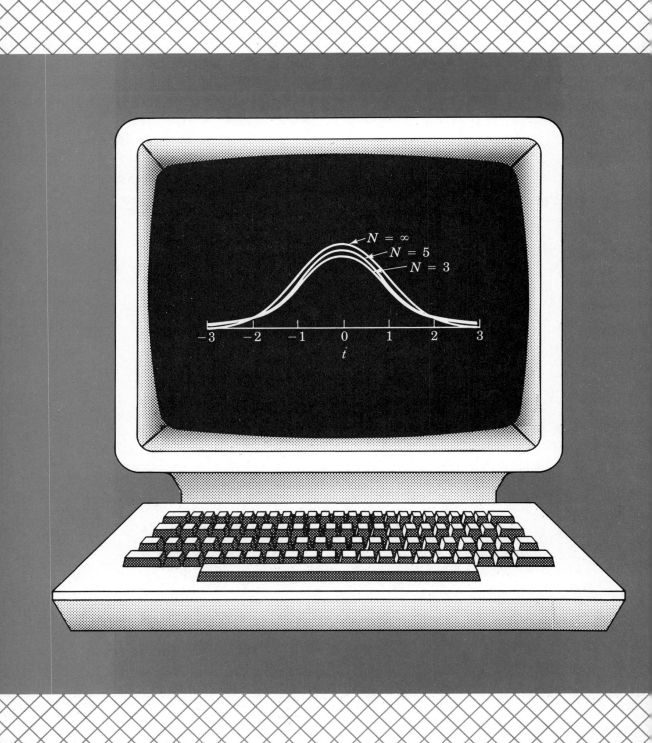

7 Inferences about Pairs of Means

T he previous chapter ended by demonstrating the need to compare
two means. This is a more typical inferential problem involving
means: the mean of one group is compared with the mean of
another. The effect of an experimental "treatment" is assessed by
comparing a group in which the treatment is given with a group in
which no treatment is given. "Treatment" could refer to a type of
behavior modification in a study of psychotherapy, or the presence of
peer pressure in a study of conformity, or the presence of a frame
of reference in a study of perception, and so forth. The classic experimental design of *transfer* is used to study the effect one kind of learning experience has on another. In one version of this design, an
experimental group is given training before a test, but the control
group is given only the test. All these examples involve comparing one
group in which a critical component is present with another in which it
is absent. In all, the experimenter is interested in whether the difference in treatment produces a difference in the behavioral measure.

Often the experimenter has a prediction about which mean will
be larger. That is, the experimenter has a *directional hypothesis*. A
nondirectional hypothesis would be that the treatment has an effect, or
that the two groups would differ, but the direction of the difference is
not specified. You should anticipate by now that either hypothesis
would be assessed by testing a null hypothesis. In this case, H_0 could be
that two groups are independently sampled from the same population
(or its logical equivalent: sampled from two populations with the same
μ and σ^2) and that any difference between the two means would be due
to chance. We have learned that an hypothesis about a statistic is
evaluated by referring to the sampling distribution of that statistic. In
the present example, the statistic is the difference between two means;
so, to test H_0, we need to know the sampling distribution of differences
between the means of two samples drawn from the same population.

We begin this chapter by examining the situation in which the
two samples are assumed to be independent. In this situation, all
subjects are assumed to be independently and randomly sampled from
the same population. An example of an experiment involving independent groups would be one in which nine subjects have been randomly
assigned to an experimental group and ten *different* subjects have been
randomly assigned to a control group. Tests between means from
independent groups are treated in Section 7.1. Another situation is one
in which the two groups of subjects, or *measures*, are related in some
way. For example, in some studies two groups are formed by matching
pairs of subjects on a measure that is expected to be related to the
dependent variable. Such groups are not likely to be independent,
since knowledge about the performance of one member of the pair is
likely to give information about the performance of the other member

of the pair. An example using related groups would be an experiment on eating behavior in which for every subject in the experimental group, there is a matched subject in the control group of approximately the same weight. Tests performed on related samples are treated in Section 7.2

7.1

t TESTS ON INDEPENDENT GROUPS

7.1.1 SAMPLING DISTRIBUTION OF MEAN DIFFERENCES

In Chapter 4, the sampling distribution of differences between two means was illustrated. In that example two independent samples were drawn from the (1, 2, 3) population, a mean of each sample was calculated, and the difference between the two means was computed. The sampling distribution, you will recall, is the set of differences obtained from all such possible pairs. The resulting sampling distribution is repeated in Fig. 7.1.

FIGURE 7.1
Example sampling distribution of $M_1 - M_2$.

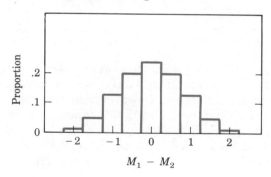

We are able to plot the sampling distribution in Fig. 7.1 because the population distribution is known. However, even when the population distribution is unknown, several statements can be made regarding the sampling distribution of $M_1 - M_2$.

First, the sampling distribution of mean differences is symmetrical. This makes sense, because if H_0 is true, M_1 is as likely to be larger than it is to be smaller than M_2, so the difference should be positive as often as negative. Second, the mean (expected value) of the distribution is zero. This too is logical because both means are based on samples from the same population. The third fact about the sampling distribution of mean differences concerns its standard deviation. Recall that the standard deviation of the sampling distribution of a statistic is called the *standard error* of that statistic. The standard error of mean differences is labelled $\sigma_{M_1 - M_1}$, and it is known that with two independent groups

$$\sigma^2_{M_1 - M_2} = \sigma^2_{M_1} + \sigma^2_{M_2} \qquad (7.1)$$

Equation 7.1 states that the square of the standard error of the difference is simply the sum of the squares of the standard errors of the respective means. The standard error of a statistic is a measure of uncertainty about that statistic. The larger its value, the greater the uncertainty. According to Equation 7.1, the greater the uncertainty about the individual means, the greater the uncertainty about their difference. We know that the standard error of the mean depends on σ^2 and N. It follows that the standard error of the difference also depends on σ^2 and N. Mathematically, since

$$\sigma^2_M = \frac{\sigma^2}{N}$$

we can rewrite Equation 7.1

$$\sigma^2_{M_1 - M_2} = \frac{\sigma^2_1}{N_1} + \frac{\sigma^2_2}{N_2}$$

We are now faced with the same problem we had in Chapter 6. Since the population variance is unknown, we have to estimate the standard error of the difference. However, adopting the same approach that was taken in Chapter 6, we obtain another sampling distribution that permits probability statements about mean differences. In general, this approach involves three steps. First, we assume the population distribution is normal. Second, we consider the ratio

$$\frac{G - E(G)}{\text{est. } \sigma_G}$$

that is, we consider the ratio of the difference between a statistic and its expected value divided by an estimate of its standard error. (You may notice the similarity between this ratio and the z score). We then consider the sampling distribution of this statistic. When the statistic G is $M_1 - M_2$, the sampling distribution is known to be t. Before the t is developed in the next section, notice what happens when $M_1 - M_2$ is substituted for the statistic G in the preceding equation. In this case, then, $E(G) = \mu_1 - \mu_2$ and est. σ_G may be written est. $\sigma_{M_1 - M_2}$. Substituting appropriately gives the ratio discussed next.

7.1.2   *t* DISTRIBUTION

When means are calculated for two samples of size N_1 and N_2, each drawn from normal populations with μ_1 and μ_2, and σ_1^2 and σ_2^2, the ratio

$$\frac{M_1 - M_2 - (\mu_1 - \mu_2)}{\text{est. } \sigma_{M_1 - M_2}}$$

is distributed as *t* with $(N_1 + N_2 - 2)$ degrees of freedom. In all applications considered here, the null hypothesis is that $\mu_1 = \mu_2$. For this reason $\mu_1 - \mu_2 = 0$, and the numerator of the ratio is simplified. Also, the estimate of the standard error of the difference is symbolized $s_{M_1 - M_2}$ so we may write

$$t = \frac{M_1 - M_2}{s_{M_1 - M_2}} \tag{7.2}$$
$$df = N_1 + N_2 - 2$$

Why is $df = N_1 + N_2 - 2$? Recall that *df* is a measure of the extent that a set of scores is free to vary and still satisfy some constraint. We saw that when a *t* test is performed on a single mean, *df* equals one less than the number of scores that go into the mean. The constraint is that $\Sigma(X_i - M) = 0$, and all scores except one are free to vary and still satisfy this constraint. Now we have two means, so we have two constraints, $\Sigma(X_i - M_1) = 0$ and $\Sigma(X_i - M_2) = 0$. Thus, two observations are "lost" from the total number, $df = N_1 + N_2 - 2$.

We next consider the problem of estimating the standard error of the difference.

7.1.3   POOLING ESTIMATES OF VARIANCE

We have already seen that the square of the standard error of the difference between two means is

$$\sigma^2_{M_1 - M_2} = \frac{\sigma_1^2}{N_1} + \frac{\sigma_2^2}{N_2}$$

Notice that under H_0, the two means are obtained from the same population. That is, under H_0: $\sigma_1^2 = \sigma_2^2 = \sigma^2$.

Looked at this way, we have two sample variances that are estimates of the same parameter. We have learned that uncertainty about the value of σ^2 can be reduced by increasing N, the number of observations used to estimate σ^2. From this perspective, we would want to somehow *combine* these two estimates to take advantage of the resulting increase in N. It might seem reasonable, therefore, that we obtain an estimate by *averaging* the two independent estimates of σ^2. That is,

$$\text{est. } \sigma^2 = \frac{s_1^2 + s_2^2}{2}$$

The problem with this particular method of combining the two estimates is that the two samples may differ in size. For example, sample 1 could have 30 observations and sample 2 could have 10. If we averaged by simply summing the two estimates and dividing by two, we would be treating the estimates as equal. What is needed is a *weighted* average, one that takes into account the difference in sample size. Here we treat the problem generally and then return to the two-sample problem.

In general, a weighted average of a set of scores is obtained by multiplying each score by its weight, adding these weighted scores, and dividing the result by the sum of the weights. The weights used to pool estimates of variance depend on sample size. More precisely, the weighted average of the variances of J samples is

$$\text{pooled est. } \sigma^2 = \frac{(N_1 - 1)s_1^2 + (N_2 - 1)s_2^2 + \ldots + (N_J - 1)s_J^2}{(N_1 - 1) + (N_2 - 1) + \ldots + (N_J - 1)} \quad (7.3)$$

A weighted average is achieved in Equation 7.3 by multiplying (weighting) each sample variance by the size of the sample, minus 1. We have seen $N - 1$ before. It has been referred to as degrees of freedom (df). Thus, each estimate is weighted by the df associated with that estimate and all are summed (as in averaging) and then divided by the total number of degrees of freedom. The weighted average is called the *pooled estimate* of σ^2. Equation 7.3 will have numerous applications in later chapters.

In the present chapter only two samples are involved ($J = 2$), so Equation 7.3 reduces to

$$\text{pooled est. } \sigma^2 = \frac{(N_1 - 1)s_1^2 + (N_2 - 1)s_2^2}{(N_1 - 1) + (N_2 - 1)}$$

or, rewriting the denominator,

$$\text{pooled est. } \sigma^2 = \frac{(N_1 - 1)s_1^2 + (N_2 - 1)s_2^2}{N_1 + N_2 - 2} \quad (7.4)$$

Equation 7.4 is important because it shows how an estimate of σ^2 is achieved by combining, or "pooling," the estimates provided by two independent sample variances. To illustrate the use of Equation 7.4, suppose that for sample 1, $N_1 = 30$, $s_1^2 = 60$, and for sample 2, $N_2 = 10$, $s_2^2 = 70$,

$$\text{pooled est. } \sigma^2 = \frac{(29)(60) + (9)(70)}{30 + 10 - 2} = \frac{2370}{38} = 62.4$$

Notice that when two samples are equal in size ($N_1 = N_2 = N$)

$$\text{pooled est. } \sigma^2 = \frac{(N - 1)s_1^2 + (N - 1)s_2^2}{2(N - 1)} = \frac{(N - 1)(s_1^2 + s_2^2)}{2(N - 1)} = \frac{s_1^2 + s_2^2}{2}$$

we *do* simply add the two estimates and divide by two.

We can now return to the problem of estimating the standard error of the difference, since we have available a pooled estimate of σ^2. We stated

$$\sigma^2_{M_1 - M_2} = \frac{\sigma_1^2}{N_1} + \frac{\sigma_2^2}{N_2}$$

But if H_0 is true, then $\sigma_1^2 = \sigma_2^2$. It follows that we can substitute our estimate in place of σ_1^2 and σ_2^2. We develop this point here with the goal, remember, of rewriting the denominator in the equation for *t*.

$$s^2_{M_1 - M_2} = \frac{1}{N_1} (\text{pooled est. } \sigma^2) + \frac{1}{N_2} (\text{pooled est. } \sigma^2)$$

$$= (\text{pooled est. } \sigma^2)\left(\frac{1}{N_1} + \frac{1}{N_2}\right)$$

Substituting from Equation 7.4

$$s^2_{M_1 - M_2} = \left[\frac{(N_1 - 1)s_1^2 + (N_2 - 1)s_2^2}{N_1 + N_2 - 2}\right]\left[\frac{1}{N_1} + \frac{1}{N_2}\right]$$

Now, the standard error of the difference, $s_{M_1 - M_2}$, is simply the square root of $s^2_{M_1 - M_2}$, or

$$s_{M_1 - M_2} = \sqrt{\left[\frac{(N_1 - 1)s_1^2 + (N_2 - 1)s_2^2}{N_1 + N_2 - 2}\right]\left[\frac{1}{N_1} + \frac{1}{N_2}\right]} \qquad (7.5)$$

and, finally, our expression for t for the difference between means of two independent samples is completed by substituting Equation 7.5 into Equation 7.2.

$$t = \frac{M_1 - M_2}{\sqrt{\left[\frac{(N_1 - 1)s_1^2 + (N_2 - 1)s_2^2}{N_1 + N_2 - 2}\right]\left[\frac{1}{N_1} + \frac{1}{N_2}\right]}}$$

$$df = N_1 + N_2 - 2$$

(7.6)

Equation 7.6 may be simplified further when the two samples are equal in size ($N_1 = N_2 = N$):

$$t = \frac{M_1 - M_2}{\sqrt{\left[\frac{(N - 1)s_1^2 + (N - 1)s_2^2}{2N - 2}\right]\left[\frac{1}{N} + \frac{1}{N}\right]}}$$

$$= \frac{M_1 - M_2}{\sqrt{\left[\frac{(N - 1)(s_1^2 + s_2^2)}{2(N - 1)}\right]\left[\frac{2}{N}\right]}}$$

$$t = \frac{M_1 - M_2}{\sqrt{\frac{S_1^2 + S_2^2}{N}}}$$

(7.7)

$$df = 2(N - 1)$$

Equations 7.6 and 7.7 are the principal equations of this section. They are to be used when the difference between the means of two independent samples is to be tested for statistical significance. Equation 7.7 is to be used when the sample sizes are equal, otherwise use Equation 7.6. However, as in any expression involving several terms, equations 7.6 and 7.7 may be written a number of ways. If you are able to calculate t from a computer program and cannot tell from the program description whether it corresponds to an equation given here, you may give the computer the data in Table 7.1 as a check. If the program does not compute the same value for t as that reported in Table 7.1, it has probably computed a t for related samples (see Section 7.2).

The use of computer statistics packages to calculate a t test on means from independent groups is discussed in sections 14.1.3, 14.2.3, and 14.3.3.

7.1.4 ILLUSTRATION OF THE COMPUTATION OF *t*

The following example illustrates an application of the *t* as a test for a difference between two means. It is based on research on memory and although the numbers do not come from a particular experiment, they are typical results. The interested reader will find a review of this area of research in Paivio (1971).

An investigator is interested in finding ways to facilitate memory. One theory states that memory for an experience can be stored in a visual memory system as well as a separate verbal memory system. When a memory is stored in both systems, according to this theory, memory performance is better than when memory is stored in only one system. To test these notions, an experiment is conducted in which subjects are given a list of 30 words to study (verbal experience). One group is told that the words will be presented one at a time and that each should be studied in preparation for a recall test to be given later. A second group is treated identically, except that the subjects are also instructed to form a mental picture of the object referred to by each word (visual experience). The experimenter randomly assigns 10 subjects to the first (control) group and 9 subjects to the second (experimental) group.

Let us review the hypotheses being considered in this experiment and the decision rules to be followed. The experimenter expects that subjects who form visual images will remember more words than those who do not. Thus, the prediction is that the experimental group will recall more words, on the average, than will the control group. The experimenter has a directional hypothesis. The countering null hypothesis would be that both groups are sampled from the same population, and any difference between the two means is due to chance alone. Suppose the experimenter sets α equal to .05. Since $df = N_1 + N_2 - 2 = 10 + 9 - 2 = 17$, we find in Table A-2 of the Appendix that the critical value for the one-tailed test is 1.74. Thus, the decision rule is to reject H_0, only if the mean number of words recalled by the experimental group is larger than the mean number of words recalled by the control group, and if *t* equals or exceeds 1.74.

The results and computation of *t* using Equation 7.6 are reported in Table 7.1 on the next page.

The experimental group recalled a mean of 20.33 words, whereas the control group recalled a mean of 14.30 words. The difference between these means is 6.03, and it is in the predicted direction. Is this difference statistically significant? The obtained *t* (2.49) exceeds the critical value (1.74), so the decision would be to reject H_0. That is, the experimenter concludes that the difference between the means is too large to have occurred by chance *if the scores came from the same*

TABLE 7.1
Example of t test on difference between two means.

Group	Number of Words Recalled by Each Subject
Experimental	12,21,30,20,26,18,16,21,19
Control	14,16,14,6,24,18,19,10,9,13

Descriptive Statistics:

Group	N	M	s^2
Experimental	9	20.33	27.750
Control	10	14.30	27.789

Experimental Hypothesis: $\mu_E > \mu_C$

$$H_0: \mu_E = \mu_C$$

$$\alpha = .05$$

$$t = \frac{M_E - M_C}{\sqrt{\left[\frac{(N_E - 1)s_E^2 + (N_C - 1)s_C^2}{N_E + N_C - 2}\right]\left[\frac{1}{N_E} + \frac{1}{N_C}\right]}}$$

$$= \frac{20.33 - 14.30}{\sqrt{\left[\frac{8(27.75) + 9(27.789)}{9 + 10 - 2}\right]\left[\frac{1}{9} + \frac{1}{10}\right]}}$$

$$t = 2.49$$

Critical value: $t(17) = 1.74$ Conclusion: Reject H_0.

population (that is, if the null hypothesis were true). Therefore, the experimenter decides to reject H_0.

Again, we distinguish between the statistical decision and claims about cause and effect. The statistical decision is to reject H_0. The experimenter may also wish to claim that the experimental group performed better because the imagery instructions produced two memories. The experimenter could be correct in the statistical decision and erroneous in the theoretical claim. For example, the claim that there are two memory systems is disputed by other theorists. We know that the probability of incorrectly rejecting H_0 (provided the assumptions of normality, etc., are valid) is equal to .05. But evaluation of the theoretical claim requires far more than the statistical decision rules discussed here.

7.1.5 POWER OF THE t TEST FOR A DIFFERENCE BETWEEN MEANS

Recall that the power of a test refers to the probability that the test will avoid a type II error, when H_0 is false. Chapter 6 reviewed the factors that influenced the power of the t test on a single mean. The factors listed there are relevant here, too. The power of the t test between two means increases with increased N and decreased variance. That is, power increases as sample size increases and σ^2 decreases. For example, in the memory experiment illustrated here, power would have been greater if $N_1 + N_2$ had been greater than 19. Power is also greater for a one-tailed test when the results are in the predicted direction. Suppose, for example, that a nondirectional test had been used in the preceding memory experiment. In this case, the critical value of t, for the same df and α level, would have been $+2.110$ (see Appendix Table A-2). In other words, a larger positive value of t would have been needed in order to be significant. In addition, the power of the t test for mean differences is influenced by the magnitude of the true difference between μ_1 and μ_2. The greater the difference, that is, the greater the treatment effect, the greater the power of t. In the imagery experiment, our best guess is that $\mu_1 - \mu_2 = 6.03$. In other words, under the conditions of the experiment, instructions to form images result in about six more words recalled than do instructions without imagery. Suppose H_0 had been false but the benefit of the imagery instruction was only a one word superiority (that is, $\mu_1 - \mu_2 = 1.0$). In this case, the likelihood of rejecting H_0 with the same sample size and variance would be considerably reduced.

7.2

t TESTS ON RELATED GROUPS

We consider here the general problem in which inferences are made about differences between *related* scores. Katahn's weight-management data is one example in which the paired scores are related because they are obtained from the same subject. In other words, one member of the pair of scores is the pretest score and the other member is the posttest score on the same subject. In general, then, the pairs of scores from a pretest-posttest experiment are related scores, related because they come from the same subjects. Often research is performed in which paired scores are related because they come from two different subjects that have been *matched* on some measure. Matching is one means of controlling variance due to extraneous variables. An experimenter wishing to manipulate one variable while controlling a second might obtain for each subject a measure on the second variable. Subjects could then be assigned to either an experimental group or a control group in such a way that for every subject in the experimental group there is a subject in the control group who has the same, or nearly the same, score on the secondary measure. The experimenter would then have pairs of subjects in the two groups that were matched on the secondary variable.

For example, if an experimenter wanted to compare the progress of a group of students receiving computer-aided instruction with the progress of a group of students receiving more standard instruction, the two groups could be matched on their Scholastic Aptitude Test (SAT) scores, which are known to predict academic performance. That is, before the subjects are assigned to the two conditions, all subjects could be given the SAT. Then, *pairs* of subjects would be identified so that each member of each pair would have nearly identical SAT scores. Finally, the experimenter would randomly assign one member of each pair to the computer-aided instruction group, leaving the other member for the standard instruction group. In this way, each subject in one condition would be matched with a subject in the other.

When the scores in two groups are related, how do we use the *t* distribution to test between the two means? There are a number of approaches to this problem, but the simplest has already been discussed in Chapter 5. Recall from Chapter 5 that we considered the use of difference scores to evaluate a treatment effect in a pretest-posttest design. We may think of this as a test between the pretest mean and the posttest mean, but an equivalent way to think of it is as a test of whether the mean difference score is zero. Here we review the use of the *t* test on the mean of difference scores with the help of a new example involving matching.

Suppose an experimenter is interested in testing a new memory technique on the elderly. The experimenter has 20 subjects available, all volunteers from a local senior citizens center, and has already tested each with an intelligence test. Since the experimenter suspects that subjects with higher IQ scores will do better on the memory task, it is decided to form two groups matched on their IQ scores. The IQ scores range from 74 to 122, and the experimenter forms matched pairs as shown in Table 7.2. The 20 subjects were assigned to groups as follows. First, the subjects were rank ordered according to their IQ scores. Then two subjects with the highest IQ were considered, with one randomly assigned to the control group, the other to the experimental group.

This procedure was followed for each successive pair in such a way that for all pairs one subject was randomly assigned to each group. The first column of the table reports the number of each of the ten pairs, then two columns are devoted to each group. The first column under each group reports the subject's number, the next column reports that subject's IQ score.

TABLE 7.2
Matched pairs formed
with 20 subjects.

	Control Group		Experimental Group	
Pair	**Subject Number**	**IQ**	**Subject Number**	**IQ**
1	14	83	8	74
2	13	86	2	88
3	19	92	4	89
4	11	93	17	102
5	7	109	3	109
6	5	110	1	112
7	6	117	16	113
8	20	119	15	120
9	12	121	18	121
10	9	121	10	122
	Mean	105.1	Mean	105.0

The experimental group is trained on the memory technique, while the control subjects are engaged in their normal activities. All subjects are then given a thirty-item memory task, and Table 7.3 reports the number of items correctly recalled by each subject.

TABLE 7.3
Data from memory ex-
periment on elderly.

Pair	Control	Experiment	D	D^2
1	14	17	3	9
2	15	24	9	81
3	16	17	1	1
4	16	15	−1	1
5	21	23	2	4
6	11	21	10	100
7	20	21	1	1
8	18	23	5	25
9	24	24	0	0
10	25	26	1	1
Sum	180	211	31	223
Mean	18.0	21.1	3.1	

As can be seen in Table 7.3, the average recall of the experimental group was 21.1 items, compared to 18.0 items by the control group. Stated another way, on the average the experimental group remembered 17% more [(21.1 − 18.0)/18.0 × 100% = 17%] than did the control group. But is this significant? To test the effect of the experimental treatment using matched pairs, the experimenter examines the difference scores for each pair (experimental minus control).* From this point, the approach is identical to that taken in Chapter 5, which examined difference scores in a pretest-posttest design. That is, the experimenter asks whether the mean of the difference scores is significantly different from zero. The mean of the difference scores is shown in Table 7.3 to be 3.1 (i.e., $M_D = 3.1$). We found in Chapter 5 that a single mean can be tested against an hypothesized μ_D by

$$t = \frac{M_D - \mu_D}{s_{M_D}}$$

with $df = N - 1$, where N equals the number of difference scores.

According to the null hypothesis, μ_D is zero (it is assumed that the difference scores are sampled from a normal population with mean equal to zero). Thus, $\mu_D = 0$ in the preceding equation for t. Next we need to calculate s_{M_D}, the standard error of the mean of the difference scores. We are doing nothing here that was not done in Chapter 5, except, of course, the numbers differ. We know that

$$s_{M_D} = \frac{s_D}{\sqrt{N}}$$

where s_D is the standard deviation of the difference scores. Our next step is to calculate this standard deviation. Recall that

$$s_D^2 = \frac{N\Sigma D^2 - (\Sigma D)^2}{N(N - 1)}$$

We can use the values of D^2 shown in Table 7.3 to calculate the standard deviation

$$s_D^2 = \frac{N\Sigma D^2 - (\Sigma D)^2}{N(N - 1)} = \frac{10(223) - (31)^2}{10(9)} = 14.1$$

$$s_D = \sqrt{s_D^2} = \sqrt{14.1} = 3.755$$

Note that N here is the number of pairs, or difference scores, not the number of subjects (which is 20), because the test is on the difference scores. Substituting these values of N and s_D into the equation for s_{M_D}:

$$s_{M_D} = \frac{s_D}{\sqrt{N}} = \frac{3.755}{\sqrt{10}} = 1.187$$

---

*The difference score could have been obtained by subtracting the experimental score from the control score. As long as the difference is taken in the same direction for each pair, the statistical decision will be the same.

We are able to calculate t

$$t = \frac{M_D - 0}{s_{M_D}} = \frac{3.1}{1.187} = 2.61$$

Is this t significant? Assume the experimenter has no directional experimental hypothesis and therefore uses a two-tailed test. Consulting Table A-2, we see that to be significant at $\alpha = .05$ and $df = 9$, a t must be 2.26 or greater. Since the obtained t is greater than the critical value, the null hypothesis is rejected and the experimenter concludes that the memory technique is an effective memory aid.

The use of computer statistics packages to calculate a t test on related groups is discussed in sections 14.1.2, 14.2.2, and 14.3.2.

7.3

SUMMARY

In this chapter, two t tests for comparing means have been considered. One is appropriate for comparing means M_1 and M_2 of two independent groups of subject size N_1 and N_2, respectively. It considers the sampling distribution of

$$\frac{M_1 - M_2}{s_{M_1 - M_2}}$$

where $s_{M_1 - M_2}$ is the standard error of the differences between two means. When the $N_1 + N_2$ scores have been independently sampled from a normal distribution, this ratio is distributed as t with $(N_1 + N_2 - 2)$ degrees of freedom. Expressions for $s_{M_1 - M_2}$ are developed and two different formulas for t are provided, depending on whether N_1 and N_2 are unequal (Equation 7.6) or equal (Equation 7.7).

The other t test is appropriate when the means are based on related groups. Pairs of scores are related when they come from the same subject or when they come from pairs of subjects matched on another measure. This t tests M_D, the mean of the differences between the members of the pairs. Its use involves the assumption that the difference scores are sampled from a normal distribution with $\mu_D = 0$. This t test is the same as that developed in Chapter 5 (Equation 5.3).

REFERENCES

Paivio, A. 1971. *Imagery and verbal processes*. New York: Holt, Rinehart & Winston.

QUESTIONS

1. Indicate which type of test would be appropriate for each: Independent (I) or Related (R) groups:
   (A) _____ Comparison of SAT scores of men and women engineering students.
   (B) _____ The effect of a time-management seminar on executives' productivity compared with other executives' productivity without the seminar.
   (C) _____ The difference in executives' productivity before and after attending a time-management seminar.
   (D) _____ The effects of a thinking skills program on reading scores of elementary school students; experimental and control students are matched on IQ.
   (E) _____ Average gas mileage of a car at 55 mph compared with its average gas mileage at 70 mph.
   (F) _____ The average number of auto sales per month in 1981 compared with average monthly sales by the same companies in 1982.
   (G) _____ Litter mates that are used to test time spent on two different kinds of activity wheels.
   (H) _____ Thirty-six undergraduates are chosen at random and divided into two equal groups.
   (I) _____ Tennis players are matched on general athletic ability in order to test different sizes of tennis racquets.
   (J) _____ Twenty-one gasoline stations are chosen at random to test the water content of the gasoline.

2. What is a benefit of using matched pairs instead of independent samples?

3. For each of the following sets of data, calculate $s_{M_1 - M_2}$:

   (A) $s_1 = 5.2$ $s_2 = 8.3$ (B) $s_1 = 36$ $s_2 = 28$
        $N_1 = 10$ $N_2 = 22$ $N_1 = 17$ $N_2 = 16$

   (C) $s_1 = 11.4$ $s_2 = 14.2$ (D) $s_1 = 56.5$ $s_2 = 62.1$
        $N_1 = 20$ $N_2 = 20$ $N_1 = 15$ $N_2 = 13$

4. For each of the following sets of data, $H_0: \mu_1 - \mu_2 = 0$. Using this information, determine whether H_0 could be rejected at the .05 level, for each of the following:

(A) $M_1 = 4.8$ $M_2 = 6.2$ (B) $M_1 = 12.3$ $M_2 = 9.8$
    $N_1 = 10$ $N_2 = 10$ $N_1 = 15$ $N_2 = 7$
        $s_{M_1 - M_2} = .95$ $s_{M_1 - M_2} = 3.9$

(C) $M_1 = 197.6$ $M_2 = 180.1$ (D) $M_1 = 3.7$ $M_2 = 4.9$
    $N_1 = 23$ $N_2 = 20$ $N_1 = 8$ $N_2 = 7$
        $s_{M_1 - M_2} = 5.2$ $s_{M_1 - M_2} = .42$

5. What is meant by $df = N_1 + N_2 - 2$? Why is the term "-2" included?

6. Even if the population distribution is unknown, several statements can be made about the sampling distribution of $M_1 - M_2$. Give three characteristics of this distribution.

7. Given two sample variances, why is a weighted combination more desirable than a simple average of the independent estimates of the variances? Provide an example. When would it be appropriate to use a simple average?

8. Name three ways in which the power of the t test for a difference between means may be increased.

9. Two book salespeople argue whether sales are better during the summer or winter months. Salesperson A says people buy more books during the summer because they have more free time; Salesperson B claims winter sales are higher because bad weather keeps people indoors. To settle the dispute, they obtain sales records for a 20-year period, and randomly select eight January and eight July periods:

Number of Books Sold ($\times$ 10,000)

January	July
4.8	5.3
6.3	4.8
5.2	6.2
6.1	4.6
4.5	5.8
6.8	4.3
5.2	5.9
5.9	6.3

(A) What is the null hypothesis you would test for these data?
(B) Compute t.
(C) What can you conclude from the results?

10. "Right-hook" Gutterman has been bowling at the "Remember to Let Go" bowling alley for several years. His average score is 180. The standard deviation is 2.6. "Righty" (as his friends call him) finally decided to try a new ball. After four games with the new ball, Righty has an average of 187. Can Righty claim there has been a significant difference in his bowling performance or could the improvement be attributed to chance? Be 95% sure of your answer. (Use a one-tailed test and a significance level of .05.)

11. D. J. Mouph is the announcer for a local small-town radio station. Often he works with commercial cassette tapes. He is concerned that tapes which are supposed to be 30 minutes in length may not play as long as the manufacturers claim. He decides to test two brands of cassette tapes and collects the following data:

Brand	Minutes				
Ellmax	29.4	31.6	32.8	30.4	34.3
Roncert	30.4	35.8	31.5	37.1	33.8

Use $\alpha = .01$ to check the claim that Roncert has a higher average tape length than Ellmax.

12. A certain drug is believed to help reduce high blood pressure. Eight participants receive a drug for one month, and their blood pressure is measured before and after. The average drop in blood pressure for this sample was 10.5 with a standard deviation of 15.5.
(A) Conduct a test to evaluate whether the decrease in blood pressure was significant.
(B) Suppose the same result had been obtained with a sample of 30 patients. What would you have concluded?

13. The Dean of the School of Engineering at State U. wondered if the grade point average (GPA) of engineering students was different from that of arts and sciences students. She took a random sample of 10 engineering students, and 10 arts and sciences students. The results are:

GPA

Engineers	Arts and Sciences
2.1	3.2
3.4	3.0
3.8	2.9
2.8	3.7
3.3	1.8
1.9	2.1
2.4	2.5
3.5	3.3
3.1	2.9
2.8	3.2

(A) What is H_0?
(B) Conduct the appropriate statistical test.
(C) Draw a conclusion from your results.

14. The volleyball coach of Spike University claimed that the players could jump higher than the players at Our Serve College. The players at S.U. jumped 16, 18, 14, 19, and 18 inches, respectively. The players at OSC jumped 15, 10, 13, 14, and 11 inches. Use $\alpha = .05$ to test the null hypothesis that there is no difference in jumping height.

15. The "Over-Haul" trucking company has been testing the claim that certain brands of diesel fuel will improve gas mileage. Sixteen trucks were used to test the fuel brands on a tank of diesel. The mileage was recorded for each truck:

Ezzon	Exico
214	201
183	176
174	175
217	212
185	182
197	195
163	164
185	185
175	169
158	157
188	184
170	167
206	204
184	178
248	236
149	145

Use $\alpha = .01$ to test the null hypothesis that there is no effect of the different brands of fuel.

16. Fifteen city grocery stores showed a mean percentage of profit of 6.59 with a variance of .94. Ten suburban grocery stores showed a mean percentage of profit of 7.28 with a variance of 1.17. Test the null hypothesis that there is no difference in profit between city and suburban stores.

17. Chemists at the Extra-Bright toothpaste laboratory developed a new formula to prevent cavities. Product testing data were collected comparing the new formula to the old. A sample of 20 people used the new formula, and 20 others used the old formula. After 6 months, the average number of cavities for the sample using the new formula was 2.3, with a standard deviation of .53. The mean number of cavities for the sample using the old formula was 2.7, with a standard deviation of .64. State the null hypothesis, conduct the appropriate test, and write a conclusion about the results.

18. A math teacher learns of a new textbook that supposedly helps students learn algebra better. The school principal is reluctant to purchase new textbooks unless some evidence exists that students really will learn more. The teacher agrees to test the new algebra book using a small sample of summer school students. Ten pairs of students, matched on mathematics achievement test scores, participate. One student of each pair uses the new textbook and the other student is given the text that is currently used. Scores on the final exam are given below:

Score of Algebra Final

Student Pair	New Text	Old Text
1	84	88
2	72	65
3	69	70
4	82	85
5	80	72
6	91	87
7	61	70
8	68	73
9	81	84
10	78	75

Should the principal authorize the purchase of the new textbook?

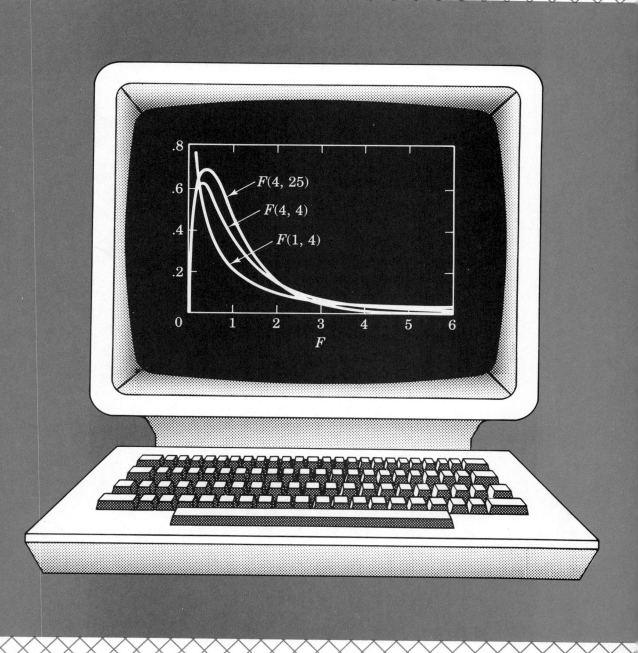

8 Inferences about Variance

his chapter deals with the problem of drawing inferences from sample variances. A sample variance, s^2, is an estimate of the population variance, σ^2. To draw inferences from s^2, we consider sampling distributions of s^2. Such distributions could be used to test hypotheses about sample variances. In practice, however, such applications are rare. Few experimenters, for example, test whether the variance of one sample is significantly different from the variance of another sample. Experimenters are usually interested in means, or averages. Nevertheless, one of the principal tools for testing among means is an analysis called the *analysis of variance*. This analysis will be covered in chapters 9 and 10. The present chapter, by considering sampling distributions of variances, introduces concepts that are important to understanding the analysis of variance.

8.1

SAMPLING DISTRIBUTION OF VARIANCES

The sampling distribution of the variance was illustrated in Chapter 4 with the (1, 2, 3) population. In that population three events, 1, 2, or 3, were possible and all were equally likely. Recall that the sampling distribution of a statistic for samples of size N is the distribution of that statistic obtained from all possible samples of size N. Fig. 8.1 shows the sampling distribution of s^2 for samples of size four from the (1, 2, 3) population.

FIGURE 8.1

Sampling distribution of s^2 for $N = 4$ from (1, 2, 3) population.

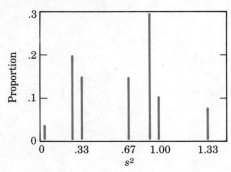

Since the population distribution is known in this case, we are able to compute the probability of any value of s^2. For example, we can compute the probability that s^2 will be 1.0 or larger to be 14/81 = .173, because 14 of the 81 (equally likely) samples have s^2 of 1.0 or larger. But how do we make inferences about s^2 when the population distribution is unknown? We know that for *any* population, the mean of the sampling distribution of s^2 (expected value of s^2) is equal to σ^2. What else can be stated about sampling distributions of variances? How do we compare variances from different samples? Do we, as with the mean, consider the sampling distribution of differences? That is, do we consider the sampling distribution of $s_1^2 - s_2^2$? As it turns out, *ratios* of sample variances provide a more productive approach to statistical inferences about variances. For two samples of size N_1 and N_2, respectively, we consider the sampling distribution of s_1^2/s_2^2. For instance, a sampling distribution of s_1^2/s_2^2 for samples of size four each ($N_1 = N_2 = 4$) from the (1, 2, 3) population could be constructed by considering all possible samples of size four (as was done to construct Fig. 8.1). The ratios of all possible pairs of sample 1 and sample 2 variances could then be calculated, and their distribution would constitute the sampling distribution of s_1^2/s_2^2 for $N_1 = N_2 = 4$ from the (1, 2, 3) population. When the population distribution is normal, the sampling distribution of s_1^2/s_2^2 has a known distribution. This new distribution, called the F distribution, is discussed in the next section.

8.2

F DISTRIBUTION

When two independent random samples of size N_1 and N_2, respectively, are drawn from a normal distribution, the ratio of their variances is distributed as F. That is,

$$F = \frac{s_1^2}{s_2^2} \tag{8.1}$$

with $df_1 = N_1 - 1$ and $df_2 = N_2 - 1$. (We saw in Chapter 5 that an estimate of variance places a constraint on the data. The df is one less than the number of observations that enter into the estimate. In this case, we have two independent estimates; hence, we have separate dfs for the numerator and denominator.) We will refer to these ratios in the following way: $F(df_1, df_2)$. The numbers in parentheses indicate the degrees of freedom associated with the F. The first number is the degrees of freedom for the numerator, the second is the degrees of freedom for the denominator.

The F distribution is illustrated in Fig. 8.2 on page 162, and as can be seen, the F is not a single distribution. Like the t, the F is a family of distributions. A particular t distribution was determined by a single value of df. The F distribution, in contrast, is determined by a pair of degrees of freedom: df_1, the degrees of freedom for the numer-

FIGURE 8.2
F distributions with
various *df*s.

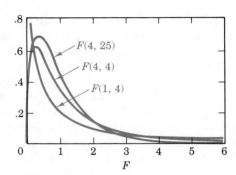

ator variance (s_1^2) and df_2, the degrees of freedom for the denominator variance (s_2^2).

8.2.1 CHARACTERISTICS OF THE *F* DISTRIBUTION

For large values of df_1 and df_2, the mean of the *F* distribution is approximately 1.0. That is, the expected value of the ratio of s_1^2/s_2^2:

$$E\left(\frac{s_1^2}{s_2^2}\right) \cong 1$$

This is logical, because both samples are from the same population. Therefore, the expected value of both s_1^2 and s_2^2 should be σ^2.

Notice other characteristics of the distribution. First, the distribution does not extend below zero. This follows from the discussion in Chapter 2 that variance cannot be negative. Second, for small values of *df* (i.e., small sample sizes), *F* tends to be positively skewed. However, as *df*s increase in size, the *F* distribution becomes more symmetrical and bell-shaped. This is shown in Fig. 8.3.

FIGURE 8.3
F distribution as *df*s
increase in size.

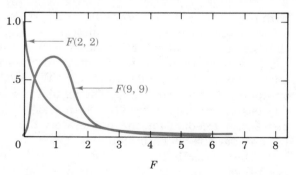

8.2.2 USE OF THE *F* TABLE

Table B in Appendix A contains critical values of *F*. A part of that table is reproduced in Table 8.1.

TABLE 8.1
Critical values of F from Appendix A, Table B.

		Degrees of Freedom for Numerator				
		1	**2**	**10**	**30**	**∞**
Degrees of Freedom for Denominator	1	161 **4052**	200 **4999**	242 **6056**	250 **6261**	254 **6366**
	10	4.96 **10.04**	4.10 **7.56**	2.97 **4.85**	2.70 **4.25**	2.54 **3.91**
	30	4.17 **7.56**	3.32 **5.39**	2.16 **2.98**	1.84 **2.38**	1.62 **2.01**
	∞	3.84 **6.64**	2.99 **4.60**	1.83 **2.32**	1.46 **1.69**	1.00 **1.00**

Source: Reproduced by permission from P. G. Hoel. 1966. *Elementary Statistics*, 2d. New York: John Wiley and Sons.

In Table 8.1 and Table B in Appendix A, the critical values are reported in regular type for $\alpha = .05$ and in boldface for $\alpha = .01$. To find the critical value for $F(df_1, df_2)$, the degrees of freedom for the numerator (df_1) are located along the top row and the degrees of freedom for the denominator (df_2) are located along the leftmost column. Consider the situation in which the df for the numerator is 2 and the df for the denominator is 10. Then, the column labelled "2 degrees of freedom for the numerator" would be entered and searched down to the row labelled "10 degrees of freedom for the denominator." There we see two entries, 4.10 in regular type and 7.56 in boldface type. The value 4.10 is the critical value for $F(2,10)$, $\alpha = .05$; **7.56** is the critical value for $F(2, 10)$, $\alpha = .01$. As another example, suppose the numerator df is 1 and the denominator df is 30. From Table 8.1, 4.17 is the critical value for $F(1, 30)$, $\alpha = .05$ and 7.56 is the critical value for $F(1, 30)$, $\alpha = .01$.

The critical values in Table B are for *one-tailed* tests. Table B is set up for the directional experimental hypothesis that σ_1^2 is larger than σ_2^2. If the experimental hypothesis is true, $\sigma_1^2 > \sigma_2^2$, and F should be larger than 1.0. For this reason, the critical value is in the tail of the distribution for large values of F, rather than in the tail near zero. To be statistically significant, therefore, F has to be *equal to* or *larger than* the critical value. For example, for $F(30, 30)$, the critical value at $\alpha = .05$ is 1.84. Five percent of $F(30, 30)$ is larger than 1.84. This means that $F(30, 30)$ must be equal to or exceed 1.84 in order to be statistically significant at the .05 level. The critical value is shown in Fig. 8.4.

The reason that Table B is prepared for one-tailed tests is that it is used primarily in conjunction with the analysis of variance. This

FIGURE 8.4
Critical value for
$F(30, 30)$, $\alpha = .05$.

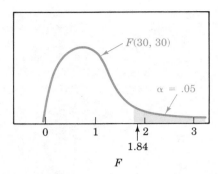

application requires one-tailed tests, as will be discussed in the next chapter.

8.3

INDEPENDENCE OF MEANS AND VARIANCES

In Chapter 3, two events were said to be independent if knowledge of the outcome of one provides no information on whether the other might occur or has occurred. Two events, A and B, are said to be independent if $P(A|B) = P(A)$. For example, when successive samples are independent, the probability that a particular element is drawn is constant across samples and unaffected by the outcome of previous samples. In the same way that we can speak of the independence of events, we can also speak of the independence of variables. Two variables are said to be independent, *unassociated*, or *unrelated*, if knowledge of the value of one variable is uninformative about the values of the other. As with events, independence between variables is a matter involving conditional probabilities. Variable Y is unrelated to variable X, if the probability distribution of Y is the same for all values of X. For example, motor dexterity and intelligence seem to be unrelated (Nunnally, 1959). Knowledge of how well a person does on a test of motor dexterity would not be predictive of how well they would do on an intelligence test.

An example of two variables that are associated, or related, would be the rate of learning by rats and magnitude of reward for correct responses. In general, the larger the reward, the faster the learning. Most experimental questions can be framed in terms of relationships among variables. For example, are attitudes about television programming related to years of education? Is longevity related to amount of cigarette smoking? Is the degree of math anxiety related to gender? Is depression related to intraversion? We have seen the t test as one method for assessing the relationship between two variables. Other methods for assessing relationships among variables are treated in later chapters.

Independence is discussed in the present chapter to explore the relationship between sample means and sample variances. The major point to be made here is that means and variances of samples from normal populations are independent. In other words, if the M of a sample were computed it would tell us nothing about the value of s^2 for that sample. This point has important implications for the development of the analysis of variance, which is treated in the next two chapters. For this reason, the relationship between sample means and variances is explored further here with the help of a specific example.

8.3.1 SAMPLING FROM A NORMAL DISTRIBUTION

To illustrate the relationship between sample means and variances, the following experiment has been conducted. With the aid of a computer, 100 random samples have been taken from a simulated normal distribution that has a mean equal to 100 and variance equal to 225. For each sample, the mean and variance were calculated. This yields 100 pairs ordered in such a way that the first member of the pair is the variance of the sample and the second member is the mean. Table 8.2 shows the first 10 of the 100 samples.

TABLE 8.2
Variances and means of the first 10 of 100 samples taken from a normal distribution with $M = 100$ and $s^2 = 225$.

Sample	s^2	M
1	117.368	105.18
2	278.231	93.07
3	214.263	101.13
4	170.030	88.68
5	157.915	97.21
6	219.141	94.41
7	174.125	92.10
8	273.898	96.57
9	162.108	103.67
10	444.741	103.44

Table 8.2 shows that sample 1 has a variance of 117.368 and a mean of 105.18, and so forth. The point of this exercise is to illustrate how M and s^2 are independent. Before pursuing this, however, we can take advantage of these samples to reiterate points made in earlier chapters. Remember, first, that $E(M) = \mu$. That is, the expected value of the sample means, the mean of their sampling distribution, is the popula-

tion mean. The mean of all 100 sample means is, in fact, 100.88. Also, the standard error of the mean, the standard deviation of the sampling distribution of means, is known to be

$$\sigma_M = \frac{\sigma}{\sqrt{N}}$$

Since our samples are size 10,

$$\sigma_M = \frac{\sqrt{225}}{\sqrt{10}} = 4.74$$

In comparison, the standard deviation of our 100 sample means is 4.68. Thus we see that, small deviations notwithstanding, our 100 samples nicely illustrate properties of the sampling distribution of the mean: the mean of our means is near μ, and their standard deviation is approximately equal to the standard error of the mean.

Now consider the joint distribution of the sample means and variances. We show this joint distribution by plotting the 100 pairs in two dimensions, each pair being represented by a single point. The value of the variance of each sample is plotted on the x-axis. The value of the mean is plotted on the y-axis. For the first sample shown in Table 8.2, the variance is 117.368 and the mean is 105.18. This pair is therefore represented at point (117.368, 105.18) on the graph. All the 100 sample pairs are similarly plotted in Fig. 8.5.

FIGURE 8.5
Scatterplot of sample variances and means.

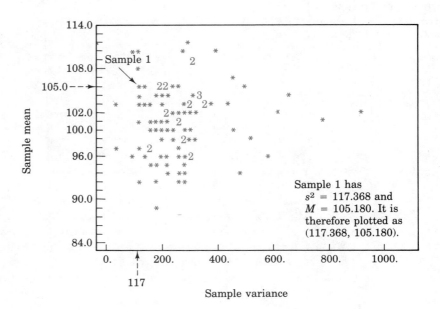

Each pair is represented by an asterisk, except where two or more points are very close. In these cases, the asterisk is replaced by a numeral that indicates the number of nearby points. This kind of graph is called a **scatterplot**—a plot of the joint distribution of two variables in which pairs of scores are shown as points in two-dimensional space. Scatterplots are discussed further in Chapter 11, which treats measures of relationships between variables.

The intent of Fig. 8.5 is to illustrate the lack of a relationship between sample means and variances that were drawn from a normal distribution. Recall that we have defined independence to mean that knowledge about the value of one variable is uninformative about the value of the other. This can be seen in Fig. 8.5 by noting that the sample means do not change systematically with different values of the sample variance. Note, for instance, the point farthest to the right in the figure. This point represents a variance of about 900 and a mean just above 100. Next consider the points in the region of a variance of 300. The points, too, have an average value of the means (y-axis) of about 100. In fact, the average value on the y-axis is generally about the same for all values of X. This is characteristic of a scatterplot that shows independence between two variables.

We have gone to some length to illustrate the independence between sample means and sample variances. We have done so in part to deepen our understanding of independence. More importantly, however, this fact of independence has permitted the development of the analysis of variance, which is a very powerful statistical tool designed to test among several means simultaneously. The analysis of variance is treated in the next two chapters.

8.4

SUMMARY

A new distribution has been introduced. F is the sampling distribution of the ratio of two estimates of σ^2. When two samples of sizes N_1 and N_2 are independently taken from a normal distribution, the ratio of their estimates is distributed as F with $N_1 = 1$ and $N_2 = 1$ degrees of freedom. We have seen that the F distribution can be used to test hypotheses about sample variances. However, the principal function of the F distribution is its role in the analysis of variance, which is a test of means.

The chapter concludes with a demonstration of the independence of means and variances of samples drawn from a normal distribution. Included is an introduction of the **scatterplot**, a plot of the joint distribution of two variables.

REFERENCES

Nunnally, J. C. 1959. *Tests and measurements*. New York: McGraw-Hill.

QUESTIONS

1. In the course of conducting experiments, it is often useful to make inferences about s^2 when the population distribution is unknown. In this case, we do not use $s_1^2 - s_2^2$. What is used?

2. Given a normally distributed population, the sampling distribution of s_1^2/s_2^2 is distributed as ——.

3. In order to determine an F and evaluate its significance, what variables must be specified?

4. How are the degrees of freedom determined for the F statistic?

5. Why is it necessary to specify two values of degrees of freedom in order to determine the critical value of F?

6. Describe three characteristics of the F distribution.

7. Why can F values not be negative?

8. If two samples are from the same population, what is the expected value of s_1^2/s_2^2? Why? What is the mean of the F distribution?

9. For each of the following, consult Table B in Appendix A to determine the critical value of F.

df	α	F
(1,5)	.05	
(10,1)	.01	
(6,6)	.05	
(20,30)	.05	
(14,12)	.01	
(20,100)	.05	

10. Using Table B in Appendix A, fill in the missing values:

α	df Numerator	df Denominator	Critical Value
.05	1	1	?
.05	∞	∞	?
.05	24	?	2.42
?	24	12	3.78
.01	?	16	3.01
.01	5	?	3.25
.01	6	11	?

11. In which case would F be significant at $\alpha = .05$?

(A) $s_1^2 = 15.8$ $s_2^2 = 4.2$ $N_1 = 12$ $N_2 = 16$

(B) $s_1^2 = 20.6$ $s_2^2 = 12.7$ $N_1 = 41$ $N_2 = 26$

(C) $s_1^2 = 4.1$ $s_2^2 = .9$ $N_1 = 10$ $N_2 = 11$

(D) $s_1^2 = 8.3$ $s_2^2 = 4.8$ $N_1 = 13$ $N_2 = 18$

(E) $s_1^2 = 17.4$ $s_2^2 = 3.6$ $N_1 = 6$ $N_2 = 22$

12. Are the critical values given in Table B for one- or two-tailed tests?

13. Complete the following, "Two variables are said to be independent. . . ."

14. If the mean of a sample is large, what can be predicted about the sample variance?

15. If we assume that a population is normal, $E(s^2) = $ _____ regardless of the value of _____.

16. Describe what is illustrated by a scatterplot.

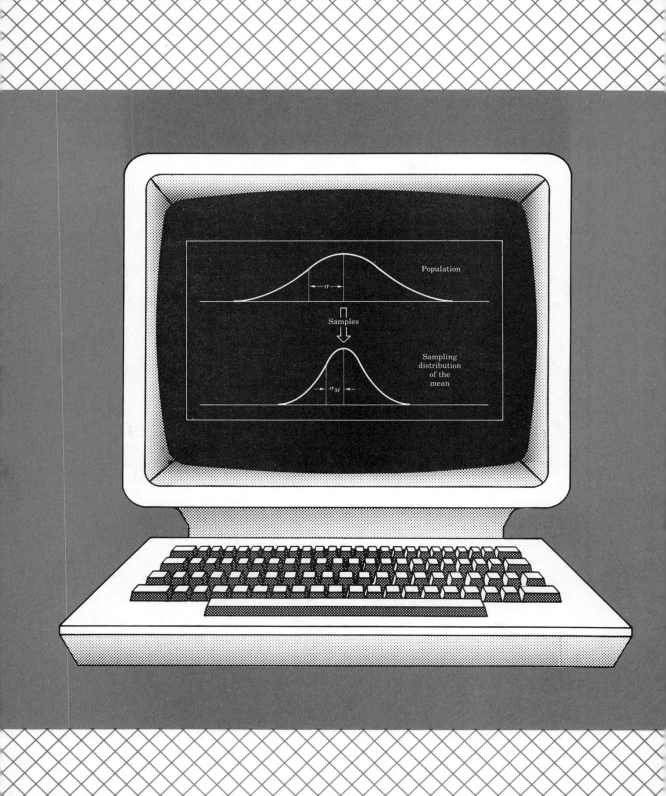

9 One-way Analysis of Variance

This chapter continues the treatment of statistical methods used to evaluate differences among means. So far we have seen that a t test can be used when two observations, each taken under different conditions, are made on the same subjects. This t test is performed on a set of difference scores, one difference score from each subject (Chapter 5), and is also appropriate for related groups (Chapter 7). We have also considered a t test involving two independent groups of subjects (Chapter 7). In this case, the t is a test of the difference between the means of two independent groups of subjects. We now turn our attention to more typical circumstances in which more than two conditions, or more than two groups, are involved. Analysis of variance is considered both in this chapter and the next. This analysis permits a simultaneous test of the differences among a set of J means ($J = 2$, or more). There are no limits on the size of J, except those placed by practical considerations. The null hypothesis is that there are no differences, except by chance, among the group means. Stated another way, H_0 is that all observations come from the same population (or its logical equivalent—from populations with the same mean and variance).

There are two kinds of t tests, one designed for two observations on each subject, another designed for a single observation on each of two groups of subjects. In a like manner, two kinds of variables influence how the analysis of variance is performed. **Within-subject variables** involve two or more observations on each subject. Each subject serves in each of the J conditions of the experiment. For example, in an experiment on perception each subject may view a display under three levels of luminance. Three perceptual measures would be obtained for each subject, one under each level. An analysis of variance on a within-subject variable is similar to the t test on difference scores (where $J = 2$), except that the analysis of variance simultaneously tests the means of all J conditions and J can exceed 2. Within-subject variables are sometimes referred to as *repeated measures* because each subject is tested, or measured, more than once. Sometimes the separate tests involve different conditions, other times they involve repetitions of the same treatment. When treatments are repeated, they are often referred to as *trials*. For instance, a subject in an experiment on motor skills would repeat a specified task a number of times. The experimenter would determine whether performance improves during training by testing the means of the successive trials.

Between-subject variables involve J groups of subjects. In this case, no subject serves in more than one group. For example, the perception study just described could be performed by having three separate groups of subjects serve under each of the luminance levels.

Often the experimenter may choose either the within-subject or the between-subject approach. Factors influencing such a choice are described in Section 9.3.4.

9.1

WHY TESTS OF MEANS ARE CALLED ANALYSES OF VARIANCE

Why is a test among means called an analysis of variance? An answer to this question offers a general introduction to the method. Computational details depend on whether variables are within- or between-subjects and on how many variables are involved, as will be shown in the next sections.

The basic test of any analysis of variance is an F test. We have seen that both the numerator and denominator of an F ratio are estimates of variance. Under H_0 both are estimates of the same σ^2, namely the variance of the population from which the scores are drawn. This is also true for the analysis of variance. Two estimates of σ^2 are obtained. These are referred to as the numerator and denominator estimates. The principal idea of analysis of variance is that a significant F implies a difference among the means. To show how this works, we first consider the numerator and denominator separately.

9.1.1 NUMERATOR ESTIMATE OF VARIANCE

In the analysis of variance, the numerator estimate of σ^2 makes use of the sampling distribution of means. Recall that the *variance* of the sampling distribution of means is the square of the standard error of the mean, σ_M^2. We have also established that

$$\sigma_M^2 = \frac{\sigma^2}{N} \tag{9.1}$$

That is, the variance of the sampling distribution of means and the variance of the population are related in a known way. The relationship is illustrated graphically on the next page in Fig. 9.1.

The point to be stressed about Fig. 9.1 is the relative size of σ and σ_M. Since $\sigma_M = \sigma/\sqrt{N}$ it follows that for samples of size 2 and larger, σ_M is smaller than σ. In this example $N = 4$, so σ_M is half the size of σ.

FIGURE 9.1
Relationship of σ^2 and
σ_M^2. Population shown
at top, sampling dis-
tribution of means on
bottom. Relative size
of σ and σ_M shown for
$N = 4$.

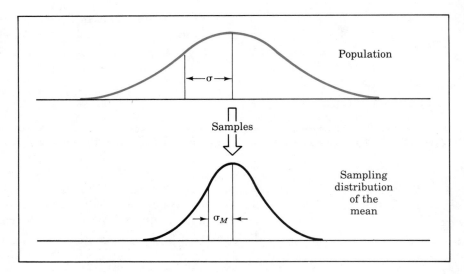

$$\sigma_M = \frac{\sigma}{\sqrt{N}} = \frac{\sigma}{\sqrt{4}} = \frac{\sigma}{2}$$

How are these facts used in the analysis of variance? Assume an experiment is conducted and J means are computed. As we have stated, it is unlikely that these means will be identical. That is, provided that $\sigma^2 > 0$, there should be some variance among these observed means. This is precisely the fact used to obtain an estimate of σ^2. How? By using the observed variance of the means as an estimate of the variance of the sampling distribution of means, σ_M^2. We label this estimate s_M^2. We then take advantage of the relation between σ_M^2 and σ^2 by substituting s_M^2 for σ_M^2 in Equation 9.1. This gives us

$$s_M^2 = \frac{\sigma^2}{N} \tag{9.1A}$$

Finally, we get an estimate of σ^2 by solving Equation 9.1A for σ^2:

$$\text{est. } \sigma^2 = N s_M^2 \tag{9.1B}$$

Suppose, for example, that three groups of eight subjects each yield the following means: 3.50, 4.25, 5.00. The mean of these means is 4.25, and the variance of these means, using Equation 2.4, is

$$s_M^2 = \frac{(3.50 - 4.25)^2 + (4.25 - 4.25)^2 + (5.00 - 4.25)^2}{3 - 1} = \frac{1.125}{2} = .5625$$

To reiterate, we have used the variance of our observed means as an estimate of the variance of the sampling distribution of the means. It is very important to notice that, in our calculation of the variance of these three means, the denominator is equal to the number of "scores" minus 1. In this case, the "scores" are three means. Recall from Chapter 2 that we distinguished between calculations of population variance and estimates of population variance. In the former case we divide by the total number of cases, in the latter by the number minus one. In the current application of the variance formula, we are *estimating* a variance, as opposed to computing a population variance, so the appropriate denominator is the number of scores minus 1, or $3 - 1 = 2$.

Now to use these three measures as an estimate of σ^2, using Equation 9.1B:

$$\text{est.}\sigma^2 = Ns_M^2 = 8(.5625) = 4.5$$

In *this* case please note that N refers to the number of observations in each group. We must be careful not to confuse our Ns. The N in Equation 9.1B refers to the number used to calculate each mean. However, when we calculate the variance of the means, it is the number of *means* that determined the denominator N. We symbolize the number of means as J.

In summary, then, our first estimate of σ^2 comes from the variance of the observed means and takes advantage of the known relation between σ^2 and σ_M^2. This estimate is the numerator estimate of σ^2 in the F ratio of the analysis of variance.

9.1.2 DENOMINATOR ESTIMATE OF VARIANCE

In general, the denominator estimate of σ^2 is a pooled estimate. You will recall from Chapter 7 (Equation 7.3) that a pooled estimate of σ^2 involves combining several independent estimates of the same σ^2. The denominator estimate of σ^2 involves similar pooling. This is most easily seen when a between-subject analysis of variance is computed. Suppose, for example, that the variances of the three groups discussed in the previous section are 1.14, 1.00, and 1.42. Under H_0 these three samples come from the same population, so each s^2 is an estimate of the same population σ^2. Since the three are based on the same number of subjects, a pooled estimate of σ^2, using Equation 7.4, would be a simple average:

$$\text{est. } \sigma^2 = \frac{1.14 + 1.00 + 1.42}{3} = \frac{3.56}{3} = 1.19$$

Notice that this estimate is smaller than the estimate based on the means.

In summary, then, two estimates of σ^2 are obtained: one based on the variance *between means* of groups or conditions, and another based on a pooled variance *within* groups or conditions. When the population distribution is normal, these two estimates are also independent, in the sense defined in Chapter 3. This comes from the fact, discussed in Chapter 8, that means and sample variances drawn from the same normal population are independent. How do we use these facts to test among means? Recall that the null hypothesis is that all means are based on samples from the same population. When H_0 is true and the population is normally distributed, it is known that the ratio of the estimates is distributed as F. The F statistic, then, becomes the test statistic for the test of H_0 in the analysis of variance.

9.1.3   WHEN THE NULL HYPOTHESIS IS FALSE

We have seen that when H_0 is true, the expected value of F (the mean of the sampling distribution of F) is approximately 1. When H_0 is not true, however, the $E(F)$ is greater than one. To see this, consider an example when H_0 is false. Figure 9.2 illustrates a situation in which a group is sampled from each of three different populations.

In this case, $\mu_1 < \mu_2 < \mu_3$ even though $\sigma_1^2 = \sigma_2^2 = \sigma_3^2 = \sigma^2$. The top part of Fig. 9.2 shows the three population distributions. The middle part shows the separate sampling distribution of means from each of these three populations for some arbitrary value of N. Notice that the expected value of each sampling distribution is the mean of its respective population distribution, and recall that the standard deviation of each sampling distribution is $\sigma/\sqrt{N}$. It is important to contrast the sampling distribution of means when H_0 is false to the sampling distribution when it is true.

Consider first that H_0 is true. If H_0 is true, only one of these distributions would be needed to show the sampling distribution of means. Suppose, for example, that H_0 is true and that all samples are drawn from the population labelled B in Fig. 9.2. The sampling distribution of means of samples taken just from B will be distributed as shown in b just below B. That b distribution has a mean equal to μ_2 and a certain variance. The major point here is that the dispersion of observed means will be larger when H_0 is false than when it is true. To show this we next consider that H_0 is false. The lower part of Fig. 9.2 shows what the sampling distribution of the means will look like when H_0 is false and the means come from A, B, and C. Notice that the variance of the means is much larger when H_0 is false (bottom distribution in Fig. 9.2) than when it is true (distribution b). It follows

FIGURE 9.2
A sampling distribution of means when H_0 is false.

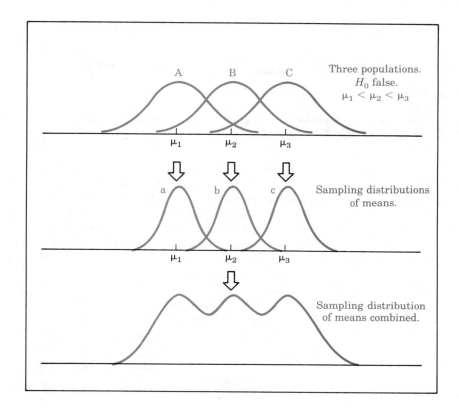

that the numerator estimate of σ^2, based on the observed variance of sample means, is expected to be larger when H_0 is false than when it is true. From this it follows that F is expected to be larger when H_0 is false than when H_0 is true. This is the reason it was stated in Chapter 8 that in the analysis of variance, the experimental hypothesis is that the numerator estimate of σ^2 will be larger than the denominator estimate of σ^2 and that the F test is a one-tailed test.

The point made with Fig. 9.2 may also be stated verbally. The variance among group or condition means may be said to have two components: variance present in the population of measurements, and variance, if any, due to or associated with the independent variable. The numerator estimate of variance is based on group means. The denominator estimate of variance, on the other hand, is not based on group means and is not influenced by the variance attributable to the independent variable. If the null hypothesis is false, the numerator of F is expected to be larger than the denominator because only the numerator is increased by the effect of the independent variable.

We will now consider the specific computational formulas for the analysis of variance. Between-subject variables will be treated first, followed by within-subject variables.

9.2

BETWEEN-SUBJECT VARIABLES

Two methods are considered here for performing an analysis of variance on J groups of scores. Both are mathematically equivalent but differ in the computational resources they require. The first is suitable when sample variances can be computed efficiently, as they can on some preprogrammed calculators that display the standard deviation of a set of previously entered scores with a single button press. The second method could be done without any calculational aids, although an easy method for summing squares would be helpful. The second method uses the raw-score computational formula commonly presented in statistical textbooks.

Although readers' computational resources vary, it is recommended that all readers review both methods. The first method offers a clearer picture of the concepts involved without getting bogged down in computational detail. On the other hand, the second method more clearly reveals how a set of scores gets transformed into a statistical test.

We begin with an example. Assume an experimenter has collected data from three groups of five subjects each. The three groups have been treated differently and the experimenter wants to know whether the different treatment produces statistically different scores. The individual data and some descriptive statistics are reported in Table 9.1.

TABLE 9.1

Scores of 15 subjects in three groups of five each.

	Group		
	1	**2**	**3**
	0	1	2
	1	2	3
	1	2	4
	3	3	5
	3	4	5
M	1.6	2.4	3.8
s^2	1.8	1.3	1.7

We see the means range from 1.6 to 3.8 and wish to determine whether any difference is statistically significant.

9.2.1 *F* RATIOS FROM VARIANCE ESTIMATES

Section 9.1 described the numerator estimate of variance as an estimate stemming from the variance of the observed means. It was also shown that

$$\text{numerator est. } \sigma^2 = N s_M^2$$

where s_M^2 is the variance of the observed J means and N is the number of observations in each sample. In the present example, the means are 1.6, 2.4, and 3.8. The variance of these three scores (using Equation 2.5, calculations not shown here) is 1.24. Substituting appropriately:

$$\text{numerator est. } \sigma^2 = N s_M^2 = (5)(1.24) = 6.20$$

The denominator estimate is a pooled estimate of σ^2. When we have a between-subject variable with J means, we may obtain a pooled estimate as described in Chapter 7. There it was shown (Equation 7.3) that

$$\text{pooled est. } \sigma^2 = \frac{(N_1 - 1)s_1^2 + (N_2 - 1)s_2^2 + \ldots + (N_J - 1)s_J^2}{(N_1 - 1) + (N_2 - 1) + \ldots + (N_J - 1)}$$

When the sample sizes are equal, as they are in the present case, i.e., $N_1 = N_2 = N_3 = N$, Equation 7.3 may be simplified:

$$\text{denominator est. } \sigma^2 = \frac{(N - 1)(s_1^2 + s_2^2 + \ldots + s_J^2)}{J(N - 1)}$$
$$= \frac{s_1^2 + s_2^2 + \ldots + s_J^2}{J}$$

In our example

$$\text{denominator est. } \sigma^2 = \frac{1.8 + 1.3 + 1.7}{3} = 1.60$$

We are now prepared to compute F.
In general,

$$F = \frac{\text{numerator est. } \sigma^2}{\text{denominator est. } \sigma^2}$$

In our particular case,

$$F = \frac{6.20}{1.60} = 3.875$$

Is this F significant? To see, we need first to consider degrees of freedom and then consult the appropriate tabled values of F.

What are the degrees of freedom for the denominator and numerator? We have learned that the degrees of freedom for a variance estimate based on N scores is $N - 1$. The numerator estimate is based on J means. Therefore,

$$df_{\text{numerator}} = J - 1$$

We have also learned that the degrees of freedom for a pooled estimate of variance is the denominator of Equation 7.3, or in the present case where $N_1 = N_2 = \ldots = N_J = N$

$$df_{\text{denominator}} = J(N - 1)$$

In our example

$$df_{\text{numerator}} = J - 1 = 3 - 1 = 2$$
$$df_{\text{denominator}} = J(N - 1) = 3(5 - 1) = 12$$

so $F(2, 12) = 3.875$.

Is this F significant at $\alpha = .05$? We enter Table B in Appendix A and find the critical value for $F(2, 12)$ to be 3.89. Since the obtained value of F is not equal to or greater than the critical value, we are unable to reject the null hypothesis of no difference. The observed differences are not statistically significant.

Before we move to raw-score formulas, a few comments about notation are in order. In the experimental literature, an estimate of variance in the analysis of variance is conventionally referred to as a **mean sum of squares**, abbreviated $\textbf{MSS}$ (or, in some texts, MS). The numerator estimate in a between-subject analysis is referred to as the **between group mean sum of squares**, abbreviated here as $\textbf{MSS}_G$. The denominator estimate is referred to as the **within group mean sum of squares**, abbreviated here as $\textbf{MSS}_{WG}$. Translating our notation:

$$\text{numerator est. } \sigma^2 = MSS_G$$
$$\text{denominator est. } \sigma^2 = MSS_{WG}$$
$$F = \frac{MSS_G}{MSS_{WG}}$$

Computational formulas in the next section will be written in this more conventional notation.

9.2.2 F RATIOS FROM RAW-SCORE FORMULAS

This section describes the computational steps needed to perform a between-subject analysis of variance on J groups of scores without using a preprogrammed calculator or a computer program. It is also applicable to the results of experiments with unequal group sizes. The calculations of F shown in the previous section require equal Ns.

Computation of MSS requires first the computation of sums of squares (SS), since in general

$$MSS = \frac{SS}{df}$$

and in particular

$$MSS_G = \frac{SS_G}{df_G}$$

$$MSS_{WG} = \frac{SS_{WG}}{df_{WG}}$$

Calculating SS requires the introduction of more new notation. Consider the following: $\Sigma\Sigma X_{ij}$. In this notation, an individual score X is subscripted both with an i and a j. The second subscript, j, refers to the *group* the score comes from ($j = 1, 2, 3, \ldots, J$) and the first subscript, i, refers to the number of the score within the group. Thus, in general, X_{ij} refers to the ith score in the jth group. For example, X_{23} refers to the second score in the third group, X_{78} refers to the seventh score in the eighth group, etc. In Table 9.1, X_{11} has value 0, X_{42} has value 3, and X_{53} has value 5, etc.

The double summation sign ($\Sigma\Sigma$) indicates that the scores are to be summed over both i and j. In other words, the scores are summed within each group and then the sums of all groups are totalled. $\Sigma\Sigma X_{ij}$, then, refers to the sum of all scores.

In general,

$$\Sigma\Sigma X_{ij} = X_{11} + X_{21} + \ldots + X_{N_11} + X_{12} + \ldots$$
$$+ X_{N_22} + \ldots + X_{1J} + \ldots + X_{N_JJ}$$

Notice that N_j is used to refer to the number of scores in group j. For the data in Table 9.1

$$\Sigma\Sigma X_{ij} = 0 + 1 + \ldots + 3 + 1 + \ldots + 4 + \ldots + 2 + \ldots + 5 = 39$$

Now that the double summation sign is introduced, it should be clear that $\Sigma\Sigma X_{ij}^2$ requires all scores to be *squared* and then summed. For the data in Table 9.1

$$\Sigma\Sigma X_{ij}^2 = 0^2 + 1^2 + \ldots + 5^2 = 133$$

Again, care should be taken not to confuse $\Sigma\Sigma X_{ij}^2$ with $(\Sigma\Sigma X_{ij})^2$. $(\Sigma\Sigma X_{ij})^2$ requires all scores to be summed and then the sum to be squared. For instance, $(\Sigma\Sigma X_{ij})^2 = (39)^2 = 1521$. But $\Sigma\Sigma X_{ij}^2 = 133$. $\sum_j[(\sum_i X_{ij})^2/N_j]$ requires special attention. The operations needed here are *first* to sum the N_1 scores within the first group; second, *square* this sum; third, *divide* this by N_1 (the number of scores that are summed); repeat this for all J groups; and finally, total across the groups. ($\sum_i$ means to sum over i, $\sum_j$ means to sum over j).

In cases where group sizes are equal, $N_1 = N_2 = \ldots = N_J = N$, this sum may be simplifed as

$$\sum_j \frac{(\sum_i X_{ij})^2}{N_j} = \frac{1}{N}\sum_j(\sum_i X_{ij})^2$$

For example, in the Table 9.1 data

For group 1 $(\sum_i X_{i1})^2 = (0 + 1 + 1 + 3 + 3)^2 = 8^2 = 64$

For group 2 $(\sum_i X_{i2})^2 = (1 + 2 + 2 + 3 + 4)^2 = 12^2 = 144$

For group 3 $(\sum_i X_{i3})^2 = (2 + 3 + 4 + 5 + 5)^2 = 19^2 = 361$

so

$$\frac{1}{N}\sum_j(\sum_i X_{ij})^2 = \frac{1}{5}(64 + 144 + 361) = \frac{569}{5} = 113.8$$

SS_T refers to *sum of squares for total*, and it may be thought of as the numerator of an estimate of σ^2, if the estimate is based on all scores. Its relation to other sums of squares is:

$$SS_T = SS_G + SS_{WG}$$

N_T is used to refer to the total number of scores, and it is the sum of the total numbers of scores across groups

$$N_T = \sum_j N_j$$

In the example data

$$N_T = \sum_j N_j = N_1 + N_2 + N_3 = 5 + 5 + 5 = 15$$

We are now ready for the computational formulas. They are presented in Table 9.2.

TABLE 9.2
Computational formulas for a one-way between-subject analysis of variance.

$$SS_T = \Sigma\Sigma X_{ij}^2 - \frac{(\Sigma\Sigma X_{ij})^2}{N_T}$$

$$SS_G = \Sigma_j \frac{(\Sigma_i X_{ij})^2}{N_j} - \frac{(\Sigma\Sigma X_{ij})^2}{N_T}$$

$$SS_{WG} = \Sigma\Sigma X_{ij}^2 - \Sigma_j \frac{(\Sigma_i X_{ij})^2}{N_j}$$

or

$$SS_{WG} = SS_T - SS_G$$

$$MSS_T = \frac{SS_T}{df_T}$$

$$MSS_G = \frac{SS_G}{df_G}$$

$$MSS_{WG} = \frac{SS_{WG}}{df_{WG}}$$

$$F(df_G, df_{WG}) = \frac{MSS_G}{MSS_{WG}}$$

$$\text{where } df_T = N_T - 1$$

$$df_G = J - 1$$

$$df_{WG} = \Sigma_j(N_j - 1) = N_T - J$$

We now apply the formulas to the data in Table 9.1.

TABLE 9.3
Analysis of variance on data in Table 9.1.

$$SS_T = \Sigma\Sigma X_{ij}^2 - \frac{(\Sigma\Sigma X_{ij})^2}{N_T} = 133 - \frac{1521}{15} = 133 - 101.4 = 31.6$$

$$SS_G = \frac{1}{N_j}\Sigma_j(\Sigma_i X_{ij})^2 - \frac{(\Sigma\Sigma X_{ij})^2}{N_T} = 113.8 - 101.4 = 12.4$$

$$SS_{WG} = 31.6 - 12.4 = 19.2$$

$$MSS_G = \frac{SS_G}{J - 1} = \frac{12.4}{2} = 6.2$$

$$MSS_{WG} = \frac{SS_{WG}}{N_T - J} = \frac{19.2}{12} = 1.6$$

$$F(df_G, df_{WG}) = \frac{MSS_G}{MSS_{WG}} = \frac{6.2}{1.6}$$

$$F(2, 12) = 3.875$$

Note that the value of F agrees with the value computed in the previous section.

When analyses of variance are reported, it is customary to report them in a summary table, as shown in Table 9.4.

TABLE 9.4
Analysis of variance summary table.

Source	SS	df	MSS	F
Between G	SS_G	df_G	MSS_G	$\dfrac{MSS_G}{MSS_{WG}}$
Within G	SS_{WG}	df_{WG}	MSS_{WG}	
Total	SS_T	df_T		

For the example data:

Source	SS	df	MSS	F
Between G	12.40	2	6.20	3.875
Within G	19.20	12	1.60	
Total	31.60	14		

The use of computer statistic packages to perform a between-subject analysis of variance is discussed in sections 14.1.4, 14.2.4, and 14.3.4.

9.3

WITHIN-SUBJECT VARIABLES

Between-subject variables involve J separate groups. Within-subject variables, in contrast, involve J observations on each subject; each subject serves under each of the J conditions. Let us reconsider, for example, the data in Table 9.1. In that table, scores are reported for five different subjects in each of three different groups. A total of 15 subjects is involved. Let us rearrange that data as if they involved a within-subject variable, as they would if a total of five subjects had served in each of the three conditions. This rearrangement is shown in Table 9.5.

Table 9.5 is arranged so that each row corresponds to a different subject. For example, the data for subject 1 is in the first row. Subject 1 obtained a score of 0 in the first condition, a 1 in the second condition, and a 2 in the third. Similarly, subject 5 received a score of 3 in the first condition, 4 in the second, and 5 in the third. Notice that the total number of observations (15) is the same as before, except that one-third as many subjects are used. (Our use of the same numbers to illustrate both the within-subject and between-subject analyses may

TABLE 9.5
Scores of five subjects each receiving three treatments.

Subject	Treatment 1	Treatment 2	Treatment 3
1	0	1	2
2	1	2	3
3	1	2	4
4	3	3	5
5	3	4	5
M	1.6	2.4	3.8
s^2	1.8	1.3	1.7

be misleading. Remember that these data are fictitious. After actual data are collected, *you cannot arbitrarily choose* which analysis to use. If different subjects serve in the different conditions, the between-subject analysis must be used. If each subject serves in all conditions, the within-subject analysis must be used.)

9.3.1 PARTITIONING SUMS OF SQUARES

The total sums of squares (SS_T) for the between-subject analysis was stated to be composed of SS_G, the between-group sum of squares, and SS_{WG}, the within-group sum of squares.

$$SS_T = SS_G + SS_{WG}$$

Breaking SS_T into these two components is sometimes referred to as *partitioning* SS_T into SS_G and SS_{WG}. The major difference between the computational formulas for between-subject and within-subject analyses of variance is in how SS_T is partitioned. SS_T for within-subject variables is partitioned into the sum of squares *between subjects*, SS_S, and the sum of squares *within subjects*, (SS_{WS}). Specifically,

$$SS_T = SS_S + SS_{WS} \qquad (9.2)$$

This is a very important difference because all analyses of within-subject variables are accomplished by further partitioning of SS_{WS}. The sums of squares contributed by differences among subjects does not play a role in the analysis of the treatment effect. It is as if the SS_S is placed aside. Thus, whatever variation there is among the subjects does not contribute to the estimates of variance used to compute the F ratio. Consider the data in Table 9.1, for example. The subjects are placed in the table in rank order of their scores and there are clear systematic differences among the subjects. Subject 1 has the lowest scores in all conditions, and subject 5 has the highest. In the computation of a within-subject analysis of variance, this systematic variation among the subjects does not contribute to the estimates of σ^2. It is as if the estimates are not "contaminated" by individual differences. Why is this important? We have shown that variance is a source of uncertainty. The larger it is, the greater the uncertainty and the less the power of the resulting statistical test. Thus, a major advantage of the within-subject analysis of variance, in comparison to the between-subject analysis of variance, is that it provides a kind of statistical control over individual differences as sources of variance. The data in Tables 9.1 and 9.5 will provide a striking example of this because, as we shall see, the same numbers lead to different conclusions about H_0 when analyzed in each of the two ways. There are, however, some disadvantages to within-subject analyses of variance, as we shall see after we examine the computational formulas for within-subject analyses of variance.

9.3.2   RAW-SCORE COMPUTATIONAL FORMULAS

The term $\Sigma\Sigma X_{ij}$ has the same meaning as before, except that now i refers to the ith subject and j refers to the jth condition. Since all subjects serve under all conditions, the numbers of observations in each condition are the same. There are N subjects and J treatments. Therefore, the total number of scores is N_T, $N_T = NJ$.

SS_T is computed exactly as before,

$$SS_T = \Sigma\Sigma X_{ij}^2 - \frac{(\Sigma\Sigma X_{ij})^2}{N_T}$$

So also is the sum of squares due to "groups," except now it is referred to as conditions, SS_C

$$SS_C = \frac{1}{N}\sum_j(\sum_i X_{ij})^2 - \frac{(\Sigma\Sigma X_{ij})^2}{N_T}$$

Our new term, *sum of squares between subjects, SS_S*, is calculated in a manner analogous to SS_C. SS_C is obtained by summing each column, squaring the sum, dividing by the number of scores, and then summing across the columns. For SS_S we perform the same operations, except the operations are performed across the *rows*. Mathematically,

$$SS_S = \frac{1}{J}\sum_i(\sum_j X_{ij})^2 - \frac{(\Sigma\Sigma X_{ij})^2}{N_T}$$

For example, for the data in Table 9.5

$$\text{For subject 1} \quad (\sum_j X_{1j})^2 = (0 + 1 + 2)^2 = 3^2 = 9$$

$$\text{For subject 2} \quad (\sum_j X_{2j})^2 = (1 + 2 + 3)^2 = 6^2 = 36$$

$$\text{For subject 3} \quad (\sum_j X_{3j})^2 = (1 + 2 + 4)^2 = 7^2 = 49$$

$$\text{For subject 4} \quad (\sum_j X_{4j})^2 = (3 + 3 + 5)^2 = 11^2 = 121$$

$$\text{For subject 5} \quad (\sum_j X_{5j})^2 = (3 + 4 + 5)^2 = 12^2 = 144$$

$$\frac{1}{J}\sum_i(\sum_j X_{ij})^2 = \frac{1}{3}(9 + 36 + 49 + 121 + 144)$$

$$\frac{1}{J}\sum_j(\sum_j X_{ij})^2 = \frac{1}{3}(359) = 119.667$$

$$SS_S = \frac{1}{J}\sum_i(\sum_j X_{ij})^2 - \frac{(\sum\sum X_{ij})^2}{N_T}$$

$$SS_S = 119.667 - \frac{1521}{15} = 18.267$$

We have stated that the SS associated with conditions is part of SS_{WS}. More precisely,

$$SS_{WS} = SS_C + SS_{\text{residual}} \tag{9.3}$$

Equation 9.3 states that the SS_{WS} has two components, one due to conditions and one due to the remainder, called *residual* and abbreviated *res*. The residual SS may be thought of as what is left after SS due to subjects and SS due to conditions are removed from the total SS. The residual MSS may then be calculated by dividing by the residual df. The residual MSS is used to estimate σ^2; it provides the denominator for the F test of condition means. That is,

$$F(df_C, df_{\text{res}}) = \frac{MSS_C}{MSS_{\text{res}}}$$

In other words, the denominator for the F test on the condition means is an estimate of σ^2 obtained by removing from SS_T all SS attributed to subjects and to conditions.

The degrees of freedom are partitioned the same way:

$$df_T = df_S + df_{WS}$$

$$df_{WS} = df_C + df_{res}$$
$$df_{res} = df_{WS} - df_C = N(J - 1) - (J - 1) = (N - 1)(J - 1)$$

These computational formulas are summarized in Table 9.6.

TABLE 9.6
Computational formulas for within-subject analysis of variance on J treatments.

$$SS_T = \Sigma\Sigma X_{ij}^2 - \frac{(\Sigma\Sigma X_{ij})^2}{N_T}$$

$$SS_S = \frac{1}{J}\Sigma_i(\Sigma_j X_{ij})^2 - \frac{(\Sigma\Sigma X_{ij})^2}{N_T}$$

$$SS_{WS} = SS_T - SS_S$$

$$SS_C = \frac{1}{N}\Sigma_j(\Sigma_i X_{ij})^2 - \frac{(\Sigma\Sigma X_{ij})^2}{N_T}$$

$$SS_{res} = SS_{WS} - SS_C$$

$$MSS_C = \frac{SS_C}{df_C}$$

$$MSS_{res} = \frac{SS_{res}}{df_{res}}$$

$$F(df_C, df_{res}) = \frac{MSS_C}{MSS_{res}}$$

$$df_T = NJ - 1$$

$$df_S = N - 1$$

$$df_{WS} = N(J - 1)$$

$$df_C = J - 1$$

$$df_{res} = (N - 1)(J - 1)$$

Table 9.7 shows the analysis of variance on the Table 9.5 data.

TABLE 9.7
Analysis of variance on the data from Table 9.5.

$$SS_T = \Sigma\Sigma X_{ij}^2 - \frac{(\Sigma\Sigma X_{ij})^2}{N_T} = 133 - \frac{1521}{15} = 133 - 101.4 = 31.6$$

$$SS_S = \frac{1}{J}\Sigma_i(\Sigma_j X_{ij})^2 - \frac{(\Sigma\Sigma X_{ij})^2}{N_T} = 119.667 - 101.4 = 18.267$$

$$SS_{WS} = SS_T - SS_S = 31.6 - 18.267 = 13.333$$

$$SS_C = \frac{1}{N}\Sigma_j(\Sigma_i X_{ij})^2 - \frac{(\Sigma\Sigma X_{ij})^2}{N_T} = 113.8 - 101.4 = 12.4$$

$$SS_{res} = SS_{WS} - SS_C = 13.333 - 12.4 = .933$$

TABLE 9.7
(continued)

$$MSS_C = \frac{SS_C}{J - 1} = \frac{12.4}{2} = 6.2$$

$$MSS_{res} = \frac{SS_{res}}{(N - 1)(J - 1)} = \frac{.933}{(4)(2)} = \frac{.933}{8} = .117$$

$$F(df_C, df_{res}) = \frac{MSS_C}{MSS_{res}}$$

$$F(2, 8) = \frac{6.2}{.117} = 53$$

The summary table is shown in Table 9.8.

TABLE 9.8
Analysis of variance summary table for within-subject analysis of variance.

Source	SS	df	MSS	F
Total	SS_T	$NJ - 1$		
Between S	SS_S	$N - 1$		
Within S	SS_{WS}	$N(J - 1)$		
Condition	SS_C	$J - 1$	MSS_C	$\dfrac{MSS_C}{MSS_{res}}$
Residual	SS_{res}	$(N - 1)(J - 1)$	MSS_{res}	
Source	**SS**	**df**	**MSS**	**F**
Total	31.6	14		
Between S	18.267	4		
Within S	13.333	10		
Condition	12.4	2	6.2	53
Residual	.933	8	.117	

We see in tables 9.7 and 9.8 that $F(2, 8) = 53$. Consulting Table B in Appendix A, we find that the critical value for $F(2, 8)$, $\alpha = .05$, is 4.46. Since the observed value exceeds the critical value, we may reject the null hypothesis of no difference.

The use of computer statistics packages to perform a within-subject analysis of variance is discussed in sections 14.1.4, 14.2.4, and 14.3.4.

9.3.4 COMPARISON OF BETWEEN-SUBJECT ANALYSES AND WITHIN-SUBJECT ANALYSES

In the example used in this chapter, the same data led to different conclusions when analyzed by between-subject and within-subject analyses. When the 15 scores were treated as if they had originated from three independent groups of subjects, the group means were not significantly different. However, when the same 15 scores were treated as three scores on each of five subjects, the means were significantly different. This illustrates a major difference between the two types of analyses. By removing individual differences as a source of variance in the denominator of the F, a test that was not significant became significant. The within-subject analysis had more *power*. This is not to suggest that a within-subject analysis will always be more powerful. The difference in power holds in the present case not only because there are individual differences but also because the data exhibit another property. If you examine Table 9.5 closely, you will see those subjects that score low under one condition, relative to the other subjects, also tend to score low on the other tests. For example, subject 1 scores lowest on all three tests. Another way to state this is to say that scores on one condition are related to scores on another. If the data did not have this property, the benefit of the within-subject analysis would disappear. The reader could verify this point by taking the same data and scrambling the scores within the conditions. The condition means would be unchanged, but the F almost certainly would be reduced. By randomly rearranging the scores within the conditions, one destroys the relation across the rows (subjects). Randomly rearranging the scores within the conditions renders the scores in the various rows independent, as they would be if different rows represented different subjects (as in the between-subject analysis).

If the scores across the conditions are *not* related, a within-subject analysis could have *less* power than a between-subject analysis that has the same number of observations. Notice that removing the between-subject sum of squares also results in reducing the number of degrees of freedom in the denominator of the F. In the present example the denominator df was 12 in the between-subject analysis of variance but was reduced to 8 in the within-subject analysis of variance. The resulting loss of power is seen by the increase in the critical value of F. For the between-subject analysis, the critical value for F was 3.89. However, for the within-subject analysis, the critical value was 4.46. An increase in the critical value is a decrease in power because a larger denominator MSS is required, and a larger denominator MSS occurs only when the differences among the means are larger. In the present case, the loss of power brought about by a reduced df was more than offset by the presence of a relation among the conditions.

In addition to these statistical considerations, there are nonstatistical factors that influence the experimenter's choice of design. For example, the within-subject design is usually less time-consuming, because fewer subjects are required for the same number of scores and often each subject can be observed under all conditions in a single experimental session. For example, in the three-condition design illustrated in the present chapter, the between-subject version required 15 subjects and the within-subject version, which resulted in the same number of observations, needed only five subjects. When subject availability is a problem, the within-subject design becomes attractive. On the other hand, the within-subject design is to be discouraged when the effects of one condition carry over to the next. Suppose, for example, an experimenter is interested in how the speed of solving a problem is influenced by different test environments. A within-subject design could not be used. Once the subject knows the problem solution, it would be pointless to present it again as required by the within-subject design. Keppel (1973) discusses the relative advantages and disadvantages of the two designs in considerable detail.

9.4

PAIRWISE COMPARISONS AMONG MEANS

If F exceeds the critical value, what exactly can be concluded? We may reject the null hypothesis that there are no differences among the means, but where, precisely, does that leave us? We conclude that a difference is present, but where is it? For example, the within-subject F was significant on a test of means with values 1.6, 2.4, and 3.8. But there are at least three differences we could consider: 1.6 versus 3.8, 1.6 versus 2.4, and 2.4 versus 3.8. The F is sometimes referred to as an "omnibus" test—it simultaneously tests all differences. This ability to test all differences at the same time is a major strength of the analysis of variance. However, a significant F is ambiguous, because it does not indicate where the difference resides.

In order to locate the source of the difference, statistical comparisons among the means are required. And, of course, there are a number of possible comparisons. In an experiment with three means, there are three possible pairwise comparisons. In an experiment with four means, there are six comparisons; with five means there are ten; and with six means there are fifteen comparisons.

The appropriate procedure for making comparisons among means depends on whether the comparisons are planned or unplanned. **Planned comparisons** are those the experimenter has considered before the experiment is conducted or before the data are examined. These are the comparisons about which experimental hypotheses have been formed and which the experimenter intends to examine in order to discuss the meaning of the results. For example, a three-group study may involve one control and two experimental groups. In this case, the experimenter may be motivated to compare the control with each of the experimental groups.

Unplanned comparisons, on the other hand, are not conceived in advance. They are often motivated *by* the results and are "after the fact." Unplanned comparisons typically occur when there are a large number of means. For example, as just pointed out, when there are six means, there are fifteen possible comparisons. It is likely that such a study is motivated by an interest to compare a few means, but after the results are in, the experimenter may wish to search for other significant effects. The reason the distinction is important is that although there is general agreement on how to proceed with planned comparisons, considerable controversy surrounds unplanned comparisons. In this book, unplanned comparisons will not be treated. Keppel (1973) provides an excellent treatment of the problem of unplanned comparisons.

In general, planned comparisons are accomplished by *t* tests. Here we consider only *t* tests between pairs of means, although more complicated comparisons are possible. For example, the mean of group 1 could be compared with the mean of groups 2 and 3 combined. Again, Keppel (1973) is an excellent source for treating these more complex comparisons.

9.4.1 BETWEEN-SUBJECT VARIABLES

Recall from Chapter 7 that the denominator for the *t* test between group means is an estimate of the standard error of the difference, and this standard error is a function of the pooled estimate of σ^2:

$$s_{M_1 - M_2} = \sqrt{(\text{pooled est. } \sigma^2)\left(\frac{1}{N_1} = \frac{1}{N_2}\right)}$$

In Chapter 7, the pooled estimate of σ^2 was based on two groups. The only difference between that *t* test and the *t* test proposed here is that the pooled estimate here is based on all *J* groups. According to H_0 all groups provide an estimate of the same σ^2, so we simply improve the quality of the estimate by taking advantage of all the data even though the *t* involves but two of the groups.

What is our pooled estimate of σ^2? The MSS_{WG}. Thus, by substitution, the appropriate *t* for a planned comparison between the mean of two groups, *A* and *B*, becomes

$$t = \frac{M_A - M_B}{\sqrt{MSS_{WG}\left(\frac{1}{N_A} = \frac{1}{N_B}\right)}} \tag{9.4}$$

with $df = N_T - J$.

When all group sizes are equal

$$t = \frac{M_A - M_B}{\sqrt{MSS_{WG} \left(\frac{2}{N}\right)}} \tag{9.5}$$

and the denominator is the same for all other t tests in this analysis.

As an example, suppose the experimenter planned to compare groups 1 and 3 in the data of Table 9.1. Table 9.9 shows the t tests computed on these means.

TABLE 9.9
t tests between two means in Table 9.1.

$$\sqrt{MSS_{WG} \left(\frac{2}{N}\right)} = \sqrt{(1.60) \left(\frac{2}{5}\right)}$$
$$= \sqrt{.64}$$
$$= .8$$

Group 1 versus group 3 $t(12) = \frac{3.8 - 1.6}{.8} = \frac{2.2}{.8} = 2.75$

Critical value of $t(12)$, $\alpha = .05$, two-tailed, is ± 2.179.

Here is an apparent paradox. The overall F was not significant, suggesting that there were no differences among the group means. But the t test between groups 1 and 3 produces a t that exceeds the critical value of 2.179. (This discrepancy should not be confused with the difference between the results of the between-subject and within-subject analyses of variance. Here we are comparing two tests, each treating the data as between-subject data.) Thus, the t supports a conclusion that H_0 is false, at least for group 1 versus group 3. The difference we observe with these data simply illustrates that the overall F and individual t may differ in power. A difference detected by the more powerful may not necessarily be detected by the less powerful. This is part of the reason that most statisticians agree that *planned t* tests between means are appropriate, even when the overall F is insignificant. In this case, then, it would be appropriate to reject the null hypothesis that groups 1 and 3 come from the same population, even though the overall F was not significant.

9.4.2 WITHIN-SUBJECT VARIABLES

Comparisons among means in within-subject designs follow the same principles. The pooled estimate of σ^2 placed in the formula for the standard error of the difference is the MSS that is used in the denominator of the F in the test for condition:

$$t = \frac{M_A - M_B}{\sqrt{MSS_{res}\left(\frac{2}{N}\right)}} \qquad (9.6)$$

Table 9.10 shows how Equation 9.6 is applied to the data in Table 9.5.

TABLE 9.10
t tests between two means in Table 9.5.

$$\sqrt{MSS_{res}\left(\frac{2}{N}\right)} = \sqrt{(.117)\left(\frac{2}{5}\right)}$$

$$= \sqrt{.047}$$

$$= .216$$

Group 1 versus group 3 $t(8) = \dfrac{3.8 - 1.6}{.21} = \dfrac{2.2}{.21} = 10.185$

Critical value of $t(8)$, $\alpha = .05$, is ± 2.306 (two-tailed).

Here the *t* and *F* tests agree.

9.5

SUMMARY

The analysis of variance is used to simultaneously test among any number of sample means. Its calculational form depends on whether the treatment variable is a within-subject variable or a between-subject variable. If all subjects serve under each of the J conditions of the experiment, the design is **within-subject**. In a **between-subject** design, no subject serves in more than one condition: J groups of subjects are used.

Why is a test among means called an analysis of variance? Because the treatment means are used to estimate σ^2 and this estimate is compared, by F test, against another, independent, estimate of σ^2. The null hypothesis is that all means are obtained on samples from the same population. If the null hypothesis is true, and if the population distribution is normal, the test statistic is distributed as F. If the null hypothesis is false, the estimate of σ^2 obtained from the treatment means is expected to be larger than the other estimate.

The between-subject analysis is based on the assumption that all subjects are randomly and independently sampled and that no subject serves in more than one condition. It is also assumed that the samples on which the means are based are taken from normally distributed populations with the same mean and variance.

The F ratio formed in a between-subject analysis of variance has a numerator estimate based on the J treatment means. This estimate is called the **between group mean sum of squares (MSS_G)** and has $(J - 1)$ degrees of freedom. The denominator is based on a pooled estimate of σ^2, using the variances within groups. This denominator is called the **within group mean sum of squares (MSS_{WG})** and has $(N_T - J)$ degrees of freedom, where N_T is the total number of subjects.

Assumptions underlying the within-subject analyses are similar except, of course, it is understood that all subjects serve under each of the J conditions. There are additional assumptions, too technical to be considered here, that can influence the form the F ratio should take in a within-subject design. These matters, some controversial, are discussed in advanced texts (e.g., Keppel, 1973) and the reader should understand that the treatment here avoids the complex issues. The F ratio formed in a within-subject analysis of variance has as numerator the same estimate of σ^2 as does the between-subject analysis of variance, namely an estimate based on the means. In this case, however, it is called the **between condition mean sum of squares (MSS_C)**, and it has $(J - 1)$ degrees of freedom. The denominator estimate of σ^2 is referred to as the **residual mean sum of squares (MSS_{res})**, and it has $(N - 1)(J - 1)$ degrees of freedom, where N refers to the number of subjects.

In designing an experiment, an experimenter often has the option of using either a between-subject or within-subject design. A within-subject design is feasible when all subjects can serve in all conditions. However, a within-subject design is to be discouraged when the effects of one treatment carry over to the next. When the within-subject design is feasible, it is attractive because it is often less time consuming than the between-subject design. In addition, the within-subject design has the advantage that it provides a means of statistical control over variance due to individual differences.

The F test in analysis of variance is an omnibus test that simultaneously tests all mean differences. To locate the source of a significant F, further statistical comparisons are needed. The appropriate procedure for making comparisons among means depends on whether the comparisons are planned or unplanned. **Planned comparisons** are those considered prior the experiment. **Unplanned comparisons** are those motivated by the results and are after the fact. Planned comparisons are accomplished by t tests. However, the denominator of the t, the standard error of the difference, uses the MSS_{WG} (between-subject design) or the MSS_{res} (within-subject design) as an estimate of σ^2.

REFERENCES

Keppel, G. 1973. *Design and analysis of experiments.* Englewood Cliffs, New Jersey: Prentice-Hall.

QUESTIONS

1. Between-subject analysis of variance is to _____ as _____ is to a *t* test for related groups.

2. What is the statistical advantage of using a within-subject design as opposed to a between-subject design? What is a *potential* disadvantage in using a within-subject design.

3. In a within-subject design, what is SS_{res}? What is its counterpart in a between-subject design? What factors contribute to SS_{res}?

4. An experiment is conducted whereby each subject participates in all the different treatment conditions. When the results are analyzed, the researcher is disappointed to find the *F* does not exceed the critical value needed to reject H_0. Consequently, it is decided to recompute the analysis as a between-subject design, because the degrees of freedom in a between-subject design are greater, therefore the critical value of *F* will be smaller. Do you agree or disagree with the researcher's decision? Why or why not?

5. An analysis of variance is a method of testing differences among means. Explain the reasoning that allows a comparison of variances to make decisions about means. Discuss how the numerator and denominator of the *F* ratio are calculated and what one expects to find under H_0.

6. A between-subject analysis of variance should be computed only if three assumptions are met. List them.

7. Express each of the following, using mathematical notation:

   (A) Sum all scores and square the total.
   (B) Square all scores and sum the total.
   (C) The mean of all squared scores.
   (D) The mean of all scores squared.
   (E) The sum of the mean squared scores.

8. An experimental psychologist is interested in four methods that may be used to improve memory for pictures. Data are collected on four memory techniques wherein three subjects are randomly assigned to each technique. Hence, there is a total of 12 subjects:

Technique A: 46, 46, 43
Technique B: 50, 43, 42
Technique C: 49, 49, 46
Technique D: 51, 52, 47

Determine whether the differences among these means are significant or whether they are attributable to chance. Use $\alpha = .05$.

9. "Right-hook" Gutterman is at it again. Ever on the lookout to improve his bowling game, "Righty" decides to try bowling with a lighter ball and a heavier ball. He decides to bowl six games with the lighter ball, six with the heavier ball, and compare the scores with six games he has rolled with his regular ball. Righty has a big tournament coming up and does not have enough time to finish. Based on the scores below, test the null hypothesis that ball weight makes no difference in Righty's scores. Plan to make comparisons among means in order to find significant differences. (For this analysis treat the observations as if they were independently sampled from the same population.)

Light Ball	Regular Ball	Heavy Ball
180	179	178
164	178	161
182	241	165
165	190	
157	186	
196	208	

10. Twelve students are randomly assigned to three conditions of a perceptual/motor experiment where ambient illumination is the independent variable and subject reaction time is the dependent variable. The scores that follow are in tenths of a second. Test the hypothesis that reaction time does not vary as a function of illumination. In other words, test H_0 that there is no difference among groups. Use $\alpha = .05$.

Bright light: 11, 10, 8, 15
Medium light: 14, 11, 16, 19
Dim light: 8, 7, 12, 13

11. The "Over-haul" trucking company is again testing fuel brands. The following are the miles per gallon that a test driver got for four tankfuls each of five brands of fuel. Use $\alpha = .05$ to test whether the differences among the five sample means can be attributed to chance.

Brand V	Brand W	Brand X	Brand Y	Brand Z
22	18	21	20	21
16	23	21	17	16
21	21	24	20	18
17	22	26	19	17

12. Speedy Gonzales Rogers is a former track star who still likes to run three miles per day. Speedy gets bored easily and has chosen four different routes which are three miles long but vary as to the type of terrain. The following are the number of minutes in which he timed himself on five different days for each route. Test the hypothesis that there are no differences among the average times needed to run each route. Use $\alpha = .05$. (For this analysis, assume each observation is independent).

    Route 1: 24, 24, 21, 22, 20
    Route 2: 27, 24, 27, 23, 20
    Route 3: 23, 20, 22, 21, 19
    Route 4: 25, 19, 19, 20, 16

13. Suppose you are interested in how geographical instruction might influence the time required to answer questions about locations. Using a computer, you collect reaction times for questions presented to 15 subjects. Subjects are randomly assigned to one of three pretest instructional conditions. Subjects either study a map for five minutes, are told to imagine a map, or are not given any instruction (a control group). Test the hypothesis that there are no differences in mean reaction times among groups. Use $\alpha = .05$. If you can reject the null hypothesis, perform comparisons to determine which means are significantly different.

                    Reaction Time (msec)
    Study map: 688, 713, 715, 723, 726
    Imagine map: 821, 796, 822, 796, 820
    No instructions: 895, 900, 894, 930, 906

14. Three groups of rats are used in this experiment. Each rat is injected with .2 mg, .8 mg, or 1.2 mg of a new tranquilizer. The following data represent the number of seconds needed to fall asleep:

    .2 mg: 18, 21, 18, 20, 16, 21, 22, 20
    .8 mg: 14, 15, 16, 17, 15, 13, 17, 15, 14, 18, 12, 14
    1.2 mg: 10, 14, 11, 8, 11, 12, 13, 9

    Use $\alpha = .05$ to test the null hypothesis that differences in dosage have no effect. If you reject H_0, determine where there are significant differences.

15. Thirty-one students were chosen at random to form five different biology courses. Each student was given the same biology test and the scores are reported as follows. Perform an analysis of variance to test the null hypothesis that the students from each class are equally knowledgeable about biology. Use $\alpha = .05$.

    Classroom A: 84, 95, 92, 64, 81, 86, 95, 72
    Classroom B: 74, 58, 94, 79, 57, 84, 70, 55, 69

Classroom C: 76, 57, 70, 88, 74, 50, 73, 79
Classroom D: 67, 61, 86, 42, 49, 63

If you can reject the null hypothesis, perform t tests to find which classroom averages are significantly different.

16. Dorf motor company has designed a new automobile suspension system. The time required to travel over a course under various road conditions was recorded in minutes. Use $\alpha = .05$ to test whether the speed of the automobile is affected by various road conditions.

Speed

Rough road conditions: 25, 22, 19, 20
Moderate road conditions: 26, 28, 21, 30, 23, 28
Smooth road conditions: 31, 29, 28, 34, 26

17. Dr. Gregor Zilstein (famous yet fictitious psychologist) has long put stock in the notion that people prefer beverages of particular colors. He suspects that people generally prefer brown (cola-colored) beverages over all others. Dr. Zilstein decided to test his theory by having 25 people taste beverages of different colors. Unknown to the subjects, Dr. Zilstein had placed tasteless food coloring into water in order to provide multicolored beverages. The scores from the questionnaire are presented beside each color group. Use $\alpha = .05$ to determine whether there is any difference in preference of beverage according to color. Remember each score represents a unique subject.

Brown: 47, 34, 38, 34, 42
Red: 39, 36, 48, 46, 36
Yellow: 37, 39, 50, 48, 37
Green: 38, 40, 52, 38, 47
Blue: 44, 41, 42, 47, 53

How could Dr. Zilstein have conducted this experiment to control for variance among subjects?

18. Nyberg and Shadick designed an experiment to examine a basic educational premise: learning by doing leads to better understanding than learning by memorizing. They assigned subjects to one of three possible conditions: in the active condition, subjects were shown a picture of a complexly-folded paper airplane. They were then given a piece of paper and told to make one. In the instructed active condition, subjects did the same thing, but they also got a set of printed instructions for the folding. In the memorize condition, subjects did not see the picture; they were given the printed instructions and told to memorize them. One week later, all participants returned and were instructed to make the plane as quickly as they could. The amount of time required to fold the paper was recorded for each subject:

Reaction Time (sec)

Active	Instructed Active	Memorize
143	121	68
132	106	75
126	93	82
98	104	61
137	115	53
118	110	97

Analyze these results. What does this evidence imply about the effectiveness of active learning in this task?

19. The Air Force wanted to study the effects of fatigue on the ability of pilots to respond to visual stimuli. Subjects sat at the controls of a simulated fighter plane and were instructed to push a button with their left hand if a green light appeared on the screen, and to push a button with their right hand if a red light appeared. Subjects remained at the controls for three hours, the data presented below is the average reaction time for the first 10 minutes of each hour.

Reaction Time (sec)

Subject	1	2	3
1	.218	.239	.407
2	.526	.510	.634
3	.447	.489	.528
4	.262	.243	.281
5	.335	.349	.487
6	.198	.213	.256
7	.429	.410	.492

Compute the appropriate F. Is reaction time affected by the amount of time the subjects are at the controls?

20. An educational psychologist was interested in the effects of different orienting instructions on students' learning of text. All students were given the same chapter from an introductory psychology text. One group was instructed to read the material and be prepared to take a test (the study group); a second group was told to read the material and be prepared to explain it to another student (the teach group); a third (control) group was given the passage without any orienting instructions. All students were given a short-answer test after they read the chapter. The psychologist predicted that the teach group would be superior to the study group, and that the study group would be superior to the control group.

Test Scores

Teach	Study	Control
57	35	37
41	28	32
39	32	29
35	30	26
48	35	39
47	29	35
42	31	20
45	40	32
48	36	29
39	37	33

(A) Analyze the results.
(B) Evaluate the two predictions made by the psychologist.

10 Two-way Analysis of Variance

The analyses described in Chapter 9 are typically used when groups or conditions differ in only one way. For example, earlier a perception study was described in which the treatment of three groups differed only in the level of luminance given. We consider in this chapter complex designs involving more than one difference. Suppose, for instance, an experiment is performed in which lesions are made in the right and left hemispheres of the brains of rats. The experimenter is interested in how memory depends on different brain structures. After training on a task, four groups receive lesions in different places. Group R receives lesions in the right hemisphere only, group L receives lesions in the left hemisphere only, group B receives lesions in both hemispheres, and group N receives lesions in neither. After recovery from surgery all rats are retrained on the original learning task and a memory score is obtained for each rat. These scores *could* be subjected to the analysis described in Chapter 9, with $J = 4$. (Obviously a between-subject analysis of variance. Here is an excellent example of a study that could *not* be done with a within-subject design!) Let us suppose that the results of this study are as shown in Table 10.1 and Fig. 10.1.

TABLE 10.1
Illustrative data. Mean error scores of four groups.

Group	Mean
Right (R)	30.0
Left (L)	40.2
Both (B)	40.4
Neither (N)	30.2

FIGURE 10.1
Mean error scores of four groups in one-way design.

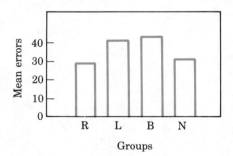

Let us further suppose that there are ten rats in each of the groups and that the pooled estimate of σ^2 is 9.912. The variance of the four means is 34.693, so if we performed a one-way analysis of variance on these data, the numerator estimate of σ^2, MSS_G, would be $N(34.693) = 10(34.693) = 346.93$ and

$$F(J - 1, N_T - J) = \frac{MSS_G}{MSS_{WG}}$$

$$F(4 - 1, 40 - 4) = \frac{346.93}{9.912}$$

$$F(3, 36) = 35$$

The critical value for $F(3, 36)$, $\alpha = .05$, is found from Table B in Appendix A to be 2.86. Since the observed value is larger than the critical value, we may reject the null hypothesis of no differences at the .05 level. Apparently, the lesions made a difference in the memory scores and we could now proceed to perform t tests among the means to locate the significant differences.

However, a far better way to consider this experiment is shown in Table 10.2.

TABLE 10.2
Four-group experiment as two-way design.

Left Hemisphere	Right Hemisphere		
		No Lesion	Lesion
	No Lesion	Group N	Group R
	Lesion	Group L	Group B

Table 10.2 arranges the four groups into two dimensions in accordance with the two ways the groups differ. The two rows represent the two ways the left hemisphere is treated: either receiving a lesion or no lesion. The two columns represent the two ways the right hemisphere is treated: lesion or no lesion. Group N, then, represents the combination of no lesion in the left hemisphere and no lesion in the right hemisphere. Group R represents the combination of no lesion in the left hemisphere and a lesion in the right hemisphere, and so forth. The data from the experiment are also rearranged in Table 10.3.

TABLE 10.3
Mean error scores in two-way experiments.

Left Hemisphere	Right Hemisphere		
		No Lesion	Lesion	Combined
	No Lesion	30.2	30.0	30.1
	Lesion	40.2	40.4	40.3
	Combined	35.2	35.2	

Table 10.3 also shows the means of groups combined. For example, the mean score for all subjects receiving no left hemisphere lesion is 30.1. Similarly, the mean for all subjects receiving no right hemisphere lesion is the same as the mean of all subjects receiving a right hemisphere lesion, 35.2. Fig. 10.2 also shows the rearranged means.

FIGURE 10.2
Mean error scores for a two-way experiment.

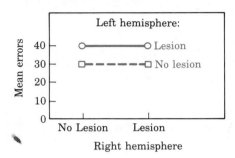

The source of the significant difference among the four groups becomes very apparent when the means are rearranged as in Table 10.3 and Fig. 10.2. Clearly the lesion in the left hemisphere had an effect. The combined mean error score was 40.3 with a left hemisphere lesion and 30.1 without a left hemisphere lesion. However, a lesion in the right hemisphere had little or no effect; the scores were about the same whether or not a lesion was made in the right hemisphere.

This experiment illustrates a design in which two variables are said to be **simultaneously manipulated**. These two variables may be referred to as factor A and factor B, and the design illustrates a *two-way factorial design*. Each factor is said to be represented by a number of *levels*. In the present example there are two levels (lesion or no lesion) of each factor (right hemisphere or left hemisphere). Factorial experiments with two levels in each factor are called 2 × 2 (two-by-two) designs. Theoretically, the number of levels of each factor is unlimited. We could have a 3 × 4 design, a 2 × 3 design, etc. The number of levels of factor A is given first. Thus, a 3 × 5 design has 3 levels of factor A and 5 levels of factor B.

In general, two-way factorial designs have J levels of factor A and K levels of factor B. In this chapter, then, we consider analysis of variance on $J \times K$ factorial experiments in which all possible combinations of J and K are represented in the design. For example, in a 3 × 4 between-subject design there are twelve groups, one for each of the combinations of 3 levels of factor A with 4 levels of factor B. It is helpful to think of the design as a table, or *matrix*, with J rows and K columns, as illustrated in Table 10.2. Each combination is referred to as a **cell** of the design. For example, level 2 of factor A combined with level 3 of factor B is cell A_2B_3.

The distinct advantage of a factorial design is that it permits the simultaneous evaluation of three effects: (1) the *main effect* of factor A, (2) the *main effect* of factor B, and (3) the *interaction* of factors A and B. **Main effect** is the effect of one variable ignoring the levels of the other. Thus, the main effect of A is the effect of the J levels of factor A combined, or "collapsed," over the K levels of factor B. To examine the main effect of A requires a comparison of the means of A_1, A_2, A_3, and so forth—all means obtained by ignoring levels of B. For example, in the lesion study we evaluate the main effect of a left hemisphere lesion. This effect is evaluated by comparing the mean error score of *all* subjects that received a lesion in the left hemisphere with the mean errors of all subjects that received no lesion in the left hemisphere. Computationally, little new is added in this chapter. The major new concept is the concept of interaction.

10.1

THE CONCEPT OF INTERACTION

Two variables are said to **interact** when the effect of one variable depends on the level of the other. Interaction may be more easily understood by first considering a result that has no interaction. Fig. 10.2 illustrates the absence of an interaction. Notice that the influence of the left hemisphere lesion is the same when no right hemisphere lesion is given as when a right hemisphere lesion is given. Stated another way, the *magnitude* of the difference between the mean of the group receiving a left hemisphere lesion and the mean of the group with no left hemisphere lesion is the same for both levels of the right hemisphere variable. Another example of a lack of an $A \times B$ interaction is shown in Fig. 10.3.

FIGURE 10.3

Example of no interaction in presence of main effects of A and B.

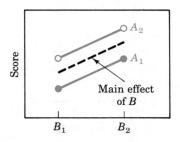

Compare figures 10.3 and 10.2. In Fig. 10.2 there was no main effect of B. In contrast, Fig. 10.3 shows a main effect of both A and B. There is a main effect of A because the mean of the cells at level 2 of factor A

(mean of A_2B_1 and A_2B_2 combined) is higher than the mean of the cells at level 1 of factor A (A_1B_1 combined with A_1B_2). Similarly, there is a main effect of B because the mean of the B_2 cells (A_1B_2 and A_2B_2) is higher than the mean of the B_1 cells (A_1B_1 and A_2B_1).

However, there is no interaction. How can we tell? Notice that the two solid lines in Fig. 10.3 are *parallel*. This occurs because the magnitude of the effect of B is the same for A_1 as it is for A_2. Another way to describe the absence of an interaction in Fig. 10.3 is to point out that the magnitude of the difference between A_2B_2 and A_1B_2 (i.e., A_2B_2 minus A_1B_2) is the same as the difference between A_2B_1 and A_1B_1.

Any significant departure from these conditions constitutes an interaction. If the function for one level of a factor is not parallel to the function for another level of that factor, there is an interaction. The analysis of variance tests for an interaction by testing the null hypothesis that there is no interaction; that is, that the functions are parallel. The data may not show exactly parallel lines even if the null hypothesis is true, in the same way that two sample means are not likely to be identical even when sampled from the same population. But the logic for testing the interaction is the same as the logic for testing means— the analysis assesses the likelihood of the obtained cell means under the hypothesis that there is no interaction.

A number of ways two variables may significantly interact are illustrated in the next three figures.

FIGURE 10.4
$A \times B$ interaction in which the magnitude of the B effect depends on the level of A.

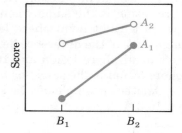

Fig. 10.4 shows a situation in which the magnitude of the effect of one variable depends on the level of the other. Here the effect of A is greater at level 1 of B than at level 2 of B. Interactions of the form illustrated in Fig. 10.4 are not uncommon when there is some limit on the size the measure can take. For example, if the measure is percent correct, the score cannot exceed 100% or be smaller than 0%. Suppose that in Fig. 10.4 factor B represents the elementary school grade level (B_1 is third grade and B_2 is seventh grade) and that factor A is sex (A_2 is female and A_1 is male). Suppose further that the measure is a test of verbal fluency. The interaction in Fig. 10.4, then, may be described as showing a relatively large superiority of females over males in the

third grade, a superiority that diminishes by the seventh grade. If the test were an easy test for seventh graders, so that most seventh graders score near 100%, the interaction between sex and grade may merely reflect an inability of the test to discriminate fluency differences in the seventh grade.

A slightly different form of interaction is shown in Fig. 10.5.

FIGURE 10.5
$A \times B$ interaction in which the B effect is eliminated at one level of A.

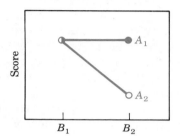

Although main effects of A and B were evident in Fig. 10.4, the best way to characterize Fig. 10.5 is simply as an interaction between A and B—whether A or B have any effect *at all* depends on the level of the other. A good example of this form of interaction comes from the study of the hippocampus. The hippocampus is a unique brain structure located near and on both sides of the midbrain in humans. Damage to the hippocampus on *both* sides of the brain has been found to produce a severe and remarkable effect on learning ability. A person with damage to the hippocampus on both sides suffers an apparent inability to learn, in a long-term way, anything new. This deficit is not produced, however, by damage to either the right or left side alone. Fig. 10.5 would illustrate this interaction, if the measure were a measure of learning ability, factor A were left hippocampal damage (A_1 = no damage, A_2 = damage), and factor B were right hippocampal damage (B_1 = no damage, B_2 = damage). Fig. 10.5 shows that only the combination of right hippocampal damage with left hippocampal damage (cell A_2B_2) shows a low score.

A *crossover* interaction is shown in Fig. 10.6.

FIGURE 10.6
A crossover interaction.

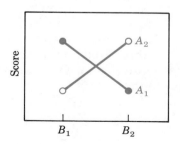

In this example there is no main effect of A, since the mean of both levels of A (combined over B_1 and B_2) is the same. Similarly, there is no main effect of B. A crossover interaction is the most dramatic form of interaction since the *direction* of an effect is reversed. For example, in Fig. 10.6 the mean scores for factor A increase, going from A_2 to A_1 when B is at level 1. When B is at level 2, however, the relation of the means on factor A is reversed. A crossover interaction may be seen in a study of gender identification. If factor A represents sex of child, factor B is sex of parent, and the measure is a measure of the degree to which the child identified with the parent, the result, at least for older children, would look like Fig. 10.6. Let A_1 be female child and A_2 be male child. Further, let B_1 be female parent, B_2 be male parent. Then the fact that cells A_1B_1 and A_2B_2 are highest reflects that girls identify with their mothers and boys identify with their fathers.

An example of the presence and absence of an interaction is a 3×4 design is shown in Fig. 10.7.

FIGURE 10.7
An interaction in a 3×4 design.

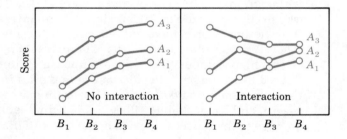

The left side of Fig. 10.7 shows no interaction; that is, all lines are parallel. Recall again, however, that data could depart from this ideal picture and still not be statistically significant or reliable. The right side of Fig. 10.7 is intended to show a reliable interaction. Note that it has a complex form, combining features of simple forms described earlier. In some ways, the effect of A diminishes as B gets larger. In other ways, (consider, for example, level 2 of A) the direction of the effect of A depends on the level of B.

In summary, an interaction between two variables is said to occur when the effect of one variable depends on the level of the other. The presence of an interaction *qualifies* the main effect of both factors. In simple forms, there are three kinds of interactions: one variable interacts with another when its effect is (1) reduced, (2) eliminated, or (3) reversed by a change in the level of the other. More complex forms, as illustrated in the right side of Fig. 10.7, involve combinations of these simple forms.

10.2

COMPUTATIONAL FORMULAS

As in the preceding chapter, the computational formulas differ somewhat depending on whether the analysis of variance is performed on data from between-subject or within-subject designs. Only those $J \times K$ factorial designs that are pure between-subject designs are discussed in this text. That is, the analyses described here are appropriate for the between-subject design with independent groups of subjects, each receiving a different combination of two factors. It is also assumed here that group sizes are equal. More complex designs are handled in Hays (1973), Keppel (1973), Kirk (1968), Meyers (1979), Winer (1971) and elsewhere.

We began this chapter by performing a one-way analysis of variance on four groups that may more appropriately be considered four groups of a 2×2 factorial design. We saw that a significant overall F on the one-way analysis appeared to be due to a main effect of factor A only. The key to understanding $J \times K$ analysis of variance is in understanding that the sum of scores (SS) for the $J \times K$ groups is calculated just as it is in the one-way analysis *but* that it has three components: SS due to factor A, SS due to factor B, and SS due to the $A \times B$ interaction. Mathematically,

$$SS_G = SS_A + SS_B + SS_{A \times B} \tag{10.1}$$

Notice that SS_G in a $J \times K$ design is calculated from the *cells* of the design. For example, in the 2×2 lesion study the sums of the cells are as shown in Table 10.4.

TABLE 10.4
Sums of scores in
2×2 lesion study.

Left Hemisphere	Right Hemisphere	
	No Lesion	Lesion
No Lesion	302	300
Lesion	402	404

Basically, as before, the first term of the sum of squares is obtained by squaring each of these sums, adding the squares, and dividing the total by the number of observations that go into each sum. Notationally, however, these operations appear somewhat more complex because of the additional factor. The operations just described are written

$$\frac{1}{N}\sum_j\sum_k(\sum_i X_i)^2$$

This states to sum over the i scores within the cell, square this sum, and repeat the operation over all j rows and k columns, adding the total. Applying this to the 2×2 lesion study

$$\frac{1}{N}\sum_j\sum_k(\sum_i X_i)^2 = \frac{1}{10}(302^2 + 300^2 + 402^2 + 404^2)$$

$$= \frac{1}{10}(506{,}024) = 50{,}602.4$$

The second term also appears somewhat more complex because of the additional factor, but the operations are the same as before. All scores are added together, the total is then squared, and the square is divided by the total number of scores: $(\sum\sum\sum X)^2/N_T$.

The complete formula for SS_G, then, is

$$SS_G = \frac{1}{N}\sum_j\sum_k(\sum_i X_i)^2 - \frac{(\sum\sum\sum X)^2}{N_T} \tag{10.2}$$

where $N_T = NJK$.

The identical principle is involved when finding SS for factors A and B, except that in each case the sums that are squared are the sums obtained by adding over (ignoring, or *collapsing* over) the other factor.

Table 10.5 shows the sums appropriate for factor A in bold type.

TABLE 10.5
Sums of scores in
2×2 lesion study,
factor A.

		Factor B		
		Level 1	Level 2	Sum
Factor A	Level 1	302	300	**602**
	Level 2	402	404	**806**

These sums, of course, are the numerators of the means being tested by the analysis of variance. Mathematically

$$SS_A = \frac{1}{NK}\sum_j(\sum_i\sum_k X_{ik})^2 - \frac{(\sum\sum\sum X)^2}{N_T} \tag{10.3}$$

Again, the first term is in principle identical to that for SS_G and SS_B. For each row a sum is found over the i subjects and k columns ($\sum_i \sum_k X_{ik}$), and this sum is squared. This entire operation is repeated over the j rows, all squares are totalled, and the total is divided by the number of observations in each row. The number of observations in each row is simply the number of observations in each cell (N) times the number of columns (K).

For the 2×2 lesion study

$$\frac{1}{NK} \sum_j (\sum_i \sum_k X_{ik})^2 = \frac{1}{10(2)} (602^2 + 806^2)$$

The SS_B is identical to SS_A except rows and columns are interchanged.

$$SS_B = \frac{1}{NJ} \sum_k (\sum_i \sum_j X_{ij})^2 - \frac{(\sum \sum \sum X)^2}{N_T} \tag{10.4}$$

The appropriate sums for factor B are shown in bold type in Table 10.6.

TABLE 10.6

Sums of scores in 2×2 lesion study, factor B.

	Factor B	
	Level 1	Level 2
Factor A — Level 1	302	300
Level 2	402	404
Sum	**704**	**704**

For these data

$$\frac{1}{NJ} \sum_k (\sum_i \sum_j X_{ij})^2 = \frac{1}{10(2)} (704^2 + 704^2)$$

The $SS_{A \times B}$ may be obtained by subtraction. Solving Equation 10.1 for $SS_{A \times B}$

$$SS_{A \times B} = SS_G - SS_A - SS_B \tag{10.5}$$

The SS_{WG} is obtained as before, since

$$SS_T = SS_G + SS_{WG}$$
$$SS_{WG} = SS_T - SS_G \tag{10.6}$$

The dfs follow the same principles. As before,

$$df_T = N_T - 1 = NJK - 1$$

The df_G is still one less than the total number of groups, except now the total number of groups is JK:

$$df_G = JK - 1$$

For factors A and B, the degrees of freedom are equal to one less than the number of means in the factor:

$$df_A = J - 1$$
$$df_B = K - 1$$

And the degrees of freedom are partitioned exactly as SS are:

$$df_T = df_G + df_{WG}$$
$$df_{WG} = df_T - df_{WG} = JKN - 1 - (JK - 1)$$
$$= JKN - 1 - JK + 1 = JKN - JK = JK(N - 1)$$

and

$$df_{A \times B} = df_G - df_A - df_B = JK - 1 - (J - 1) - (K - 1)$$
$$= JK - 1 - J + 1 - K + 1 = JK - J - K + 1$$
$$= (J - 1)(K - 1)$$

10.2.1 SUMMARY OF FORMULAS

The computational formulas derived so far are summarized in Table 10.7.

TABLE 10.7
Computational formulas for two-way between-subject analysis of variance.

$$SS_T = \Sigma\Sigma\Sigma X_{ijk}^2 - \frac{(\Sigma\Sigma\Sigma X)^2}{N_T}$$

$$SS_G = \frac{1}{N}\sum_j\sum_k(\sum_i X_i)^2 - \frac{(\Sigma\Sigma\Sigma X)^2}{N_T}$$

$$SS_{WG} = SS_T - SS_G$$

$$SS_A = \frac{1}{NK}\sum_j(\sum_i\sum_k X_{ik})^2 - \frac{(\Sigma\Sigma\Sigma X)^2}{N_T}$$

TABLE 10.7
(continued)

$$SS_B = \frac{1}{NJ}\sum_k(\sum_i\sum_j X_{ij})^2 - \frac{(\sum\sum\sum X)^2}{N_T}$$

$$SS_{A \times B} = SS_G - SS_A - SS_B$$

$$MSS_A = \frac{SS_A}{df_A}$$

$$MSS_B = \frac{SS_B}{df_B}$$

$$MSS_{A \times B} = \frac{SS_{A \times B}}{df_{A \times B}}$$

$$MSS_{WG} = \frac{SS_{WG}}{df_{WG}}$$

$$df_T = N_T - 1$$

$$df_G = JK - 1$$

$$df_{WG} = JK(N - 1)$$

$$df_A = J - 1$$

$$df_B = K - 1$$

$$df_{A \times B} = (J - 1)(K - 1)$$

$$F(df_A, df_{WG}) = \frac{MSS_A}{MSS_{WG}} \text{ for factor } A$$

$$F(df_B, df_{WG}) = \frac{MSS_B}{MSS_{WG}} \text{ for factor } B$$

$$F(df_{A \times B}, df_{WG}) = \frac{MSS_{A \times B}}{MSS_{WG}} \text{ for } A \times B \text{ interaction}$$

In summary, the three effects are (1) the main effect of A, which has J means; (2) the main effect of B, which has K means; and (3) the interaction of A and B, which has JK means. Each effect is tested by using the relevant means to estimate σ^2. In each case, the effect is tested by the F test using the within group mean sum of squares (MSS_{WG}) as the denominator estimate of σ^2. Each is tested separately with a critical value of F determined by the degrees of freedom for the effect and the degrees of freedom for MSS_{WG}. The numerator df for factor A is $J - 1$, for B is $K - 1$, for $A \times B$ is $(J - 1)(K - 1)$, and for the denominator the df is equal to $JK(N - 1)$.

10.2.2 COMPUTATIONAL EXAMPLE

An example is given in the next three tables.

TABLE 10.8
Scores from a 2 × 3 factorial design with eight subjects per cell.

	B_1	B_2	B_3	Total
	24	32	36	
	14	40	43	
	23	30	38	
	28	32	35	
A_1	25	21	33	
	21	25	35	
	20	26	25	
	34	29	29	
Sum	189	235	274	698
	14	27	42	
	9	33	32	
	15	22	36	
	12	22	40	
A_2	17	18	41	
	20	24	38	
	24	29	46	
	19	25	50	
Sum	130	200	325	655
Total	319	435	599	1353

Cell means:

	B_1	B_2	B_3
A_1	23.625	29.375	34.250
A_2	16.250	25.000	40.625

TABLE 10.9
Analysis of variance on data in Table 10.8.

$$SS_T = \Sigma\Sigma\Sigma X_{ijk}^2 - \frac{(\Sigma\Sigma\Sigma X)^2}{N_T}$$

$$= (24^2 + 14^2 + \ldots + 46^2 + 50^2) - \frac{(1353)^2}{48}$$

$$= 42{,}281 - 38{,}137.688$$

$$= 4143.3125$$

$$SS_G = \frac{1}{N}\Sigma_j\Sigma_k(\Sigma_i X_i)^2 - \frac{(\Sigma\Sigma\Sigma X)^2}{N_T}$$

$$= \frac{1}{8}(189^2 + 235^2 + \ldots + 325^2) - 38{,}137.688$$

$$= 2930.687$$

$$SS_{WG} = SS_T - SS_G = 4143.3125 - 2930.687 = 1212.626$$

$$SS_A = \frac{1}{NK}\sum_j(\sum_i\sum_k X_{ik})^2 - \frac{(\sum\sum\sum X)^2}{N_T}$$

$$= \frac{1}{8(3)}(698^2 + 655^2) - 38{,}137.688 = 38.52$$

$$SS_B = \frac{1}{NJ}\sum_k(\sum_i\sum_j X_{ij})^2 - \frac{(\sum\sum\sum X)^2}{N_T}$$

$$= \frac{1}{8(2)}(319^2 + 435^2 + 599^2) - 38{,}137.688$$

$$= 2474$$

$$SS_{A\times B} = SS_G - SS_A - SS_B$$

$$= 2930.687 - 38.520 - 2474$$

$$= 418.167$$

$$df_T = N_T - 1 = 48 - 1 = 47$$

$$df_G = JK - 1 = (2)(3) - 1 = 5$$

$$df_{WG} = JK(N - 1) = (2)(3)(8 - 1) = 42$$

$$df_A = J - 1 = 2 - 1 = 1$$

$$df_B = K - 1 = 3 - 1 = 2$$

$$df_{A\times B} = (J - 1)(K - 1) = (1)(2) = 2$$

$$MSS_A = \frac{SS_A}{df_A} = \frac{38.52}{1} = 38.52$$

$$MSS_B = \frac{SS_B}{df_B} = \frac{2474}{2} = 1237$$

$$MSS_{A\times B} = \frac{SS_{A\times B}}{df_{A\times B}} = \frac{418.167}{2} = 209.084$$

$$MSS_{WG} = \frac{SS_{WG}}{df_{WG}} = \frac{1212.626}{42} = 28.872$$

$$F(1, 42) = \frac{MSS_A}{MSS_{WG}} = \frac{38.52}{28.872} = 1.334 \text{ for factor } A$$

$$F(2, 42) = \frac{MSS_B}{MSS_{WG}} = \frac{1237}{28.872} = 42.84 \text{ for factor } B$$

$$F(2, 42) = \frac{MSS_{A\times B}}{MSS_{WG}} = \frac{209.084}{28.872} = 7.242 \text{ for factor } A \times B$$

TABLE 10.10
Summary table of
analysis of variance
on data in Table 10.7.

Source	SS	df	MSS	F
Total	4143.3125	47		
Groups	2930.687	5		
A	38.52	1	38.52	1.33
B	2474	2	1237	42.84
A × B	418.167	2	209.084	7.24
Within group	1212.626	42	28.872	

The critical value for $F(1, 42)$, $\alpha = .05$, is found in Table B (Appendix A) to be 4.07. Since the F for factor A is smaller than the critical value, we cannot reject the null hypothesis at the conventional level of confidence in the case of factor A. However, the critical value for $F(2, 42)$ is 3.22, and the Fs for both the main effect of B and $A \times B$ exceed this value. For both, then, we may reject the hypothesis for no difference at the .05 level.

The cell means are plotted in Fig. 10.8 to show the significant interaction.

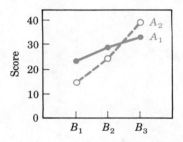

FIGURE 10.8
Plot of means in sig-
nificant interaction of
data in Table 10.7.

Although the main effect of B is significant, it appears from Fig. 10.8 that the best way to characterize the results is as a crossover interaction. We will return to these data in the next section.

The use of computer statistics packages to perform a two-way between-subject analysis of variance is discussed in sections 14.1.5, 14.2.5, and 14.3.5.

10.3

PAIRWISE COMPARISONS AMONG MEANS

Planned tests between pairs of means in $J \times K$ between-subject designs are conducted in the same way as in one-way between-subject designs. The MSS_{WG} is used as the pooled estimate of σ^2 and inserted into the formula for the standard error of the difference between means. Since group sizes are equal, the appropriate formula is Equation 9.5.

As an example, assume that the experimenter for the data shown in Table 10.8 and graphed in Fig. 10.8 had predicted a crossover interaction. As can be seen in Table 10.10, the $A \times B$ interaction is statistically significant, but a significant F provides no information about the form of the interaction. The form can only be determined by examining the graph of the interaction and, in some cases, by testing between selected pairs of means. For instance, although a crossover interaction is apparent in Fig. 10.8, we cannot tell without further analysis whether the crossover is significant. In this particular example, the mean of cell A_1B_3 may not be significantly different from the mean of cell A_2B_3. Stated another way, the two cells may have been sampled from the same population and the observed difference may be due to chance. Since the experimenter predicted the difference in the direction that was obtained, a t test is in order.

We see in Table 10.8 that the two means are 40.625 and 34.250. In Table 10.10, the MSS_{WG} is found to be 28.872. Substituting these values into Equation 9.5:

$$t = \frac{M_A - M_B}{\sqrt{MSS_{WG}\left(\frac{2}{N}\right)}} = \frac{40.625 - 34.250}{\sqrt{28.872\left(\frac{2}{8}\right)}} = \frac{6.375}{2.687} = 2.37$$

with $df = df_{WG} = 42$

Table A-2 does not contain the critical value for t with 42 df. However, we see that for $df = 40$, $\alpha = .05$, for a one-tailed test, the critical value of t is 1.684. Since the critical value for a larger df must be smaller than 1.684, and since the observed value is larger, the experimenter may conclude that the difference is reliable. Focusing now on the other end of the interaction, at level 1 of factor A, we see that the difference between the mean of A_1B_1 and the mean of A_2B_1 is even larger than the difference just tested. We could perform the t on this difference, but since the denominator would be the same as the t just computed, we know that the resulting t has to be larger. The combination of these two tests, then, confirms that the interaction is a crossover interaction.

10.4

SUMMARY

Two-way factorial designs have J levels of factor A and K levels of factor B. In this chapter we consider analysis of variance on the results of $J \times K$ factorial experiments in which all possible combinations of J and K are represented by different groups. In such designs, A and B are said to be **simultaneously manipulated**, and each of the JK conditions are referred to as **cells** of the design.

Such designs permit the simultaneous evaluation of (1) the main effect of A, (2) the main effect of B, and (3) the interaction of A and B. A **main effect** is the effect of one variable, while ignoring the levels of the other. An **interaction** between variables is said to occur when the effect of one depends on the level of the other. Each of the three effects is tested by F test. Each F has the same denominator, which is the **within group mean sum of squares (MSS_{WG})**, an estimate of σ^2 based on variances within the JK groups. It has $JK(N - 1)$ degrees of freedom, where N is the number of subjects in each of the JK groups. The numerators for the Fs are estimates of σ^2 based on the relevant means. For factor A the numerator is MSS_A, based on the variance of the J means of factor A. It has $(J - 1)$ degrees of freedom. For factor B the numerator is MSS_B, based on the variance of the K means of factor B. It has $(K - 1)$ degrees of freedom. For the $A \times B$ interaction, the numerator of F is based on the variance of JK means of the cells of the design. It has $(J - 1)(K - 1)$ degrees of freedom.

REFERENCES

Hays, W. L. 1973. *Statistics for the social sciences*. New York: Holt, Rinehart & Winston.

Keppel, G. 1973. *Design and analysis of experiments*. Englewood Cliffs, New Jersey: Prentice-Hall.

Kirk, R. E. 1968. *Experimental design: procedures for the behavioral sciences*. Monterey, California: Brooks/Cole.

Myers, J. L. 1979. *Fundamentals of experimental design*. (3d ed.) Boston, Mass.: Allyn & Bacon, Inc.

Winer, B. J. 1971. *Statistical principles in experimental design*. (2d ed.), New York: McGraw-Hill.

QUESTIONS

1. The distinct advantage of a $J \times K$ factorial design is that it permits the simultaneous evaluation of three effects. Name these effects.

2. Experiments with N-way factorial designs are such that each factor is said to be manipulated _____. The conditions in such a design are referred to as _____ of the design. The factors are said to be represented by a number of _____.

3. (A) Define the concept of interaction.
   (B) Graphically, how can you tell if an interaction exists?
   (C) Draw the results of a 2×3 experiment where there is:
   (I) No interaction.
   (II) A crossover interaction.
   (III) An interaction that does not cross over.

4.

	B_1	B_2
A_1	32	61
A_2	58	29

   If the average of B_1 is 45 and the average of B_2 is 45, can we conclude that factor B has no effect? Justify your answer.

5. In order to investigate the "grease effectiveness" of three brands of dishwashing liquid in three different water temperatures, 18 dishes were dipped in grease, washed, and given a "greasiness" rating according to two expert dishwashers:

	Hot	Warm	Cold
Detergent A	50, 52	48, 53	60, 61
Detergent B	42, 48	45, 49	61, 65
Detergent C	47, 45	49, 42	51, 48

   Perform a two-way analysis of variance (ANOVA) using $\alpha = .05$. What can you conclude about the null hypothesis that there is no effect of detergent brand in either water temperatures?

6. Referring to Question 12 in Chapter 9, suppose that Speedy has run an extra day and noticed that for half the days the weather was sunny and for the other half it was raining. The data may now be represented:

	Sunny	Rainy
Route 1	24, 24, 24	21, 22, 20
Route 2	27, 24, 25	27, 23, 20
Route 3	23, 20, 22	22, 21, 19
Route 4	25, 19, 22	19, 20, 16

   Perform a two-way analysis of variance using $\alpha = .05$. What can you conclude?

7. Suppose that in Question 11 in Chapter 9, the "Over-Haul" trucking company had used two trucks instead of one. Use $\alpha = .05$ to perform a two-way analysis of variance. What can you now conclude?

	Brand V	Brand W	Brand X	Brand Y	Brand Z
Truck 1	22	18	21	20	21
	16	23	21	17	16
Truck 2	21	21	24	20	18
	17	22	26	19	17

8. A local sporting goods store is interested in the strength of two brands of fishing line on four kinds of rods. Fishing lines are tested according to pounds of force required to break the line. Perform a two-way analysis of variance to test whether there are differences in breaking points of the line and whether there are differences in strength as a function of type of rod. Use $\alpha = .05$.

	Line 1	Line 2
Bamboo	26.1, 30.7, 30.6	30.2, 32.4, 34.1
Fiberglass	26.2, 28.9, 27.0	32.0, 31.9, 29.3
Aluminum	25.5, 27.1, 29.1	32.6, 30.6, 33.8
Birch wood	26.9, 29.4, 30.7	29.8, 30.2, 32.6

9. A mnemonic aid is a technique that is often used to remember lists. Three techniques were taught to three groups of college undergraduates: psychology majors, engineering majors, and art majors. The mean recall score for each student is presented below. Perform a two-way analysis of variance using $\alpha = .05$. Is it reasonable to consider the three techniques as equivalent?

		Major	
Technique	Psychology	Engineering	Art
A	15.27	12.55	10.73
	16.73	12.73	10.18
	16.00	13.10	9.64
B	14.70	12.60	11.30
	13.68	11.43	12.40
	13.20	13.25	12.52
C	18.19	14.40	15.20
	17.63	14.95	15.58
	16.95	15.17	16.75

10. Suppose that in the experiment of Question 5, three dishes were washed with each combination of dishwashing liquid and water temperature. The new results are as follows:

	Hot	Warm	Cold
Detergent A	50, 44, 51	48, 51, 46	60, 53, 58
Detergent B	42, 37, 48	45, 42, 51	61, 56, 58
Detergent C	47, 47, 51	49, 50, 43	51, 54, 47

Use $\alpha = .05$ to test for difference due to water temperature, dishwashing liquid, and an interaction.

11. Two baseball pitchers have been matched to test four radar guns and three types of baseballs. Each pitcher threw a fastball using each ball while being clocked by the different radar guns. The results are in miles per hour. In each case, the first figure is the speed of the ball thrown by the first pitcher and the second the speed of the ball thrown by the second pitcher:

	Gun 1	Gun 2	Gun 3	Gun 4
Brand A	91, 87	95, 93	99, 93	98, 92
Brand B	93, 86	96, 98	99, 95	91, 86
Brand C	100, 93	94, 96	92, 88	95, 91

Use $\alpha = .05$ to analyze this experiment. What can you conclude about each factor? Each interaction?

12. A cognitive psychologist was studying a phenomenon called "transfer appropriate processing." She designed an experiment in which students were taught a lesson on binomial probability. Half of the students received instruction which focused on the concepts underlying the topic, and half of the students received instruction which stressed the computation of binomial probability. Following this instruction, students either received a test on concepts or a test requiring computation. The test scores are given below:

		Test	
		Concepts	Computation
Teach	Concepts	45	32
		38	28
		43	21
		47	31
		44	24
	Computation	26	41
		30	49
		22	39
		25	42
		29	43

(A) Compute a two-way analysis of variance.
(B) Draw a graph of the means.
(C) What can you conclude?

13. A college physical education instructor wanted to determine the best way to increase students' understanding of baseball game strategies. He contacted his favorite local psychologist (Dr. Gregor Zilstein), and together they designed this experiment. One factor in the design was

whether students got to watch videotapes of baseball games, or if they just heard announcers' descriptions of those games. A second factor manipulated the learning activities of the students: half of the students were simply told to try to remember the events (passive), and half were asked questions about the events which were expected to help them better understand the game strategies (active). Following instruction, all students were given a test on baseball strategies. A total of 40 students were randomly assigned to one of the four conditions.

		Method of Presentation	
		Video	Audio
	Active	95, 82	82, 80
		88, 89	85, 81
		80, 92	81, 78
		89, 94	75, 80
		94, 90	86, 83
Learning Activities	Passive	69, 62	52, 45
		65, 59	49, 55
		60, 65	47, 48
		64, 63	53, 52
		75, 67	50, 48

(A) Analyze these data.

(B) Graph the results.

(C) Discuss your findings.

14. Aptitude-treatment interactions (ATIs) have been observed in a variety of educational settings: some instructional techniques work well with certain types of students but do not work well with others. Suppose a study was conducted with sixth grade students, where self-paced science instruction was compared with the traditional teacher-paced instruction. Ten low-achieving students and ten high-achieving students were assigned at random to one or the other instructional condition. Scores on the end-of-year science achievement test are as follows:

		Instruction	
		Self-paced	Traditional
	High	87, 95	80, 85
		82, 97	75, 90
		80	81
Achievement Level	Low	85, 76	42, 51
		79, 80	53, 60
		68	41

(A) Is there evidence for an ATI here? Compute an analysis of variance, graph your results, and comment.

(B) Is there a significant difference between high- and low-achievement students when instruction is paced by the teacher?

(C) Do high-achievement students learn better when instruction is self-paced than when it is teacher-paced?

(D) What would your advice be to a teacher who was considering using these self-paced science materials?

15. Suppose a researcher wanted to evaluate three weight-loss plans. She compared a plan that combined dieting and exercise with just dieting or just exercising. Fifteen men and fifteen women participated and were assigned at random to the three groups. Subjects were weighed before and after the six-month program. The following scores are the number of pounds *lost* (negative scores indicate weight *gains*).

	Exercise	Dieting	Exercise and Dieting
Male	1.8, 5.1 4.9, −.7 3.2	4.6, 2.9 3.4, 3.8 5.2	10.2, 14.7 12.5, 11.8 5.1
Female	2.1, −1.6 −4.8, 1.9 .9	6.5, 4.2 9.1, 5 2.3	12.4, 8.7 10.9, 14.2 15.6

(A) Analyze the data, graph the means, and discuss your results.

(B) Do the weight-management plans affect men and women differently?

16. A physician is studying the phenomenon of insomnia. He wishes to test two new soporific (sleep-inducing) drugs. In addition, he believes that relaxation training can help reduce insomnia, especially in combination with medication. A total of 24 chronic insomniacs participate and are assigned at random to one of four conditions. The number of minutes taken for each subject to fall asleep is recorded. The results of the experiment are:

	Drug X	Drug Y
Relaxation Training	143, 127 150, 133 161, 135	90, 125 103, 128 110, 131
No Relaxation Training	178, 151 162, 183 170, 168	134, 123 152, 140 162, 147

(A) Conduct a test, graph the means, and discuss the results.

(B) What appears to be the most promising method for the physician to pursue?

17. What is the null hypothesis for the test of main effects? What is the null hypothesis for the test of the interaction?

18. A comparison was made of the running speeds of boys and girls at two grade levels: 3rd and 7th. The results for 40 children running the 50-yard dash are as given (in seconds).

	Grade	
	3	7
Male	14.2, 13.5	6.8, 6.1
	15.1, 14.8	8.3, 6.7
	13, 12.1	7.4, 7
	12.9, 11	5.9, 7.3
	12.2, 10.7	5.5, 6.2
Female	12.3, 13.7	9.8, 8.3
	13.1, 13.3	7.4, 8.6
	12, 12.1	8.2, 7.7
	11.9, 14.1	10.1, 7.2
	10.8, 12.4	9.5, 6.8

(A) Analyze the data, graph the means, and comment on any main effects or interaction.

(B) Is the difference between male and female significant at grade 3? Is it significant at grade 7?

19. A social psychologist was studying attribution theory (the effects of background knowledge on the way we perceive other people and events). He wanted to see if people evaluate the quality of others' writing differently when told beforehand that the author was male or female. Twelve men and twelve women were presented an essay on economic theory. Half of the men and women were told the author was male; half were told the author was female. All subjects read the passage and rated its quality on a scale of 1 (low) to 10 (high). Analyze and graph the results of the data presented below. Does it make a difference if subjects are told the author was male or female? Are men and women affected differently by the experimenter's manipulation?

		Told Author Was	
		Male	Female
Rater	Male	10, 9	8, 10
		10, 9	9, 10
		9, 9	10, 9
	Female	7, 6	8, 6
		8, 6	7, 6
		8, 6	8, 8

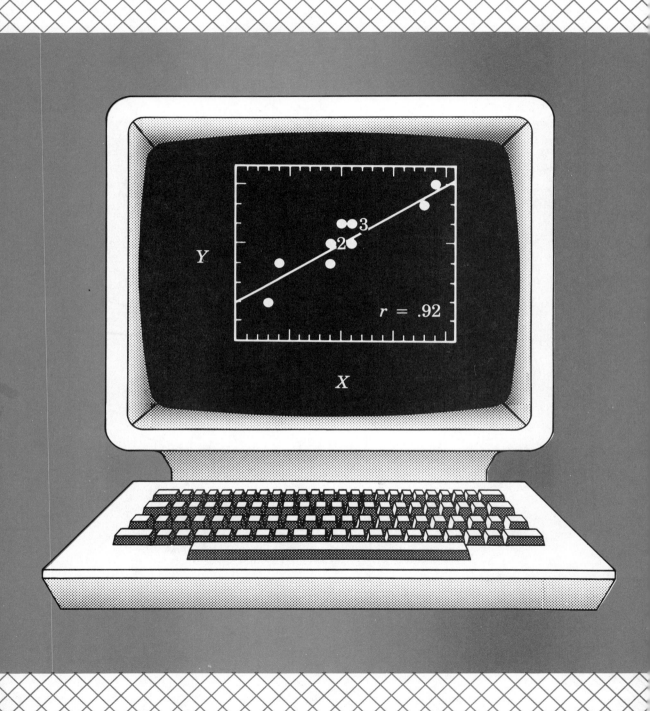

11 Regression and Correlation

In chapters 8 and 10, two variables were stated to be unrelated, if knowledge about the value of one provides no information about the values of the other. Statistical techniques have been described in chapters 5, 7, 8, and 9 for assessing whether a relationship exists between a variable manipulated by the experimenter and some measure of subject performance. In this chapter we expand our treatment of relations between variables in two ways. First, we discuss the problem of predicting the value of one variable from knowledge of the value of the other. That is, we go beyond the issue of whether two variables are related to the problem of using information about one to predict the other. This problem, known as the problem of *regression*, leads to the more subtle issue of determining the *form* of the relationship between the two variables. We are concerned here with quantitative, or numerical, variables, and we therefore consider the *mathematical* relationship between variables.

Second, based partly on developments in our treatment of regression, we introduce a measure of the *strength* of relationships. In doing so we also manage to enlarge the extent of our coverage of research designs. So far we have considered designs used to assess the relationship between measures on subjects and variables manipulated by the experimenter. These two variables have been referred to as *dependent* and *independent* variables, respectively. For example, in the perception study the independent variable was illumination level and the dependent variable was perceptual performance. In this chapter we consider how relationships are assessed when neither variable is controlled by the experimenter; that is, when both are measures on the same subjects, and it is difficult to state which of the two is the independent variable. As an example, consider the scores on the Scholastic Aptitude Test (SAT) and school grades. The relationship between these two variables could be assessed by obtaining both an SAT and a grade point average for each of a number of students. The technique described in this chapter operates on these *pairs* of scores to obtain a measure of the strength of the relationship. This measure is a measure of co-relation, or correlation, and is called the *correlation coefficient*. The pairs of scores obtained from N subjects are assumed to be sampled from two populations, a distribution of X and a distribution of Y. Since we are considering the correlation of X and Y, it is necessary to consider the *joint* distribution of X and Y.

11.1

THE JOINT DISTRIBUTION OF X AND Y

The joint distribution of X and Y may be thought of as the collection of all (X, Y) pairs. Consider how a joint distribution might be graphed.

When a distribution on a *single* variable is graphed, the horizontal axis represents the variable and the vertical axis is used to represent frequency. However, when a joint distribution is graphed, three dimensions are necessary, one dimension for each of the two variables, and a third dimension to represent frequency. Fig. 11.1 illustrates a graph of a joint distribution drawn in three dimensions. In this graph, *X* and *Y* appear tilted back in order to represent a third dimension. This third dimension, frequency, is represented by "height."

FIGURE 11.1
One example of the joint distribution of *X* and *Y*. Here *X* and *Y* are moderately related.

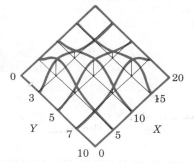

The surface in Fig. 11.1 has its greatest height at *X* = 10, *Y* = 5. Notice the shape of the joint distribution. This example illustrates a situation in which *X* and *Y* are moderately related. Pairs that have high scores on the *X* variable tend to have high scores on the *Y* variable. For instance, notice that when *Y* is 3, the mode of *X* is approximately 5. But when *Y* is 7, the mode of *X* is approximately 15.

Suppose that Fig. 11.1 illustrates the relationship between performance on a test of spatial ability (*X* variable) and a test of geometry proficiency (*Y* variable). The figure would represent the joint distribution on some large population. Now consider a sample from this population. Suppose a sample of 14 subjects is drawn and each subject is tested on the two measures. Table 11.1 illustrates such a sample. In this table, 14 pairs of scores are shown. These results are also plotted in Fig. 11.2.

TABLE 11.1
Sample of pairs taken from population shown in Fig. 11.1.

Pair	X	Y
1	9	4
2	10	5
3	4	4
4	12	6
5	9	5
6	3	2
7	19	8
8	10	6

TABLE 11.1
(*continued*)

Pair	X	Y
9	12	6
10	10	5
11	11	5
12	12	6
13	18	7
14	11	6

FIGURE 11.2
Scatterplot of data in
Table 11.1.

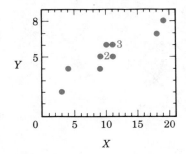

11.1.1 SCATTERPLOTS

Scatterplots, introduced in Chapter 8, are graphic representations of sample pairs obtained by plotting each pair as a point in two-dimensional space. For example, pair number 7 in Table 11.1 is plotted as point (19, 8) in the upper right-hand corner of Fig. 11.2. When more than one pair has the same value, the point is replaced by the number of such pairs. For example, in our data three pairs (#4, #9, and #12) have the values (12, 6). These pairs are represented in Fig. 11.2 by the number 3.

Fig. 11.3 shows the population and one sample in which X and Y are related in a different way.

FIGURE 11.3
Another example of
the joint distribution
of X and Y.

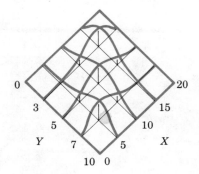

Notice that when Y is 3, the mode of X is approximately 15, but when Y is 7, the mode of X is approximately 5. A sample from this distribution is illustrated in the next table and figure.

TABLE 11.2
Sample of pairs from joint distribution in Fig. 11.3.

Pair	X	Y
1	10	4
2	15	2
3	12	6
4	20	1
5	10	5
6	6	6
7	11	7
8	10	5
9	18	3
10	7	5
11	2	8
12	2	7
13	12	4
14	9	4

FIGURE 11.4
Scatterplot of data in Table 11.2.

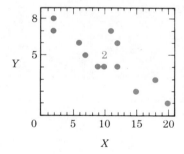

Here the pairs with high values of X tend to have low values of Y. Fig. 11.4 illustrates data one might obtain from 14 subjects on a task in which both speed and accuracy are measured. Assume X is accuracy and Y is speed. The figure shows that subjects who respond quickly (have high values of Y) in this task tend to be less accurate (have low values of X).

Fig. 11.5, on the next page, illustrates the joint distribution of two unrelated variables.

Notice that the mode of X remains 10 for all values of Y. A sample from this distribution is illustrated in Table 11.3.

FIGURE 11.5
An example of the
joint distribution of
two unrelated vari-
ables.

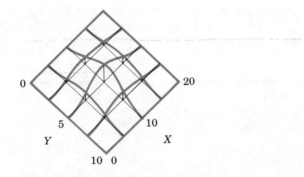

TABLE 11.3
Sample from distribu-
tion in Fig. 11.5.

Pair	X	Y
1	8	6
2	15	4
3	4	5
4	12	8
5	10	7
6	8	8
7	10	5
8	7	3
9	12	5
10	11	2
11	12	5
12	10	5
13	12	4
14	16	6

FIGURE 11.6
Scatterplot of data in
Table 11.3.

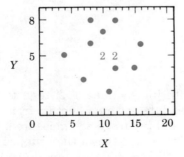

Notice that there is no systematic change in the value of Y as X
changes. We have already seen a scatterplot of two unrelated variables. In Chapter 8, pairs of sample means and variances are plotted to
illustrate their independence. That graph (Fig. 8.5) resembles Fig. 11.6
with the slight exception that the X variable in Fig. 8.5 is positively
skewed, whereas here X is normally distributed.

These figures have also demonstrated that in a joint distribution the means of X and Y need not agree. In all three figures the mean of the X population is 10 and the mean of Y is 5. The variances differ as well. To reflect these possibilities, we use the terms μ_X and μ_Y to refer to the mean of the population distribution of X and Y, respectively. M_X and M_Y refer to the mean of samples from X and Y, respectively. Similarly, σ^2 and s^2 are subscripted to denote the appropriate population or sample; σ_X^2 and σ_Y^2 refer to the variance of X and Y, respectively, and s_X^2 and s_Y^2 denote the variance of samples from these two distributions.

11.1.2 THE PROBLEM OF PREDICTION

We have characterized the problem of regression as the problem of predicting the value of one variable from knowledge of the value of the other; for example, predicting Y from X or predicting X from Y. In the absence of information concerning a related variable, the best prediction for a score from a distribution is the mean of that distribution. For instance, suppose you knew the mean grade of your class on the last test and were asked to predict the performance of a classmate. If you knew nothing about the person, the best prediction you could make is the mean class grade. This is so because, as we have seen, the sum of the signed values of the deviations from the mean is equal to zero, and the sum of the squared deviations is at a minimum.

But suppose Y is related to X. Now μ_Y will be different for different values of X. To see this, Fig. 11.1 is redrawn in Fig. 11.7 to show how μ_Y changes with different values of X.

FIGURE 11.7
Joint distribution of X and Y. M plotted across X.

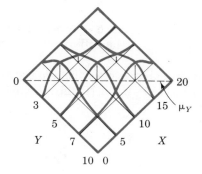

Notice that μ_Y is shown as a line.* When X equals 5, μ_Y equals 3; when X equals 15, μ_Y equals 7, etc. Another point that can be made with reference to Fig. 11.7 is that if Y is predicted from knowledge of X by

---

*To be precise, the line in Fig. 11.7 should be labelled $\mu_{Y|X}$, the mean of Y *given* X, to indicate that the mean of Y is a function of X.

using μ_Y there will still be some error in the prediction. That is, there is variance about μ_Y for all values of X. One way to state the problem of **regression**, then, is that it attempts to find a prediction of Y from X (or X from Y) in such a way as to minimize the deviations from the predictions. Returning to our class-test example, you would very likely improve your prediction of your classmate's performance if you knew his grade on the last test in this class, his grade in a related class, or even his overall grade point average. Each of these are measures that are probably related to performance on the last test and could be used to reduce error in your prediction, relative to the error made from a prediction that the score is equal to the mean score. We will review specific techniques for generating such a prediction after we obtain an overview of correlation.

11.1.3   USING CORRELATION TO ACCOUNT FOR VARIANCE

Reducing uncertainty about an inference concerning a statistic was described in Chapter 6 as a matter of controlling variance. Various methods were discussed for controlling, reducing, or *accounting for*, variance. Correlation may be described as a statistical technique for controlling variance. Stated another way, if two variables are related, the variance in one is *reduced* by knowledge of the value of the other. To express this difference in variance when X is known, versus when X is unknown, we define the quantity $\sigma^2_{Y|X}$, the variance of Y *given* X, and distinguish it from σ^2_Y as used to refer to the variance in Y (X not known). The difference between the two

$$\sigma^2_Y - \sigma^2_{Y|X}$$

reflects the decrease in variance that occurs when X is known. If $\sigma^2_{Y|X}$ has the same value as σ^2_Y, the difference is zero and no reduction in variance occurs.

We are now prepared to describe a measure of the strength of the relation between X and Y. In particular, the measure is the *correlation ratio*, ρ^2:

$$\rho^2 = \frac{\sigma^2_Y - \sigma^2_{Y|X}}{\sigma^2_Y} \tag{11.1}$$

The greek letter ρ is written out as *rho* and pronounced *row*. Notice that the correlation ratio is one in which the numerator is $\sigma^2_Y - \sigma^2_{Y|X}$ and the denominator is σ^2_Y. The numerator is the difference between the variance of Y and the variance of Y given X. When $\sigma^2_{Y|X} = \sigma^2_Y$, uncertainty about Y is not reduced by knowledge of X, the numerator is equal to zero, and therefore ρ^2 is equal to zero. Stated another way, ρ^2 is zero, when X and Y are unrelated. On the other hand, when variance in Y is *totally eliminated* by knowledge of X, then $\sigma^2_{Y|X} = 0$. In this case, the numerator of ρ^2 becomes σ^2_Y and the value of ρ^2 is 1.0. Thus, the correlation ratio is a measure of association strength that varies between zero (when the two variables are unrelated) and 1.0 (when the two are perfectly related).

The fact that the correlation is written in terms of population parameters means that it itself is a population parameter. It is a measure of the association between X and Y in their joint distribution. In figures 11.1 and 11.3, ρ^2 is approximately .80. In Fig. 11.5, ρ^2 is equal to zero. The *statistic* corresponding to ρ^2 is r^2, the square of the **Pearson product-moment correlation**. Pearson's r will be developed further after the following treatment of regression.

This chapter has two major sections. In Section 11.2, the problem of regression will be treated. This is the problem of predicting the value of one variable from knowledge of the other. Section 11.2 includes a review of the equation for a straight line. Also introduced is the concept of the *standard error of estimate*, a measure of the amount of variance in one variable about its predicted value. Correlation is treated in Section 11.3. Computational formulas for r are also provided. The problem of making inferences from r is then discussed. These inferences are based on the assumptions that X and Y are normally distributed and linearly related. Indices of strength of correlation that do not depend on these assumptions are discussed in Chapter 13.

11.2

REGRESSION

Predicting performance on one measure from knowledge of performance on another is a common problem in education and industry. The admissions office of an undergraduate college attempts to predict college performance from secondary or preparatory school grades. An admissions decision by a graduate or professional school usually involves a measure of undergraduate performance. The personnel office in an industrial setting tries to predict future performance on the job. Typically, more than one measure is used in these applications. For example, graduate school application includes undergraduate grades, scores on the Graduate Record Examination (GRE), and possibly other relevant test scores. In this chapter, we consider the relatively simple task of predicting one score from knowledge of only one other. Predictions in education and industry are also typically limited to categorical decisions. That is, the prediction of interest is whether the candidate will be successful or fail. Thus, in these applications, one variable (the predictor) is quantitative and the other (the criterion) is categorical. Nevertheless, the tools that are used, which are reviewed here, are appropriate when both variables are quantitative and the relationship between the two may be described mathematically. We consider a simple mathematical form, namely a *linear relationship*. That is, we explore predictions made when the relationship between two variables may be described by a straight line.

11.2.1 FORMULA FOR THE STRAIGHT LINE

The general expression for a straight line is

$$Y = a + bX \tag{11.2}$$

where a is a constant determining the y-intercept and b determines the slope of the line (how rapidly Y changes with changes in X). If $X = 0$, then

$$Y = a$$

This is what is meant when a is described as a y-intercept; it is the value that Y takes on when $X = 0$. Examples of straight lines having the same slope but different intercepts are shown in Fig. 11.8.

FIGURE 11.8
Examples of straight lines with the same slope but different y-intercepts.

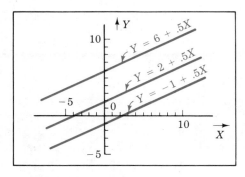

To see the effect b has on the straight line, set $a = 0$. Now the equation becomes

$$Y = 0 + bX = bX$$

If $b = 1$, then

$$Y = X$$

the simplest form of the equation. In this case, Y has the same value of X, for all values of X. But if $b = 2$, then

$$Y = 2X$$

Each value of Y has twice the value of X, and the slope of the line is twice that of the line $Y = X$. An important result occurs when b is negative. For example, suppose $b = -1$ (a remaining equal to 0). Then

$$Y = -X$$

Under these circumstances, values of Y decrease with increases in X. These three lines are graphed in Fig. 11.9.

FIGURE 11.9
Straight lines with equal y-intercepts but different slopes.

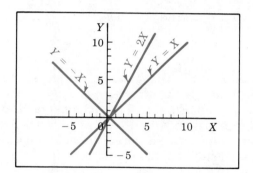

To anticipate matters a bit, let us compare Fig. 11.9 with Fig. 11.7. In Fig. 11.7, we saw one relationship in which values of Y increased, in general, with increased values of X, and another relationship in which Y decreased as X increased. It may come as no surprise that the former is called a *positive* relationship and the latter is called a *negative* relationship. The slope determines whether the relationship is positive or negative.

There is another important way that the graph of straight lines in Fig. 11.9 differs from the joint distributions graphed in Fig. 11.7. In the graph of a straight line there is only one value of Y for each value of X. However, in the joint distribution, at least as illustrated, there is dispersion, or variance, in the value of Y for each value of X (and vice versa). As we will now begin to develop, it is the *magnitude* of this variability (in Y and X) that determines the *magnitude* of the *correlation ratio*.

11.2.2 REGRESSION LINES

Fig. 11.10 shows two lines imposed on the scatterplot from Fig. 11.2.

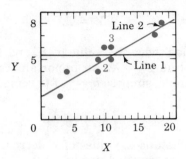

Line 2 is obviously a better "fit" than line 1. Line 1 is a straight line with $b = 0$. In fact, its equation is

$$Y = M_Y$$

Lines 1 and 2 may be thought of as two candidates to use for predicting Y from X. Line 1 is the prediction of Y using the mean of Y. The point to be gained from Fig. 11.10 is that when a relation exists between X and Y, the prediction $Y = M_Y$ can be bettered. Stated another way, if X and Y are related, the best predictor of Y is not the mean of Y. Rather, the best predictor of Y takes into account the value of X. But what do we use as an index of quality of fit? We use the magnitude of the deviation of Y from its prediction. Fig. 11.11 shows well how these deviations differ for lines 1 and 2.

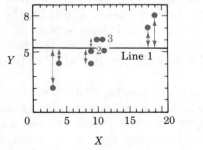
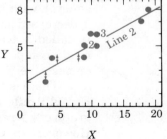

We see in the figure on the left that many points, particularly those with relatively large or small values of X, are far away from line 1. The distance of the point from the line, the difference between the point and its predicted value, is the *deviation* of the point from the line. Fig. 11.11 shows that the points deviate more from line 1 than from line 2.

We will refer to the *prediction* of Y as $\tilde{Y}$ (Y "tilde") and consider $\Sigma(Y - \tilde{Y})^2$, the sum of the squares of the deviations of Y from the predicted values of Y. Assuming that a straight line is used to predict Y from X, we can consider what a and b must be in order to *minimize* $\Sigma(Y - \tilde{Y})^2$. This is a problem for the calculus and its solution is well known. The solution may be expressed in terms of $(Y - \tilde{Y})$, but it is more convenient computationally to write the solution in terms of the "raw" scores X and Y. The raw-score form of the solution for b in order to minimize $\Sigma(Y - \tilde{Y})^2$ is

$$b = \frac{N(\Sigma XY) - (\Sigma X)(\Sigma Y)}{N\Sigma X^2 - (\Sigma X)^2} \tag{11.3}$$

Examine Equation 11.3 closely. Most terms are familiar: N represents the number of points (or pairs, or subjects, *not* the number of scores which equal $2N$), ΣX is the sum of X scores, ΣY is the sum of Y scores, ΣX^2 is the sum of all X scores squared, and so forth. The new term in Equation 11.3 is ΣXY. To determine ΣXY we find the product of X times Y (the "cross product") for each pair and sum across all pairs. We illustrate this operation with a small sample in Table 11.4.

TABLE 11.4
A demonstration of the calculation of b.

Pair	X	Y	X²	XY
1	5	9	25	45
2	8	7	64	56
3	1	4	1	4
4	5	4	25	20
5	7	8	49	56
6	10	10	100	100
Sum	36	42	264	281

$$b = \frac{N(\Sigma XY) - (\Sigma X)(\Sigma Y)}{N\Sigma X^2 - (\Sigma X)^2} = \frac{6(281) - (36)(42)}{6(264) - (36)^2} = \frac{174}{288} = .6042$$

$$a = M_Y - bM_X = \frac{42}{6} - (.6042)\frac{36}{6} = 3.375$$

$$\tilde{Y} = 3.375 + (.6042)X$$

Table 11.4 also shows that a is obtained by

$$a = M_Y - bM_X \tag{11.4}$$

The resulting line is plotted against the data in Fig. 11.12.

FIGURE 11.12
Scatterplot of six pairs
showing best fitting
straight line.

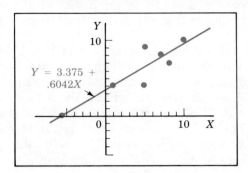

Remember that the line shown in Fig. 11.12 is obtained when a and b are chosen in such a way as to minimize $\Sigma(Y - \tilde{Y})^2$. This line is referred to as the *regression* of Y on X. The regression of X on Y can also be calculated. This line is defined in the same way as the regression of Y on X, except that X is predicted from Y and $\Sigma(X - \tilde{X})^2$ is minimized. These two **regression lines** coincide only when all points fall precisely on a straight line of nonzero slope. We are now prepared to discuss variance about the regression of Y on X.

11.2.3   STANDARD ERROR OF ESTIMATE

The reader should not overlook the powerful similarity between $\Sigma(Y - \tilde{Y})^2$ and $\Sigma(Y - M_Y)^2$. The former is the sum of squared deviations in Y from a prediction in Y based on a straight-line relation between Y and X. The latter, $\Sigma(Y - M_Y)^2$, is the sum of squared deviation in Y from a prediction that Y is equal to M_Y. But $\Sigma(Y - M_Y)^2$ should also be recognized as the numerator of the definitional equation for s_Y^2. That is, we know that a measure of variance about M_Y is obtained by dividing the sum of squared deviations of Y from the mean of Y by the degrees of freedom. The same logic may be applied to $\Sigma(Y - \tilde{Y})^2$. A measure of variance about the *predicted* value of Y is obtained by dividing $\Sigma(Y - \tilde{Y})^2$ by the appropriate degrees of freedom. The appropriate degrees of freedom for the variance, you will recall, is $N - 1$. The variance of a sample, then, written in our new terms, is

$$s_Y^2 = \frac{\Sigma(Y - M_Y)^2}{N - 1}$$

In contrast, the degrees of freedom for the variance about the predicted value is $N - 2$.

In other words,

$$s^2_{Y|X} = \frac{\Sigma(Y - \tilde{Y})^2}{N - 2}$$

is defined as the variance in Y about a predicted value based on the assumption that X and Y are linearly related. The positive square root of this newly defined variance, $s_{Y|X}$, is known as the **standard error of estimate**. As $s_{Y|X}$ increases in magnitude, there is increasing uncertainty about the prediction ("estimate") of Y from X. When $s_{Y|X}$ is zero, there is no error of estimate—all points fall on the predicted line.

The alert reader may have noticed the similarity between $s^2_{Y|X}$, the square of the standard error of estimate, and $\sigma^2_{Y|X}$, the variance of Y given X found in the formula for ρ^2. In fact, $s^2_{Y|X}$ is an estimate of $\sigma^2_{Y|X}$ (provided X and Y are normal, linearly related, and $\sigma^2_{Y|X}$ is the same for all values of X). Having now established a critical line between regression and correlation, we return to the development of Pearson's r.

11.3

CORRELATION

We have defined ρ^2, the correlation ratio, as

$$\rho^2 = \frac{\sigma^2_Y - \sigma^2_{Y|X}}{\sigma^2_Y}$$

Recall that ρ^2 is a population parameter, a measure of association between X and Y in the population. Its components are the population parameters, σ^2_Y and $\sigma^2_{Y|X}$, which are typically unknown. However, we have assembled estimates of these parameters. Our estimate of σ^2_Y is s^2_Y and our estimate of $\sigma^2_{Y|X}$ is $[(N - 2)/(N - 1)]s^2_{Y|X}$. We are now prepared to define the *statistic* equivalent to ρ^2, by substituting our estimates for the parameters. The resulting correlation coefficient is the square of Pearson's r:

$$r^2 = \frac{s^2_Y - \text{est. } \sigma^2_{Y|X}}{s^2_Y} \tag{11.5}$$

Equation 11.5 is not very convenient for computation because est. $\sigma^2_{Y|X}$ is defined in terms of $(Y - \tilde{Y})$. Pearson's r^2 may be written in more convenient forms, but the form of choice depends on the user's computational resources. The computational form offered here is:

$$r^2 = \frac{[N\Sigma XY - (\Sigma X)(\Sigma Y)]^2}{[N\Sigma X^2 - (\Sigma X)^2][N\Sigma Y^2 - (\Sigma Y)^2]} \qquad (11.6A)$$

$$r = \frac{N\Sigma XY - (\Sigma X)(\Sigma Y)}{\sqrt{[N\Sigma X^2 - (\Sigma X)^2][N\Sigma Y^2 - (\Sigma Y)^2]}} \qquad (11.6B)$$

Equation 11.6 is useful when the cross-products (X times Y) are easily obtained and summed. However, the formulas for r^2 and r can be written in a variety of equivalent ways. Some are written in terms of raw scores, as is Equation 11.6, but others are written in terms of s_Y^2, s_X^2, and b. Still others are written in terms of z scores. If you find that the formula for correlation provided with your computer or calculator program is not identical to Equation 11.6, simply give the program the data presented in the following example as a check on equivalence.

11.3.1   A COMPUTATIONAL EXAMPLE

The data in Table 11.4 are repeated here in Table 11.5 along with the computations needed to illustrate the definitional and computational formulas for r.

TABLE 11.5
An example of the computations of r.

Pair	X	Y	X²	Y²	XY	$\tilde{Y}$	$Y - \tilde{Y}$	$(Y - \tilde{Y})^2$
1	5	9	25	81	45	6.396	2.604	6.7808
2	8	7	64	49	56	8.208	−1.208	1.4593
3	1	4	1	16	4	3.979	.021	.0004
4	5	4	25	16	20	6.396	−2.396	5.7408
5	7	8	49	64	56	7.604	.396	.1568
6	10	10	100	100	100	9.417	.583	.3399
Sum	36	42	264	326	281			14.4780

$$b = \frac{N\Sigma XY - (\Sigma X)(\Sigma Y)}{N\Sigma X^2 - (\Sigma X)^2} = \frac{6(281) - 36(42)}{6(264) - (36)^2} = \frac{174}{288} = .6042$$

$$a = M_Y - bM_X = \frac{42}{6} - (.6042)\frac{36}{6} = 7 - (.6042)(6) = 3.375$$

$$\tilde{Y} = a + bX = 3.375 + (.6042)X$$

$$s_{Y|X}^2 = \frac{\Sigma(Y - \tilde{Y})^2}{N - 2} = \frac{14.478}{6 - 2} = 3.6195$$

TABLE 11.5
(*continued*)

Definitional Formula for r

$$r^2 = \frac{s_Y^2 - \text{est. } \sigma_{Y|X}^2}{s_Y^2} \qquad \text{where est. } \sigma_{Y|X}^2 = \frac{N-2}{N-1}s_{Y|X}^2$$

$$r^2 = \frac{s_Y^2 - \frac{(N-2)}{(N-1)}s_{Y|X}^2}{s_Y^2} = \frac{6.4 - \frac{4}{5}(3.6195)}{6.4} = .548$$

$$r = \sqrt{.548} = .74$$

Computational Formula for r

$$r = \frac{N\Sigma XY - (\Sigma X)(\Sigma Y)}{\sqrt{[N\Sigma X^2 - (\Sigma X)^2][N\Sigma Y^2 - (\Sigma Y)^2}} = \frac{6(281) - (36)(42)}{\sqrt{[6(264) - (36)^2][6(326) - (42)^2]}}$$

$$r = .74$$

The use of computer statistics packages to compute r is discussed in sections 14.1.6, 14.2.6, and 14.3.6.

11.3.2 MAKING INFERENCES FROM r

As always, we make inferences about a statistic by considering the sampling distribution of that statistic. What, then, is the sampling distribution of r? That is, if we assumed that $\rho^2 = 0$ and drew samples from the joint distribution of X and Y, each sample having N pairs, how would r be distributed? There is a fortunate relationship between r and t that can be used to make inferences about r. In particular, provided that X and Y are normal, it can be shown that

$$t = \frac{r}{\sqrt{1 - r^2}}\sqrt{N - 2}$$

with $(N - 2)$ degrees of freedom. Why is $df = N - 2$? Recall that the formula defining r involves estimating two variances, σ_Y^2 and $\sigma_{Y|X}^2$. This means that the N pairs used to calculate r are constrained in two ways. First, determining s_Y^2, the estimate of σ_Y^2, requires that $\Sigma(Y - M_Y) = 0$, as in the previous discussion. In addition, however, recall that the estimate of $\sigma_{Y|X}^2$ involves $\Sigma(Y - \tilde{Y})^2$, and that this is minimized by estimating a and b in $\tilde{Y} = a + bX$. Together these two constraints result in $df = N - 2$.

It is possible to substitute various values of N and tabled critical values of t into the equation relating r to t in order to solve for critical values of r at various levels of α. This has been done and the results are presented in Table C in Appendix A. That is, Table C can be used to test the statistical significance of an obtained sample value of r. Stated still another way, Table C can be consulted in order to use the obtained r as a test against the null hypothesis of no relation between X and Y.

As an example, let us test the r computed in Table 11.5. This r is based on six pairs, so the $df = 6 - 2 = 4$. As we examine Table C for the critical value of $r(4\ df)$, we note that the table gives critical values for one-tailed tests and two-tailed tests. As before, one-tailed tests are appropriate when H_0 is countered by a directional hypothesis. H_0, of course, is that there is no relation between X and Y, that ρ^2 is equal to zero. In this case, a directional experimental hypothesis would be one that specified the direction of the relation, that is, whether positive or negative. A two-tailed test is appropriate when there is no experimental hypothesis about the sign of r. Let us suppose that the data in Table 11.5 is based on a study in which a positive relation is predicted, i.e., in which it is expected that increased values of X will be accompanied by increased values of Y. In this case, a one-tailed test is used and we find in Table C that the critical value for $r(4)$, one-tailed, $\alpha = .05$, is equal to .729. As with t, we reject the null hypothesis when the observed value is *larger* than the critical value. This is so, because the expected value of r is zero under the null hypothesis. In the example, the observed value is .74. We therefore reject the null hypothesis. We conclude that the probability of obtaining the observed value is too small to retain the hypothesis of no relation.

Notice in Table C that for small values of N, large values of r will be required to reject H_0. As N increases, small values of r will permit the rejection of H_0. This is not to say, remember, that the probability of incorrectly rejecting a true H_0 (making a type I error) increases as N increases. Recall that the probability of making a type I error depends on the α set by the experimenter and is unrelated to the size of the sample. However, as N (the number of pairs) increases, the ability to detect moderate relationships does increase; that is, power increases with increased N.

Also, do not confuse the magnitude of r with the *importance* of the relationship. Pearson's r is a measure of the strength of the relationship, not a measure of its importance. An r of .90 from one study is not necessarily more important than a statistically significant r of .06 in another. The r of .06 might form the initial basis for the discovery of a new cure for cancer, and the r of .90 could reflect the trivial relationship between height and weight.

11.3.3   FACTORS INFLUENCING THE SIZE OF *r*

It should be clear by now that *r* depends on the magnitude of the variance about the predicted value of *Y*. To emphasize this point graphically, Fig. 11.13 shows two scatterplots in which *b* is held constant, but the variance about *Y* is varied.

FIGURE 11.13
Two scatterplots that differ in the standard error of estimate.

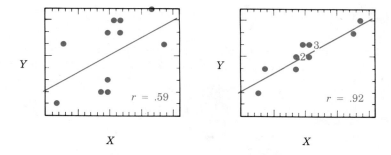

In the scatterplot on the left side of the figure, most points deviate considerably from the regression line. However, the points on the right side are packed tightly about the regression line. The relative size of the corresponding *r*s, as shown in the figure, reflect this difference.

Fig. 11.13 should be helpful in reiterating another point about correlation. Whereas the variance in *Y* about $\tilde{Y}$ influences the magnitude of *r*, the slope of *Y* against *X* (the sign of *b*) determines whether *r* is positive or negative. If all values of *Y* fell precisely on $\tilde{Y} = a + bX$, *r* would be equal to 1.0 regardless of *a* and the magnitude of *b* (provided that *b* is nonzero). However, if *b* is negative, so also is *r*.

What problems can arise when H_0 cannot be rejected? Of course, the possibility exists that the test lacks power (*N* is small) and retention of H_0 is a type II error. However, in the study of correlation there are two other sources of a zero correlation. One possibility is that there is a relation between the variables, but the range of one of the variables is restricted in the sample. The other source of a nonsignificant *r* is that a relation exists, but it is nonlinear. We develop these points after gaining some practice at estimating the size of *r* from a visual inspection of the scatterplot. To begin this practice, examine Fig. 11.14 on the next page.

The general outlines of five scatterplots are depicted. These represent *r* ranging from high positive to high negative. The figure is intended to show that the general form of the scatterplot can be used to estimate the size and sign of *r*. If the points fall nearly on a straight line, the correlation is near ± 1. As the scatterplot becomes more circular, the size of the correlation approaches zero.

FIGURE 11.14
Outlines of scatterplots.

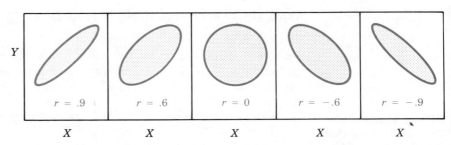

Now let us consider the impact of a restricted range on the magnitude of *r*. To illustrate this effect, consider Fig. 11.15.

FIGURE 11.15
An example of the effect of restricted range on the magnitude of *r*.

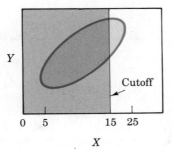

This figure shows a moderately high correlation between *X* and *Y*. Notice that *X* ranges roughly between 5 and 25. Now suppose a study of the relation of *X* and *Y* is conducted in such a way that no subjects are sampled if their score on *X* is below 15. A cutoff of 15 is shown in the figure, and the aspect to be noticed is the shape of the scatterplot to the right of the cutoff. The scatterplot to the right of the cutoff clearly does not have the oval shape characteristic of the overall shape. In fact, the scatterplot to the right of the cutoff appears generally circular, as occurs when *r* is zero. If a sample were drawn from this restricted range of *X*, the value of *r* would be substantially reduced.

But why would an investigator conduct a study in which the range was restricted and thereby increase the chance of a type II error? Since the investigator does not control the variables in a correlational study, such a restriction may be unavoidable. Consider a study of the relation between performance on the Graduate Record Examination (GRE) and success in graduate school. Often the correlation between these two variables is essentially zero. This point is sometimes used as an argument against the use of the GRE in admissions decisions. Yet, these studies are often performed on students in graduate school that were selected on the basis of their GRE scores. In other words, students with low GREs were not likely to be subjects in the studies. The result *could* be similar to the use of the 15 cutoff in Fig. 11.15. On the other hand, the conclusion of no relation could be correct. The point is that the conclusion about the absence of a correlation is ambiguous as long as performance on one of the variables is used initially in the decision to include subjects.

As another example of the role of restricted range in the study of correlation, imagine that a study was performed examining the relation between performance on a measure of intelligence and a laboratory measure of learning rate, e.g., speed of learning verbal lists in the laboratory. Suppose the study finds a nonsignificant correlation. Suppose further that the study was performed on college students. By using college students as subjects, the study has seriously restricted the range of possible intelligence test scores. Thus, while it may be safe to conclude that there is no correlation between the two measures when the population is college students, this would be an unsafe generalization for the entire population.

The second consideration when a nonsignificant correlation is observed is the possibility that a relation exists, but it is nonlinear. Fig. 11.16 illustrates the situation in which a strong relation exists between X and Y, but the resulting r is zero.

FIGURE 11.16

Example of curvilinear relation that produces $r = 0$.

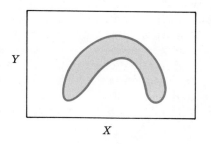

In this figure the relation is strong but curvilinear. The relation is "strong," because if X is known the value of Y can be predicted with some confidence. As X increases from a small to moderate value, Y also increases. But as X continues to increase, the value of Y begins to decrease. Y is largest at some intermediate value of X. A relation resembling Fig. 11.16 might be seen in a study of the accuracy of a tennis serve and the speed or force of the serve. If the speed is either too soft or too hard, the serve may be all over the court. Optimal accuracy is probably found at some moderate serving speed.

In general, if r is near zero, the possibility of a nonlinear relation should be assessed. This can be accomplished by inspecting the scatterplot. There are also statistical tests that do not depend, as Pearson's r does, on the assumption of a linear relation. Treatment of these tests may be found in advanced texts (e.g., Hays, 1973).

11.3.4   CORRELATION AND CAUSATION

A distinction was made in Chapter 6 between statistical decisions and explanations. There it was cautioned that a null hypothesis could be correctly rejected but given an erroneous explanation. This caution is especially appropriate when correlational techniques are used, since the experimenter has no control over the values of the variables. Thus, if a significant correlation between two variables is found, it is hazardous to infer a *causal* relationship. For example, a finding that school grades are negatively related to coffee drinking does not permit us to conclude that either causes the other. Among nations, it is often pointed out, Sweden has the highest rate of suicide and also an advanced social welfare program. Any conclusion from this observation is ludicrous. In the first place, a relationship cannot be inferred from a single point. The fact that a single pair has high values for both X and Y does not itself tell us anything about other pairs. Second, even if such a correlation existed, it would be unclear whether social welfare causes suicides, or whether conversely, a high suicide rate initiates social welfare programs.

In all these examples, of course, a conclusion about causation is ambiguous because there are three possible conclusions. Either X causes Y, Y causes X, or both are caused by another unidentified variable. In the absence of additional information, there is no way to decide among the alternatives. (Even with additional information, conclusions can be difficult. For example, the attempt to account for the cause of Sweden's suicide rate by pointing to other related variables in culture and geography is stymied by the observation that Norway, which borders Sweden, has a suicide rate among the world's lowest.)

11.4
SUMMARY

The problem of **regression** concerns predicting the value of one variable from knowledge of another, for instance, predicting Y from X. The **correlation ratio, ρ^2**, measures the strength of the relationship between two variables, X and Y. It is expressed in terms of the proportion of variance in one variable that is reduced by knowledge of the value of the other. The statistic associated with ρ^2 is r^2, the square of the **Pearson product-moment correlation**.

Y is predicted from X by the **regression line**,

$$\tilde{Y} = a + bX$$

which is selected in such a way as to minimize the squared deviations of $\tilde{Y}$ from Y. The variance in Y about $\tilde{Y}$ is a measure of the uncertainty of the prediction. The positive square root of this variance is the **standard error of estimate**.

The relation of r to t is shown to illustrate how inferences are made from r. These inferences are based on the assumptions that X and Y are normally distributed and linearly related. The magnitude of r was shown to depend on several factors. In general, r is smaller the larger the standard error of estimate. It can be decreased by reductions in the range of X values sampled and decreased when the relation between X and Y is nonlinear.

Finally, the observation was made that statistically significant correlations need not connote a causal relationship.

REFERENCES

Hays, W. L. 1973. *Statistics for the social sciences*. New York: Holt, Rinehart & Winston.

QUESTIONS

1. (A) If $\sigma_Y^2 = \sigma_{Y|X}^2$, what is the relationship between X and Y?

   (B) If $\dfrac{\sigma_Y^2 - \sigma_{Y|X}^2}{\sigma_Y^2} = 1$, what is the relationship between X and Y?

2. How is ρ^2 different from r^2?

3. Graph the following equations. State whether the variables in each are positively or negatively related.
   (A) $y = 2x + 5$
   (B) $y = \dfrac{3}{5}x - 3$
   (C) $y = -1.5x + 2$
   (D) $y = 6x$
   (E) $y = -\dfrac{1}{3}x - 1$

4. Correlation does not imply causation. Explain this statement, and give an example.

5. The following data represent the number of years applicants for a chauffeur's license have been driving and the scores they received on a driver's license test.

Number of Years	Grade on Test
5	77
4	50
6	75
7	86
5	59
6	80
6	74
4	60
7	91
5	65

(A) Calculate a and b for the regression equation $y = a + bX$.

(B) Plot the line obtained in part (A) and use it to "read off" the predicted grade of someone who has been driving for 5 years.

(C) Use the regression equation to predict the grade of someone who has been driving for 5 years.

6. Traffic safety officials have collected the following data on blood alcohol level (BAL) and traffic fatalities:

BAL	Fatalities
.05	10,000
.10	20,000
.15	25,000
.20	35,000
.25	37,000
.30	30,000
.35	28,000

(A) Calculate the regression equation, plot the points, and sketch a line from the scatterplot that will allow one to predict accidents from BAL.

(B) Use the regression equation obtained in (A) and the scatterplot to predict traffic fatalities when BAL = .175.

7. The following are data of the average annual yield of corn (in bushels) for a given county and the annual rainfall (in inches) per annum.

Bushels of Corn	Rainfall
34.5	7.6
37.4	9.0
64.2	14.7
47.3	11.9

55.4	11.7
21.6	6.0
45.1	10.1
73.5	17.4

(A) Calculate the regression equation and use it to sketch a line where average rainfall may be used to predict the average annual corn yield.

(B) Use the equation in (A) to predict the average annual corn yield, if the average rainfall is 7.8 inches.

8. A psychologist interested in the "speed/accuracy trade-off" has collected the following data. Average reaction measures were calculated from subjects' ability to discriminate pattern stimuli. Accuracy is represented as the average errors per trial session.

Reaction Time (msec)	Number of Errors
.9976	15
1.3482	13
.7692	19
1.5400	12
1.4580	13
.5656	22
2.0101	9
1.9765	10
.4397	25
.6012	21
1.8070	11

(A) Solve for the regression equation and sketch a scatterplot of the pairs of points given above. Draw in the regression line.

(B) What number of errors can be predicted if a subject's average reaction time = 1.000?

9. A statistics professor investigating the possible relationship between the number of hours spent studying statistics and grades achieved on a statistics test collected the following data:

Hours	Grade
0	55
9	100
8	100
.5	72
1	73
2	80
3	90
4	95

(A) Compute the regression equation and sketch a scatterplot.

(B) What would you predict the grade to be if a student spent 2.5 hours studying statistics?

10. High Tech University wanted to determine a way of awarding financial aid to their most promising incoming students. One measure under consideration was a test of mechanical reasoning. A preliminary study was conducted to assess the predictive strength of this test for success in the first-year engineering program. A sample of 12 freshmen was given the mechanical reasoning test in September, and at the end of May these students' scores were correlated with first-year GPA. The results were:

Mechanical Reasoning	GPA
16	1.7
49	3.1
55	2.3
47	1.8
59	3.7
60	2.4
18	3.0
25	2.2
57	3.1
44	2.8
56	3.2
35	2.7

(A) Draw a scatterplot of the data.

(B) Calculate r and discuss your results.

(C) Compute a regression equation.

(D) What would you predict a person's GPA to be, if he or she scored exactly at the mean of the mechanical reasoning test?

(E) If the school wanted only to give financial aid to those who were most likely to maintain a minimum GPA of 2.8, what is the lowest mechanical reasoning score a student could make to qualify for aid?

11. For each of the following, state whether you would expect a positive, negative, or no correlation.

(A) The number of hours that an archer practices and his score at an archery meet.

(B) The IQs of husbands and wives.

(C) Eye color and hair color.

(D) Levels of smog and the sale of eye drops.

(E) Hat size and amount of income tax paid yearly.

(F) Shoe size and sense of humor.

(G) Company profits and price of shares.

(H) Number of hours typing practice and number of typing errors.

12. Suppose you were asked to calculate r for the following two sets of data. Would you be surprised to find $r = 1$ and $r = -1$, respectively? Why?

X	Y	X	Y
42	9	19	8
57	13	22	12

13. Calculate r for Question 8 and test for significance at the .05 level.

14. The following are the number of mistakes that 14 supermarket clerks made at their cash registers during the first hour at work and during the last hour of their shift.

First Hour	Last Hour
3	5
4	5
2	1
2	2
4	4
3	2
5	6
0	3
4	3
5	2
1	4
2	3
3	2
4	3

Calculate r and test the null hypothesis that there is no relationship at the .05 level of significance.

15. A sociologist wanted to know if job satisfaction was related to income. He surveyed 10 people between the ages of 35 and 40, and asked them to rate their overall satisfaction with their work. A scale of 1 (very dissatisfied) to 10 (very satisfied) was used. The results are:

Income	Satisfaction Rating
40,000	5
82,000	9
31,000	2
12,000	5
50,000	3
63,000	7
71,000	4
42,000	7
18,000	9
69,000	1

Compute r. What is the relationship between income and job satisfaction?

16. Based on this sample of eight married couples, what can be concluded about the relationship between marital harmony and duration of the marriage?

Years Married	Number of Arguments/Month
10	4
6	6
4	2
18	2
37	1
26	4
2	5
1	9

17. Based on the data that follow, is there a significant correlation between the age at which a woman marries and the age at which her mother married?

Daughter	Mother
27	21
32	29
27	34
27	28
36	29
23	24
20	18
18	26
25	25
33	24
26	25

18. A concern that is often heard about children watching too much television is that they will spend less time reading and therefore fail to gain proficiency in reading skills.
   (A) Do the following (fictitious) data support such a concern? Calculate r and test it for significance against the null hypothesis.
   (B) Using the formula in Section 11.3.2, conduct a t test and compare your result with that obtained in (A).

Number of Hours of Television Viewing/Day	Comprehension Scores, (%)
1	97
1	50
1	62
2	70
2	45
2	91
3	62

3	50
4	60
4	33
5	25
5	60
6	50

19. Suppose one brown die and one white die are rolled five times and the following results are obtained:

Brown Die	White Die
1	2
3	1
1	3
3	1
6	4

Calculate r for these data. Although r is rather high, can we claim the values of the dice are significantly related?

20. The manager of the Savoir-Faire Restaurant wanted to screen applicants so that the waiters and waitresses hired would make the fewest mistakes on the job. A written test was designed and given to 35 applicants. The six highest scoring applicants were hired, and the following is a record of any errors they made during their first month at work.

Test Score	Number of Errors
95	4
92	12
89	10
93	18
91	5
93	11

(A) Analyze the data for the six employees and discuss your results.

(B) Can you conclude that the manager's test was helpful in screening applicants? What might have been a better way to assess the test's validity? (*Hint*: Who was hired?)

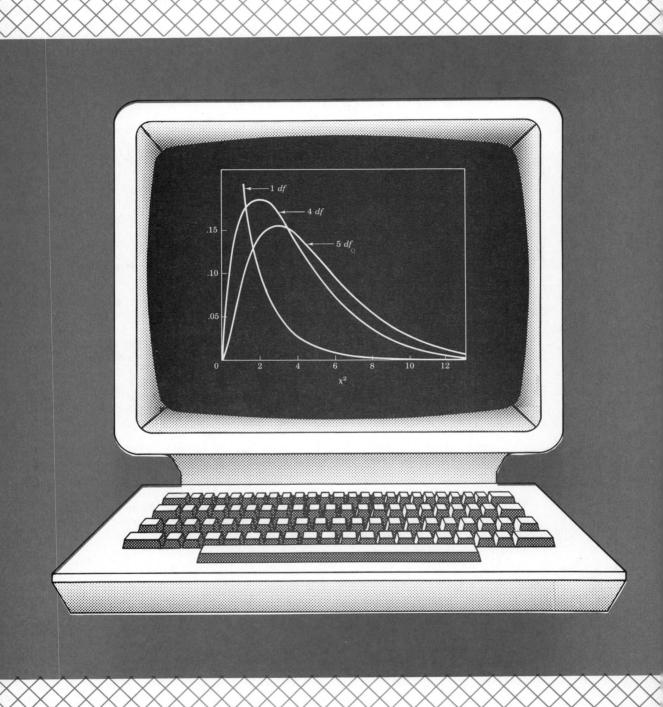

12 Analysis of Categorical Data

The statistical tests described in the preceding chapters were for use with numerical data. The present chapter covers analyses of categorical data. Analyses of categorical and numerical data differ because numerical measures assign *numbers* to subjects, whereas categorical measures assign *categories* to subjects. For example, questions on surveys and personal inventories are typically of the form: "Do you agree or disagree with the following statement?" The answer is used to place the respondent in one of two or three categories: agree, disagree, and (sometimes) undecided. Other examples of categorical variables are sex, occupation, and place of birth. In addition, subjects may be categorized in terms of their experimental treatment. For example, subjects in the experimental group are placed in one category and subjects in the control group are placed in another. Analyses of categorical data, then, are performed on the numbers, or "frequencies," of subjects in each category. Notice that these analyses are still performed on numbers. The difference is that the numbers are *frequencies* of subjects in categories, not numbers assigned to the subject.

Sometimes categorical frequencies are transformed to percentages—percents of subjects that fall into each category: "Eighty percent preferred our product." What the investigator hopes to do with such data, of course, is to generalize from some specific sample to some population at large. When an advertiser claims that "9 out of 10 doctors recommend our brand," the hope is that the reader will conclude that 90% of *all doctors* feel the same way. There are a host of potential problems with such generalizations, and in this chapter we review statistical tests that are appropriate when attempts are made to generalize from categorical data. We begin by defining the *contingency table*, which is a means of organizing categorical data.

12.1

CONTINGENCY TABLES

We anticipated this discussion when, in Chapter 3, we defined independence. There we considered populations that were partitioned, or categorized, according to two variables *A* and *B*. An example involved race and voting behavior, and data were presented that suggested that blacks were less likely to vote in the 1976 national election than were whites; that is, that race and voting behavior were associated (not independent). A contingency table is used when a sample is drawn and a statistical test is made of independence. In general, a contingency table shows the observed frequencies in the joint distribution of two categorical variables. Suppose, for example, that 100 males and 60 females were asked whether they support a constitutional amendment prohibiting abortion. Table 12.1 illustrates, with totally fabricated data, how such a poll might turn out.

TABLE 12.1
Example of 2 × 2 contingency table. Responses by males and females to question: "Do you agree that a constitutional amendment should be passed prohibiting abortion?"

Sex	Agree	Disagree	Total
Male	60	40	100
Female	20	40	60
Total	80	80	160

Table 12.1 shows that 60 of the 100 males sampled (60%) favored the amendment but only 20 of the 60 females (33%) were in favor. A statistical test of these data would be a test of independence. That is, the null hypothesis would be that the proportion of males in the population favoring the amendment would be the same as the proportion of females favoring the amendment. Notice that the test is *not* a test of the absolute magnitude of the overall proportion; it is a test of whether the proportion is the same for both sexes.

Table 12.1 illustrates a 2 × 2 contingency table, that is, one with two rows and two columns. In general, an **R × C contingency table** has R rows and C columns; the row variable has R categories and the column variable has C categories. If the poll reported in the table had permitted a category for undecided, the table would be 2 × 3.

12.1.1 AN EXAMPLE OF ALL POSSIBLE SAMPLES

To see how inferences might be made from data in contingency tables, we consider a simple example in which only two males and two females are polled on the amendment issue. For purposes of this example, assume that males and females do *not* differ on the issue and that 50% of the population favors the amendment. Table 12.2, on the next page, lists 16 possible results (samples) of our poll.

Sample 1 shows the outcome in which both males and both females disagree with the statement. Sample 2 shows the outcome in which both males disagree with the statement but one female agrees and the other female disagrees. Notice that samples 2 and 3 appear the same. They are listed separately, however, to reflect the two possible ways the females could respond differently (first female agrees but the second disagrees, and vice versa). Similarly, samples 5 and 9 look the same but reflect the two ways the two males could split on the question. These 16 outcomes show all possible results of a poll of two men and two women. Under the assumption that the population proportion

TABLE 12.2
All possible results of poll of two men and two women on the amendment issue. Agree = A, Disagree = D, Male = M, Female = F.

Sample 1	A	D
M	0	2
F	0	2

Sample 2	A	D
M	0	2
F	1	1

Sample 3	A	D
M	0	2
F	1	1

Sample 4	A	D
M	0	2
F	2	0

Sample 5	A	D
M	1	1
F	0	2

Sample 6	A	D
M	1	1
F	1	1

Sample 7	A	D
M	1	1
F	1	1

Sample 8	A	D
M	1	1
F	2	0

Sample 9	A	D
M	1	1
F	0	2

Sample 10	A	D
M	1	1
F	1	1

Sample 11	A	D
M	1	1
F	1	1

Sample 12	A	D
M	1	1
F	2	0

Sample 13	A	D
M	2	0
F	0	2

Sample 14	A	D
M	2	0
F	1	1

Sample 15	A	D
M	2	0
F	1	1

Sample 16	A	D
M	2	0
F	2	0

favoring the statement is 50% for both men and women, each of these outcomes is equally likely—each has probability 1/16 of occurring. We could use these facts to calculate a number of possibilities. For instance, consider the event in which the proportion of males agreeing with the statement is the same as the proportion of females agreeing with the statement. The probability of this event is 6/16 (six outcomes satisfy the event: samples 1, 6, 7, 10, 11, and 16; and there are 16 possible outcomes). Notice also that there are only four samples that have the same proportion as the population (samples 6, 7, 10, and 11). This means that the probability that the sample proportions will be the same as the population proportions is only 4/16 = .25.

This exercise is designed to show how the probability of all possible outcomes in a contingency table can be calculated, when an assumption is made about the population proportion and about independence. These facts could be used to calculate the likelihood of data such as those given in Table 12.1 and to form the basis of a statistical test of independence. *Fisher's exact test* is based on an approach similar to that outlined here. However, rather substantial computational resources are needed in some applications of the test. We were able to calculate event probabilities from Table 12.2 because only four subjects are considered. Fisher's test can become very laborious when the total number of subjects becomes large (i.e., exceeds 30).

Fortunately, however, an *approximation* is available that is not only suitable for data in a 2 × 2 contingency table but also can be generalized to any $R \times C$ table. This test is **Pearson's χ^2 (chi-square) test**. Pearson's test is used to test the null hypothesis that the data in a $R \times C$ contingency table are sampled from a population in which the proportion falling into the various C categories is the same for all R categories. Applied to the data in Table 12.1, Pearson's χ^2 would test the null hypothesis that the proportion of males agreeing with the abortion statement is the same as the proportion of females agreeing with the statement. In other words, Pearson's χ^2 test tests the null hypothesis that the variables of sex and attitude toward the abortion amendment are independent. We will describe the test and then apply it to the data in Table 12.1.

12.1.2 EXPECTED FREQUENCIES

A critical concept to understanding the χ^2 test of independence is *expected frequency*. Expected frequency should not be confused with *expected value*, defined in Chapter 4 as the mean of a sampling distribution. **Expected frequency**, in contrast, is the frequency that would be expected in a contingency table, if the null hypothesis of independence were true. For example, consider again the data in Table 12.1. There we see that 60 of the 100 males agreed with the amendment statement but only 20 of the 60 females agreed with the statement. What would these frequencies be if the same proportion of males and females agreed with the statement? Notice that Table 12.1 not only reports the frequencies for each of the four cells of the contingency table, but the totals for the rows and columns are listed as well. These totals are called **marginal frequencies** and play a role in the computation of the expected frequencies. Summing across the two sexes we see that 80 of the 160 total subjects agreed with the statement. That is, ignoring sex, half of the subjects (80/160) agreed with the statement. It is this marginal frequency, or more correctly *marginal proportion*, that is used to compute the expected frequencies within the table.

The logic is as follows: if the sexes do not differ on the amendment issue, the fact that the sample proportions differ must be due to chance. Under the circumstances the best estimate of the proportion of the total population (male and female) agreeing with the statement is the proportion of all sampled subjects that agree. The reasoning is not unlike that used to pool estimates of population variance; the more scores used, the better the estimate of the population parameter. In this case, 50% of the total sample agrees with the statement, so our best estimate is that 50% of the population agrees. The expected frequency of males agreeing, then, is 50% of the 160, or 80 males. Similarly, the expected frequency of females agreeing is 50% of 60, or 30 females. These are reported in Table 12.3.

TABLE 12.3
Expected frequencies for data in Table 12.1.

Sex	Agree	Disagree	Total
Male	50	50	100
Female	30	30	60
Total	80	80	160

Note that the marginal frequencies are unchanged. The only differences between tables 12.1 and 12.3 are in the frequencies in the cells within the tables. Recall that the expected frequency is calculated in such a way that the proportion of the column categories within a row is the same as the marginal proportion. In a 2 × 2 table, only one expected frequency need be calculated from the proportions; the remainder of the cells can be determined by subtraction. After it is determined that the expected frequency for females to agree is 30, it follows that 50 males (80 − 30) are expected to agree, and so forth.

In summary, then, to compute the expected frequency for the cell in row i, column j, divide the marginal total of column j by the total and multiply this result by the marginal total of row i.

In order to gain further practice at calculating expected frequencies, let us consider new data. Suppose that a sample of 300 high school seniors who plan to attend college are asked two questions: "Do you enjoy video games?" (Yes or No), and "Do you plan to major in engineering or computer science?" (Yes, No, or Undecided). Their answers permit the investigator to place each senior into one of six cells of a 2 × 3 contingency table. Illustrative data are shown in Table 12.4.

TABLE 12.4
Classification of 300 seniors on basis of answers to two questions.

Enjoy Video Games?	Plan to Major in Engineering or Computer Science?			
	Yes	No	Undecided	Total
Yes	42	129	41	212
No	10	62	16	88
Total	52	191	57	300

We see that of 212 seniors who enjoy video games, 42 (19.8%) plan to major in engineering or computer science. In contrast, only 10 of the 88 (11.4%) seniors who do not enjoy video games plan to major in engineering or computer science. What would the expected frequency be if the answers to these questions were independent; that is, if we were unable to predict the answer to one from knowledge of the other? The rule states that for cell i column j, the expected frequency is obtained by dividing the marginal total of column j by the overall total and multiplying this result by the marginal total of row i. Applying this rule for the cell in row 1 column 1 (row "Yes," column "Yes"),

$$\frac{52}{300}(212) = (.173)(212) = 36.75$$

Notice that the expected frequency for this cell is smaller than the observed frequency (42). Next, the expected frequency of the cell in row 1 column 2 (students responding "Yes" to video games and "No" to engineering),

$$\frac{191}{300}(212) = (.637)(212) = 134.97$$

We could use this rule to calculate the expected frequency for each cell in the contingency table, but a simpler procedure is possible since the marginal totals must remain unchanged. For example, since the total of column 1 (the total of the students planning to major in engineering and computer science) must remain 52, and the expected frequency of those 52 that like video games is 36.75, it follows that the expected frequency that do not like video games is $52 - 36.75 = 15.25$. Similarly, since 134.97 of the 191 seniors not planning to major in engineering are expected to answer "Yes" to video games, then $191 - 134.97 = 56.03$ are expected to answer "No." The remaining cells are determined similarly.

TABLE 12.5
Expected frequencies for the data in Table 12.4.

Enjoy Video Games?	Plan to Major in Engineering or Computer Science?			
	Yes	No	Undecided	Total
Yes	36.75	134.97	40.28	212.00
No	15.25	56.03	16.72	88.00
Total	52	191	57	300

We are now prepared to define a new statistic:

$$\Sigma\Sigma \frac{(O_{ij} - E_{ij})^2}{E_{ij}}$$

The term O_{ij} in this expression refers to the **observed frequency** in the ith row and jth column of an $R \times C$ contingency table. For example, in Table 12.4, O_{12} is 129. The term E_{ij} refers to the **expected frequency** in the ith row and jth column under the null hypothesis of no relation between the two variables. For example, E_{12} in Table 12.5 is 134.97. The specified operations are: (1) find the differences between the observed and expected frequencies, (2) square the differences, (3) divide the differences by the expected frequencies, and finally, (4) total all the resulting ratios (totalling over rows and columns, hence the double summation sign). This statistic is approximately distributed by a well-known probability distribution. This distribution, the χ^2 distribution, is covered in a general way in the next section. After the distribution is discussed, we will return to its application in Pearson's chi-square test of independence.

12.2

CHI-SQUARE DISTRIBUTION

The chi-square distribution is illustrated in Fig. 12.1.

FIGURE 12.1
Chi-square distribution with various degrees of freedom.

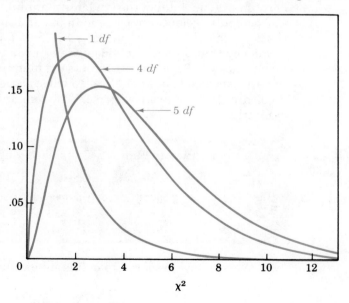

As with t and F, we see that χ^2 is actually a family of curves, each specified by a value of df. Three values are graphed in Fig. 12.1, which shows χ^2 to be positively skewed. It is severely skewed (J-shaped) for $df = 1$, and we find that as df increases, the amount of skew decreases and the mode of the distribution increases. In fact, the mean, or expected value, of each χ^2 distribution is equal to its degree of freedom.

12.2.1 HOW TO READ CHI-SQUARE TABLES

Critical values of χ^2 are reported in Table D in Appendix A and reproduced in part in Table 12.6.

TABLE 12.6
Critical values for the chi-square distribution.

df	5%	1%
1	3.84	6.64
2	5.99	9.21
4	9.49	13.28
5	11.07	15.09

Table 12.6 reports two values for each *df*. One value (in the column labelled 5%) is that value of χ^2 above which 5% of the distribution falls. For example, for $df = 1$, we see that 5% of the distribution falls above 3.84. For $df = 1$, 1% of the distribution falls above 6.64.

Fig. 12.2 illustrates the critical values for $\alpha = .05$ for $df = 1, 4,$ and 5.

FIGURE 12.2
Chi-square distributions showing critical values for $\alpha = .05$.

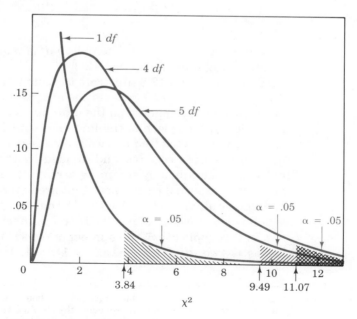

The critical value for $\chi^2 (df = 5)$, $\alpha = .05$, is found in Table 12.7 to be 11.07. In Fig. 12.2 the area above the critical value is shaded as ▨.

That is, the area shaded ▨ is the upper 5% of the χ^2 ($df = 5$). Similarly, for χ^2 ($df = 4$), $\alpha = .05$, the critical value is 9.49 and the area above that value is shaded as ▨. Fig. 12.2 also shows the critical value for χ^2 ($df = 1$), $\alpha = .05$.

In summary, the values reported in Table D in Appendix A and Table 12.6 are critical values of χ^2 for $\alpha = .05$ and $\alpha = .01$. The entry in the table for a given df is the value of χ^2 above which 5% or 1% of the distribution falls. The table may therefore be used to evaluate the likelihood of a statistical outcome. If a statistic is distributed as χ^2, the probability that it will exceed the tabled critical value is less than or equal to the associated α. For example, if a statistic is distributed as χ^2 ($df = 1$), the probability that it will exceed 3.84 is .05.

12.2.2   COMPUTATIONAL EXAMPLES

We have stated that the previously defined statistic $\sum\sum\dfrac{(O_{ij} - E_{ij})^2}{E_{ij}}$ is distributed *approximately* as χ^2. The approximation is improved when the absolute values of the differences are reduced by .5. Equation 12.1 shows the Pearson χ^2 test corrected* in this way.

$$\chi^2(df) = \sum\sum\frac{(|O_{ij} - E_{ij}| - .5)^2}{E_{ij}} \tag{12.1}$$

where $df = (R - 1)(C - 1)$. The reason df is related to R and C in this manner is explained in Section 12.3.

Tables 12.7 and 12.8 illustrate the computation of χ^2 on the data in Table 12.1. Table 12.7 repeats the observed and expected frequencies, and the χ^2 is computed on the differences as shown in Table 12.8.

Notice that for each of the four cells, the difference between the observed and expected frequencies is obtained, and then reduced by .5. This result is then squared and divided by the expected frequency. Finally, χ^2 is obtained by summing across all cells. Be certain to note that it is the expected frequencies that are entered into the denominators. In general, $df = (R - 1)(C - 1)$; that is, df equals the product of the number of rows less one *times* the number of columns less one. Since, in this application, the number of rows is two and the number of columns is two, $df = (2 - 1)(2 - 1) = 1$. If we now compare the

---

*There is some disagreement among statisticians about the conditions under which χ^2 should be corrected. This correction (reducing the absolute value of each difference by .5) is known as a *correction for continuity*. It is often suggested that the correction is unnecessary when df is larger than one.

TABLE 12.7
Responses by males and females to question: "Do you agree that a constitutional amendment should be passed prohibiting abortion?"

Sex	Observed Frequencies		
	Agree	Disagree	Total
Male	60	40	100
Female	20	40	60
Total	80	80	160

Sex	Expected Frequencies		
	Agree	Disagree	Total
Male	50	50	100
Female	30	30	60
Total	80	80	160

TABLE 12.8
Computation of chi-square statistic for data in Table 12.7.

$$\chi^2(R-1)(C-1) = \sum\sum \frac{(|O_{ij} - E_{ij}| - .5)^2}{E_{ij}}$$

$$\chi^2(2-1)(2-1) = \frac{(|60-50|-.5)^2}{50} + \frac{(|40-50|-.5)^2}{50} + \frac{(|20-30|-.5)^2}{30} + \frac{(|40-30|^2-.5)^2}{30}$$

$$\chi^2(1) = \frac{(10-.5)^2}{50} + \frac{(10-.5)^2}{50} + \frac{(10-.5)^2}{30} + \frac{(10-.5)^2}{30}$$

$$= 1.805 + 1.805 + 3.008 + 3.008 = 9.626$$

obtained χ^2 with Table 12.6, $df = 1$, we find that the obtained values exceed the tabled value of χ^2 (1), $\alpha = .05$. What do we conclude? If males and females did not differ in their expressed attitude toward the amendment statement, the probability of obtaining a χ^2 of this magnitude is less than .05. We may therefore reject the null hypothesis that sex and expressed attitude toward the abortion amendment are independent in favor of the hypothesis that the sexes differ in this regard. Stated another way, knowing a person's sex improves one's ability to predict how they will respond to the abortion amendment poll.

You may have noticed that the test used was *one-tailed*. That is, the critical values for the χ^2 statistic were placed at only one end of the distribution. Pearson's χ^2 test of independence is a one-tailed test in the same way that the F test in an analysis of variance is one-tailed. If

the null hypothesis of no difference is false, the statistic should be larger than the expected value regardless of the direction of the difference. For example, we *might* have entertained the experimental hypothesis that males are more likely to support the amendment statement than are females. But the χ^2 value is insensitive to the direction of the difference. To see this, suppose the agree-disagree frequencies in Table 12.1 were reversed so that 40 females had agreed and 20 disagreed and 60 males agreed but 40 disagreed. The value of the resulting χ^2 would be unchanged. It would be identical to that already calculated, and the null hypothesis would be rejected.

The use of computer statistics packages to calculate a χ^2 test of independence is discussed in sections 14.1.7, 14.2.7, and 14.3.7.

As a second example, we now test independence in the data reported in Table 12.4. Using the expected frequencies in Table 12.5, the statistic is calculated in Table 12.9.

TABLE 12.9
Computation of chi-square statistic on data in Table 12.4.

$$\chi^2(R - 1)(C - 1) = \sum\sum \frac{(|O_{ij} - E_{ij}| - .5)^2}{E_{ij}}$$

$$\chi^2(2 - 1)(3 - 1) = \frac{(|42 - 36.75| - .5)^2}{36.75} + \frac{(|129 - 134.97| - .5)^2}{134.97} + \frac{(|41 - 40.28| - .5)^2}{40.28} + \frac{(|10 - 15.25| - .5)^2}{15.25}$$

$$+ \frac{(|62 - 56.03| - .5)^2}{56.03} + \frac{(|16 - 16.72| - .5)^2}{16.72}$$

$$\chi^2(1)(2) = \frac{(4.75)^2}{36.75} + \frac{(5.47)^2}{134.97} + \frac{(.22)^2}{40.28} + \frac{(4.75)^2}{15.25} + \frac{(5.47)^2}{56.03} + \frac{(.22)^2}{16.72}$$

$$\chi^2(2) = 2.853$$

Comparing the observed value of χ^2 with the critical value of χ^2 (2), $\alpha = .05$, we see that the observed value is smaller than the critical value (5.99). In this case, we cannot reject the null hypothesis of independence. Recall that these data involved the answers to two questions, one about video games, another about planned college major. The χ^2 test does not permit us to conclude that answers to the two questions are related.

12.2.3 ASSUMPTIONS UNDERLYING THE USE OF CHI SQUARE AS A TEST OF INDEPENDENCE

In contrast to the t and F tests, which are based on the assumption that the samples are taken from normal distributions, Pearson's test of independence does not depend on a assumption about the form of the population distribution. Statistical tests that have this property are called **distribution-free** statistics. Additional distribution-free tests are described in the next chapter.

However, the χ^2 test is based on the assumption that every observation is independent of the other. Observations are independent, you will recall, when knowledge of one does not aid in the prediction of the other. Pearson's χ^2 test would be inappropriate if this assumption were violated, which would occur if an individual subject contributed more than one observation. Suppose, for example, that 20 subjects were given a four-item test. The resulting 80 observations (categorized pass-fail, for example) should not be submitted to Pearson's test, because the different responses made by the same subject are not likely to be independent. It is important to stress, in other words, that the N observations in a contingency table must be single observations from N independently sampled subjects.

Another limitation on the use of Pearson's test arises from the fact that it is only approximately distributed as χ^2. As we have seen, the approximation is improved when .5 is subtracted from each of the differences between observed and expected frequencies. Nevertheless, under certain conditions even the corrected statistic departs too greatly from χ^2 to be useful. These circumstances arise when the expected frequencies are small. Although there is some disagreement among statisticians, a useful rule is that when $df = 1$, all expected frequencies should be greater than 10. When $df = 1$ and any *expected* frequency is less than 10, Fisher's exact test is recommended. Useful tables for Fisher's exact test for sample sizes of 30 and less may be found in Siegel (1956), and Pearson and Hartley (1966). When df is larger than 1 and any expected frequency is less than 5, Pearson's statistic is not well approximated. Often this problem can be met by collapsing categories, although the reader is advised to consult Hays (1973) before pursuing this option.

12.3

THE NUMBERS OF DEGREES OF FREEDOM

Recall that df is a measure of the extent to which the values of the observations are free to vary and still satisfy some constraint. In a contingency table, the constraint is that the marginal frequencies are fixed. Now, given that condition, how much are the cell frequencies free to vary? Suppose the marginal frequencies are:

		20
		30
25	25	50

How many cells are free to vary? Suppose the cell in the first row, first column is 8.

8		20
		30
25	25	50

We see immediately that once a cell, any cell, is fixed, the rest of the cells are determined. Given that the marginal frequencies must remain unchanged, if an 8 is placed in row 1 column 1, then row 1 column 2 must be 12, because the first row must sum to 20; and row 2 column 1 must equal 17, because the first column must total 25. Once these are determined, we see that row 2 column 2 must equal 13 to satisfy the marginal totals of row 2 and column 2.

8	12	20
17	13	30
25	25	50

We have seen, then, that in a 2 × 2 contingency table the value of one cell is free to vary; that is, there is one degree of freedom, once the marginal frequencies are fixed.

Now consider a 2 × 3 contingency table. Suppose we have

			60
			40
35	33	32	100

How many degrees of freedom are there? Suppose 15 is in the first row, first column:

15			60
			40
35	33	32	100

Since the marginal totals are fixed, the value of row 2 column 1 must now be 20:

15			60
20			40
35	33	32	100

but no other cell is determined. We have more than one *df* here. Consider what remains. We are left with an empty 2 × 2. As before, once any remaining cell is fixed the rest are determined. Placing 18 as follows:

15	18		60
20			40
35	33	32	100

forces:

15	18	27	60
20	15	5	40
35	33	32	100

In this case we have 2 *df*. Any two cells are free to vary before the rest are determined by the marginal frequencies. Recall that when the expected frequencies were calculated for the 2 × 3 contingency table in Table 12.4, we noted that after two cells were computed the rest could be determined by subtraction. This reflects the fact that two cells are free to vary, but once determined, the rest are fixed. The same logic would apply in extending the treatment to any $R \times C$ table. In general, there are $(R - 1)(C - 1)$ *df* because once $(R - 1)(C - 1)$ cells are determined, the rest are fixed.

12.4

SUMMARY

Categorical measures are those in which subjects are assigned to categories. Analyses of categorical data are performed on the frequencies with which categories occur. When subjects are assigned to categories according to two variables simultaneously, they are placed into a contingency table. An **R × C contingency table** has R categories for the row variable and C categories for the column variable. The entries of a contingency table are the frequencies of subjects assigned to each of the $R \times C$ combinations of the two variables. **Pearson's χ^2 (chi-square) test** is a test of independence between the row and column variables which uses the **observed frequencies, O_{ij},** in a contingency table. **Expected frequencies, E_{ij},** are frequencies in a contingency table that would be expected if the two variables were independent. **Marginal frequencies** are row and column totals.

Pearson's χ^2 test is an example of a **distribution-free test**, a statistical test that does not depend on an assumption about the form of the population distribution. However, Pearson's test does depend on the assumption that each subject is independently sampled. The number of degrees of freedom (df) for the test equals $(R - 1)(C - 1)$.

REFERENCES

Hays, W. L. 1973. *Statistics for the social sciences*. New York: Holt, Rinehart & Winston.

Pearson, E. S., and Hartley, H. O., eds. 1966. *Biometrika tables for statisticians*, vol. I. London: Cambridge University Press.

Siegel, S. 1956. *Nonparametric tests for the behavioral sciences*. New York: McGraw-Hill.

QUESTIONS

1. What is a distribution-free test?

2. Other than being "distribution-free," what assumption is associated with a chi-square test?

3. When performing analyses on categorical data, we use computations involving sample numbers. Explain the difference between the numbers that are used in analyses of numerical data versus those used in analyses of categorical data.

4. Describe the difference between *analyses* done with categorical and numerical data.

5. Define *expected frequency*. How is an expected frequency obtained?

6. Describe the null hypothesis that is tested by Pearson's χ^2 test.

7. Could χ^2 be used for a two-tailed test? Why or why not?

8. A sample survey, designed to investigate whether persons living in different parts of the country like to play video games, yielded the following data:

	East	South	Mid-South	Mid-West
Yes	424	405	386	385
No	244	257	294	265
Never Played	247	253	235	265

Test the null hypothesis that enjoying video games is independent of the geographic area of the country. Use $\alpha = .05$.

9. The following data pertain to shipments received by a car dealership from three manufacturers:

	Rejected	Flawed But Accepted	Accepted
Manufacturer 1	34	45	111
Manufacturer 2	30	34	84
Manufacturer 3	43	52	141

Use $\alpha = .05$ to test whether the three manufacturers ship cars of equal quality.

10. Consider the following data as a study of whether graduates with a particular major acquired the job they preferred:

	Preferred	Not Preferred	No Job
Humanities	49	65	39
Engineering	77	81	34
Social Sciences	45	46	41
Physical Sciences	48	49	53
All Other Fields	56	84	58

Use $\alpha = .05$ to test the null hypothesis that the true proportions of students attaining preferred, not preferred, or no jobs are the same for all majors.

11. Suppose the data collected from the 300 seniors described earlier in this chapter had resulted as in the following table:

Major in Engineering or Computer Science?

Enjoy Video Games?	Yes	No	Undecided
Yes	47	160	49
No	5	31	8

Using $\alpha = .05$, what can you conclude about the data?

12. A social psychologist interested in whether there is a relationship between perceived attractiveness and a person's hair color collected the following data:

	Attractive	Average	Homely
Redhead	72	115	68
Brunette	67	100	65
Blonde	80	144	79

Use $\alpha = .05$ to test the null hypothesis that there is no relationship.

13. Use the following data from a random sample of 1140 people to determine if level of education and amount of television viewing are independent.

Average Hours of Daily Television Viewing

Educational Level	≤ 1	1–2	2–3	2–4	>4
High School or Less	22	49	128	115	60
1–4 yrs. College	62	84	157	109	25
More Than B.A.	169	271	341	256	103

14. A psychologist hypothesized that birth order was a variable related to school achievement. Test this hypothesis against the null hypothesis of independence using the data below:

	Youngest	Middle	Oldest
High Achievement	11	19	25
Low Achievement	20	20	5

15. Two hundred students attending a lecture were asked the month of the year in which they were born. The frequencies follow. Devise a test to see whether the frequencies differ significantly across months. (That is, do there seem to be more births some months than others? *Note:* The degrees of freedom for this problem equal the number of categories minus one.)

Month	Number	Month	Number
Jan.	16	July	16
Feb.	17	Aug.	18

March	16	Sept.	16
April	13	Oct.	18
May	19	Nov.	15
June	20	Dec.	16

16. A recent city election included a one-cent gas tax referendum to help finance the mass transit system. The following data are a sample of results from three districts: district 1 is a low income area, district 2 is predominantly middle class, and district 3 is an upper class area.

District	For	Against
1	1142	682
2	1517	1752
3	1150	1974

(A) Were there significant differences in voting patterns among the three districts?

(B) If voting behavior was independent of district, how many people in district 1 would you have expected to vote against the referendum?

17. An industrial psychologist wanted to know if the need for physical privacy in the workplace differed for different types of employees. Employees of a large company who worked in open offices (large rooms with no partitions among workspaces) were asked if they felt they had enough privacy in their work environment. Their responses were

	Yes	No
Clerical Workers	45	15
Managers	12	23

Compute a χ^2 and decide if managers and clerical workers have the same need for privacy.

18. A gas station attendant was of the opinion that women drivers are more reluctant to pump their own gas. To test his idea, he observed 225 customers at his station and noted which service lane they chose. Evaluate the data and decide if the attendant is correct.

	Self Serve	Full Service
Men	81	58
Women	39	47

13 Nonparametric Statistics

The inferential statistics described in chapters 5–11 are all *parametric* statistics that involve estimates of the parameters μ and σ^2. The statistics discussed in the present chapter do not involve estimates of μ and σ^2, and they are called *nonparametric* statistics. They also have the property of being *distribution free*, by which is meant that significance levels do not depend on an assumption that the population distribution has a particular form. Inferences made about t, F, and r depend on the assumption that the population is normally distributed. The statistical tests described in this chapter do not depend on this assumption.

There are two kinds of nonparametric statistics, those based on *counting* and those based on *ordering*. Statistics based on counting simply involve determining the frequencies with which certain categories occur. We have seen an example of this operation in Pearson's χ^2 (chi-square) test. Statistics based on ordering, on the other hand, involve arranging or *ranking* measures in relative order on some dimension. Suppose, for example, we were to rank three people in order of height. Assigning a "1" to a person would mean that person is tallest, giving another a "2" would indicate a middle height, and the shortest would then be assigned "3." We have used an example in which a quantitative measure (e.g., in inches) is transformed into a rank-order measure. Often, however, a rank-order measure is used because a quantitative measure is not feasible or appropriate.

Measures of preferences are good examples of rank-order measures. If asked to indicate your preferences among a set of possible careers (air traffic controller, architect, archeologist, forest ranger, teacher, TV game show host), you may find it easier to rank order them than to assign a number to each. Similarly, many experimental measures are obtained by having trained judges rank order subjects or stimuli on some dimension. For example, an investigation of social dominance in gorillas may involve assessing a "pecking order" in a troop of gorillas. Or a study of reading may have teachers rank order kindergarten children on "reading readiness."

Measures based on rank ordering are referred to as *ordinal* measures. An important feature of an ordinal scale is that it has no zero. Thus, a set of rank-ordered subjects (1, 2, 3, . . .) could all be near the high end of the measured dimension or near the low end, and we cannot tell which. The ordinal measure provides only relative information about the members of the set being ranked.

13.1

COMPARING PARAMETRIC AND NONPARAMETRIC TESTS

Although the normality assumption is lacking in nonparametric tests, the logic guiding statistical inference is the same as for parametric tests. First, a rule is stated so that some number, G, can be assigned to any set of data. Second, a null hypothesis and assumptions are made so that the sampling distribution of G can be determined. From this distribution, critical values associated with $\alpha = .01$ and $\alpha = .05$ can be computed and tabled. Finally, the observed value of G is compared with the critical value and H_0 is rejected or not. Consider briefly, for example, the *sign test*, which involves a test on pairs of observations. The sign test might be used instead of the t test on difference scores in the pretest-posttest design. To illustrate this, Katahn's weight-management data are repeated in Table 13.1.

TABLE 13.1

Katahn's weight-management data.

Subject	Pretest	Posttest	D
1	261.50	256.50	+5.00
2	212.75	214.00	−1.25
3	218.25	215.00	+3.25
4	185.75	184.75	+1.00
5	204.25	203.75	+.50
6	152.00	151.75	+.25
7	207.00	204.50	+2.50
8	293.75	290.50	+3.25
		Sum	14.50

The rule applied to calculate the statistic used in a sign test is simple: count the number of positive difference scores. That is, count the number of times the first member of the pair is larger than the second member. Table 13.1 reveals seven positive scores in the eight pairs. The null hypothesis would be that each difference score is as likely to be positive as it is to be negative. Assuming that each score is independent, the resulting sampling distribution is a classic distribution in probability theory. From this distribution we can calculate the likelihood of different numbers of positive and negative scores from a set of N difference scores. We will explore the sign test further shortly. The point emphasized here is that although the distributions used in nonparametric statistics may differ from those arising from the normality assumption, the logic guiding the inference is the same.

In the sections that follow, several nonparametric tests are described. Each test may be thought of as an alternative to a particular parametric test. For example, the sign test is an alternative to the t test on a mean of difference scores. If the experimental measures are ordinal measures, the appropriate test is the nonparametric test. If, however, the measure is neither ordinal nor categorical, the experimenter has to choose between a parametric test or a nonparametric test. One advantage of selecting the nonparametric alternative is that the nonparametric test does not depend on the assumption of normality. Thus, if an investigator finds the data depart substantially from the normal distribution, the nonparametric alternative may be chosen. Another advantage of nonparametric tests is that they are often easier to compute than their parametric counterparts. This advantage disappears, however, when computer statistical packages, such as those described in Chapter 14, are available.

Nonparametric tests are typically less *powerful* than their parametric counterparts. That is, the parametric test usually has a greater probability of rejecting an untrue H_0 than does the nonparametric test, especially when the distribution is normal. We saw in Chapter 6 that the power of a test is often a dominant concern of the experimenter. If a difference between conditions is present, we want to be able to detect it. For this reason, when a choice is available, parametric tests are often preferred over nonparametric ones.

In the following sections, nonparametric tests for a variety of statistical problems are discussed. Their order of coverage is the same as the order of their parametric counterparts in chapters 5–12. We also illustrate many of the tests with the identical data discussed in these earlier chapters. This helps us compare the two kinds of tests. The reader is reminded, however, that when the measures are ordinal or categorical, the appropriate test is a nonparametric test.

The use of computer statistics packages to perform these nonparametric tests is discussed in sections 14.1.8, 14.2.8, and 14.3.8.

13.2

ALTERNATIVES TO A t TEST ON A MEAN OF DIFFERENCES

In chapters 5 and 7 we saw examples in which two groups of related, or matched, subjects are observed and the parametric t test was used to test a null hypothesis about the mean score. For instance, in Katahn's study eight subjects are weighed twice in a pretest-posttest evaluation of a weight-management program. A set of eight difference scores was submitted to a *t* test of the null hypothesis that the mean difference score is zero. In another example, two groups of subjects are matched and a *t* test is performed on the mean of the differences between the matched pairs. A nonparametric alternative could have been used in either example. In fact, at least two nonparametric tests are available to test differences between matched or related pairs. Each tests a slightly different null hypothesis. One H_0 is that the *sign* of any difference score is as likely to be positive as negative. This is the approach taken by the sign test. Another way to nonparametrically test the effect of the manipulated variable on difference scores would be to use the *Wilcoxon test*, which will be described after the sign test.

13.2.1 SIGN TEST

As stated, the **sign test** is performed on a set of difference scores obtained from measures on related or matched pairs. Only the signs of the differences are assessed and the critical statistic is the number of positive differences (or the number of negative differences). In Katahn's data there are seven positive and one negative difference scores. Whether the sum of positive or the sum of negative differences are to be used depends on the experimental hypothesis. A directional experimental hypothesis specifies whether more positive or more negative difference scores are expected. For example, in Katahn's experiment the subjects are expected to lose weight, so subtracting postweights from preweights should produce more positive than negative difference scores. In general, the number to be compared with the critical value is the number (positive total or negative total) predicted to be the larger by the experimental hypothesis. This is, of course, a one-tailed test. Thus, in Katahn's data the number of positive difference scores is to be compared with the critical value.

If, however, a nondirectional hypothesis is involved (that is, a two-tailed test is used), the number to be compared with the critical value is whichever is the larger. For example, if there were eighteen negative and three positive scores, the eighteen should be used in a two-tailed sign test.

Table E in Appendix A reports critical values computed from the sampling distribution to be used with the sign test. Part of Table E is repeated on the next page as Table 13.2.

The entries in Table E and Table 13.2 are the numbers of positive (or negative) scores out of N difference scores required for the sign test to be significant at $\alpha = .05$ and $\alpha = .01$. The tables report critical values for one-tailed and two-tailed tests. First, find in the first column of the table the number of difference scores involved (N). The critical values for each N are found in one of the next four columns on that row.

TABLE 13.2

Critical values for the sign test. Selected values from Table E in Appendix A.

N	One-tailed .01	One-tailed .05	Two-tailed .01	Two-tailed .05
7	7	7	—	7
8	8	7	8	8
9	9	8	9	8
10	10	9	10	9
11	10	9	11	10
12	11	10	11	10
13	12	10	12	11

In order to be significant, the obtained value must be *equal to* or *greater than* the tabled value. Suppose, for example, there are thirteen difference scores ($N = 13$). We see from the table that for a one-tailed sign test to be significant at the .05 level, the number of difference scores that are in the predicted direction should equal 10 (the tabled critical value) or greater. Remember that for a one-tailed test, the critical value is to be compared with the number of difference scores that have a sign in the predicted direction. However, for a two-tailed test (nondirectional hypothesis), the critical value is to be compared with the number of positive scores or the number of negative scores, whichever is larger. For example, if $N = 13$ and a two-tailed test is used, Table 13.2 tells us that to reject H_0 at the .05 level of confidence, there must either be 11 or more positive scores *or* 11 or more negative scores.

Katahn's data produced a significant t when a one-tailed test was used with $\alpha = .05$. What happens when a sign test is used on the same data? We see in Table 13.2 that for $\alpha = .05$, using a one-tailed test, the critical value is 7. Since the number of positive difference scores is equal to 7, we may reject H_0. Notice, incidentally, that if the test had been two-tailed, 8 of the difference scores would have had to be positive (or negative). In this case, H_0 would not have been rejected. This outcome is different from the conclusion that would be reached with a t test. A two-tailed t test would have permitted rejecting H_0 at the .05 level. This example is consistent with what we know about the relative power of the t and sign tests. Here we see a situation in which t permits the rejection of H_0 but the sign test does not. If H_0 is in fact false, the application of a sign test would have led to a type II error, but the t test would have led to the correct decision. That is, in this example the t test has sufficient power to avoid the type II error.

In summary, the **sign test** is a nonparametric test between two matched or related sets of scores. To perform the sign test, the differences between the pairs are determined and the statistic used is the number of positive (or negative, depending on the experimental hypothesis) differences. This number is compared to the critical values reported in Table E of Appendix A. If the critical value is exceeded, the null hypothesis of no difference may be rejected.

13.2.2 WILCOXON TEST

Like the sign test, the **Wilcoxon test** is applied to difference scores from matched pairs. Unlike the sign test, however, the Wilcoxon does not ignore the magnitude of the differences. In the Wilcoxon test, these magnitudes are rank ordered in terms of their *absolute* magnitude; that is, regardless of sign. This operation is illustrated with the Katahn data in Table 13.3.

TABLE 13.3
Rank order of absolute magnitudes of Katahn difference scores.

Subject	D	Rank Order
1	+5.00	8
2	−1.25	4
3	+3.25	6.5
4	+1.00	3
5	+.50	2
6	+.25	1
7	+2.50	5
8	+3.25	6.5

Notice that the rule is to assign rank order "1" to the *smallest* value, assign "2" to the next smallest, and so forth. Note also that the differences are ranked in order *regardless* of sign. For this reason, the score of −1.25 is ranked fourth between +1.00 (ranked third) and +2.50 (ranked fifth). The next step is to identify the sign (positive or negative) that is the less frequent. In Katahn's data the negative sign, which occurs one time, is less frequent than the positive sign, which occurs seven times.

The statistic W, then, is defined as the *sum of the ranks of scores of the less frequent sign*. For example, if there were 20 scores and four were negative, W would equal the sum of the ranks of the four negative difference scores. In Katahn's data there is only one negative score. Its value is 4, so the value of W is 4 in this case.

Critical values from the sampling distribution of W are reported in Table F of Appendix A and repeated in part in Table 13.4.

To use Table F, first find the row labelled with the appropriate N, the number of difference scores. For each N two critical values are given, one for $\alpha = .05$, another for $\alpha = .01$. (A dash, —, in the table indicates that no decision is possible at the stated level of significance.)

N	Level of Significance for One-tailed Test	
	.05	.01
6	2	—
7	3	0
8	5	1
9	8	3

The obtained value must be *equal to* or *smaller than* the critical value in order to be significant at the indicated level of α. For example, for $N = 9$, W must equal 8 or less to be significant at the .05 level.

The value of W obtained from the Katahn data is 4, and the tabled critical value for $N = 8$ and $\alpha = .05$ is $W = 5$. Since the obtained value is smaller than the tabled value, we may reject H_0. This, of course, is the same decision reached with the t and sign tests on these data.

Table F reports critical values of W for $N = 25$ or smaller. When sample sizes are larger, we can take advantage of the fact that the sampling distribution of W is approximately normal, with known mean

$$E(W) = \frac{N(N + 1)}{4}$$

and known standard deviation

$$\sigma_W = \sqrt{\frac{N(N + 1)(2N + 1)}{24}}$$

This permits us to use the normal distribution to make inferences about obtained values of W when sample sizes are large. We do this by transforming the obtained value of W into a *standard normal score*. You will recall from Chapter 2 that in general a score X from a distribution with mean μ and standard deviation σ is transformed into a standard score z by

$$z = \frac{X - \mu}{\sigma}$$

In the current application, the relevant distribution is the sampling distribution of W, which is approximately normal with mean $E(W)$ and standard deviation σ_W. Therefore, to transform an obtained W into a standard score, we substitute appropriately:

$$z_W = \frac{W - E(W)}{\sigma_W}$$

$$z_W = \frac{W - \dfrac{N(N + 1)}{4}}{\sqrt{\dfrac{N(N + 1)(2N + 1)}{24}}} \tag{13.1}$$

There is a very important result of all this. Since the standardized normal is tabled, when Table F is useless because $N > 25$, we may transform W into z_W. We then determine the likelihood of z_W by consulting tabled values of the standardized normal. These are reported in Table 13.5.

TABLE 13.5
Critical values for the standardized normal distribution.

Type of Test	.05	.01
One-tailed test	1.645	2.326
Two-tailed test	1.960	2.576

To be significant, the obtained value must equal or exceed the tabled value.

Suppose that an experiment is performed testing a nondirectional hypothesis. Suppose, further, that data are collected from 30 pairs of matched subjects and the resulting W is computed to be 330. To test the significance of this observed value of W, we insert it along with the value of N into Equation 13.1 and compare the obtained value of z_W with the critical value given in Table 13.5.

$$z_W = \frac{W - \dfrac{N(N + 1)}{4}}{\sqrt{\dfrac{N(N + 1)(2N + 1)}{24}}} = \frac{330 - \dfrac{(30)(31)}{4}}{\sqrt{\dfrac{(30)(31)(61)}{24}}}$$

$$= \frac{330 - 232.5}{48.618} = 2.005$$

Using a two-tailed test with $\alpha = .05$, we find that the observed value of z_W is larger than the critical value of 1.960. We may therefore reject the null hypothesis of no difference between the sets of paired scores at the .05 level of confidence. That is, there is a statistically significant difference.

In summary, the **Wilcoxon test** is a test between two related or matched groups. It first requires that the magnitude of the difference scores be rank ordered. The statistic W is defined as the sum of the ranks of scores of the less frequent sign. Determination of the critical values of W depends on whether N, the number of difference scores, is larger or smaller than 25. When N is 25 or less, W is compared to critical values in Table F. When N is larger than 25, W is transformed to z_W, according to Equation 13.1, and compared against critical values of the standardized normal distribution, reported in Table 13.5.

13.2.3　A NOTE ON TIES

There are two types of "ties," or equalities, that can influence how the sign and Wilcoxon tests are applied. First, there may be a tie between the matched scores. That is, a difference score may be equal to zero. When this type of tie occurs, the zero scores are placed aside and N is reduced by the number of zero differences. For example, if two of Katahn's eight subjects had neither gained nor lost weight, the sign and Wilcoxon tests would have been performed on the remaining subjects with $N = 8 - 2 = 6$.

A second type of tie is that between ranked scores. That is, two or more scores may tie for the same rank. This type of tie is illustrated in Table 13.3, where it can be seen that two subjects had the same D score ($+3.25$) and were tied for sixth and seventh. When a tie occurs in rank orders, the tied values are assigned the *average* rank for the tied values. Thus, in Table 13.3 the values tied for 6 and 7 were assigned 6.5. If three scores were tied for 13th, 14th and 15th, they all would receive a rank of 14. Remember, in such cases the next score assigned a rank continues with the rank beyond 15, which is 16. Thus, in this example, no score would have a rank of either 13 or 15.

This rule of averaging tied ranks can be applied to the Wilcoxon test as well as to the Mann-Whitney, Friedman, and Kruskal-Wallis tests described next. However, these tests all assume that the underlying dimension is continuous and that no ties would exist if measurements were precise enough. For this reason, correction formulas have been suggested for the special condition existing when ties are present and large sample approximations are used. Such corrections are usually negligible, however, and are not discussed in this text. See Bradley (1968) for advanced discussion of these issues.

13.3

ALTERNATIVE TO A *t* TEST ON TWO MEANS

The **Mann-Whitney test** provides a test of the difference between two independent groups. However, it does not specifically test between two means, as does the *t* test. In contrast, the Mann-Whitney tests the null hypothesis that the two groups are sampled from populations with identical distributions. Nevertheless, it is a popular nonparametric alternative to the *t* test and is used to compare two independent groups.

13.3.1 CALCULATING THE MANN-WHITNEY *U*

Let N_1 and N_2 refer to the numbers of subjects in two groups. (Either group may be arbitrarily assigned as the group with N_1 subjects.) To calculate the Mann-Whitney U, the two groups are combined and rank ordered. As with the Wilcoxon, the rank "1" is assigned to the smallest score. The sum of the ranks is determined for the group with N_1 subjects. This sum is called R_1. The value of R_1 is then inserted into the formula for U:

$$U = N_1 N_2 + \frac{N_1(N_1 + 1)}{2} - R_1 \qquad (13.2)$$

The calculation of U is illustrated using the data from Table 7.1. These data are from an experimental group ($N_1 = 9$) and a control group ($N_2 = 10$). Table 13.6, on the next page, first repeats the scores from Table 7.1, then combines the two groups and rank orders all 19 subjects.

Incidentally, Table 13.6 also shows the computation of R_2, which is the sum of ranks for group 2, in this case, the control group. There is a valuable property of the sum of R_1 and R_2 that can be used to check your computations when they are done by hand. Specifically, if we let the total number of subjects (ranks) be N_T

$$N_T = N_1 + N_2$$

Then it can be shown that

$$R_1 + R_2 = \frac{N_T(N_T + 1)}{2}$$

In Table 13.6

$$R_1 + R_2 = 117.5 + 72.5 = 190$$

To check our calculations we determine that

$$\frac{N_T(N_T + 1)}{2} = \frac{19(20)}{2} = 190$$

We see that the two values agree, as they should.

TABLE 13.6
An example of the calculation of the Mann-Whitney U.

Group	Scores	
Experimental	12, 21, 30, 20, 26, 18, 16, 21, 19	$N_1 = 9$
Control	14, 16, 14, 6, 24, 18, 19, 10, 9, 13	$N_2 = 10$

Rank Ordering

Experimental		Control	
Score	Rank	Score	Rank
30	19		
26	18		
		24	17
21	15.5		
21	15.5		
20	14		
19	12.5	19	12.5
18	10.5	18	10.5
16	8.5	16	8.5
		14	6.5
		14	6.5
		13	5
12	4		
		10	3
		9	2
		6	1
Sum	$R_1 = 117.5$		$R_2 = 72.5$

$$U = N_1 N_2 + \frac{N_1(N_1 + 1)}{2} - R_1$$

$$U = (9)(10) + \frac{9(10)}{2} - 117.5$$

$$U = 17.5$$

13.3.2   USING THE TABLES OF U

Critical values for U are presented in Table G in Appendix A. Table G differs from other appendix tables in that it is composed of four separate tables, one each for the combinations of two levels of α (.05 and .01) with two types of tests (one-tailed and two-tailed). Table 13.7 repeats part of the table associated with $\alpha = .05$, one-tailed.

The columns of the table are devoted to different values of N_1, the rows to the values of N_2. To find the critical values for your test, find the value of N_2 along the left side of the table and the value of N_1 along

TABLE 13.7

Critical values for the Mann-Whitney U test, $\alpha = .05$, one-tailed test. Selected values from Table G in Appendix A.

			N_1		
		7	8	9	10
N_2	7	4, 45	6, 50	7, 56	9, 61
	8	6, 50	7, 57	9, 63	11, 69
	9	7, 56	9, 63	11, 70	13, 77
	10	9, 61	11, 69	13, 77	16, 84

the top. Notice that for each combination of N_1 and N_2, *two* critical values are reported. For example, for $N_1 = N_2 = 7$, the critical values are 4 and 45 ($\alpha = .05$, one-tailed). If the observed value falls *between* these two values, H_0 cannot be rejected at the stated level of significance. If the test is two-tailed and the observed value does not fall between the two critical values, H_0 may be rejected. If the test is one-tailed, however, two conditions must be met in order to reject H_0: (1) the observed value of U cannot fall between the tabled critical values, and (2) the direction of the difference must be in the direction predicted by the experimental hypothesis.

As an example, consider the value of U calculated on the data in Table 13.6. In this case, $N_1 = 9$ and $N_2 = 10$. Inspecting Table 13.7 for a one-tailed test at $\alpha = .05$, we find the critical values are 13 and 77. Since the obtained value, calculated in Table 13.6, is found to be 17.5 we are unable to reject H_0 with this test.

13.3.3 LARGE SAMPLE SIZES

Table G reports critical values for the Mann-Whitney for sample sizes up to $N_1 = N_2 = 20$. For larger samples we can take advantage of the fact that as Ns increase, the sampling distribution of U approaches the normal distribution with mean

$$E(U) = \frac{N_1 N_2}{2}$$

and standard deviation

$$\sigma_U = \sqrt{\frac{N_1 N_2 (N_1 + N_2 + 1)}{12}}$$

Thus, for sample sizes of 20 or larger we can use the standardized normal distribution to test the significance of an observed value of U:

$$z_U = \frac{U - E(U)}{\sigma_U}$$

$$z_U = \frac{U - \dfrac{N_1 N_2}{2}}{\sqrt{\dfrac{N_1 N_2 (N_1 + N_2 + 1)}{12}}} \tag{13.3}$$

The observed value of U is therefore transformed into z_U in accordance with Equation 13.3 and the result compared with the critical values reported in Table 13.5.

In summary, the **Mann-Whitney test** is a nonparametric test between two independent groups of sizes N_1 and N_2. It is performed on ordinal data, or on numerical data that is rank ordered. All scores are combined and rank ordered. Then the ranks of all scores in the group with N_1 scores are summed and entered into Equation 13.2 to determine the magnitude of U. If N_1 and N_2 are each less than 21, the observed value of U is compared with critical values reported in Table G. For larger sample sizes, U can be transformed into z_U and compared with critical values for the standardized normal, reported in Table 13.5.

13.4

ALTERNATIVE TO A WITHIN-SUBJECT ANALYSIS OF VARIANCE

The **Friedman test** is a nonparametric alternative to a within-subject analysis of variance. It is appropriately applied whenever the scores on the N subjects are ranked *across* the J conditions. That is, for each subject the scores are ranked $1, 2, \ldots, J$. We then define R_j as the sum of ranks for condition j (that is, summed across subjects). When this is accomplished, we calculate

$$\chi_R^2 = \frac{12}{NJ(J + 1)} \Sigma R_j^2 - 3N(J + 1) \tag{13.4}$$

Notice that Equation 13.4 calls for the sum of ranks to be determined for each condition, each of these sums to be squared and totalled across conditions (this yields ΣR_j). This quantity is then multiplied by $12/NJ(J + 1)$ and finally reduced by $3N(J + 1)$.

How is χ_R^2 evaluated for significance? It is distributed approximately as χ^2 with $(J - 1)$ degrees of freedom. Therefore, Table D in Appendix A for χ^2 can be used to determine critical values.

An example of the computation of χ^2 is shown in Table 13.8. The data are from an experiment involving three conditions (A, B, and C). The table displays the original scores on the left side and the ranks on the right side. In some experiments, of course, the scores may be ranks originally.

TABLE 13.8
An example of the Friedman test.

Subject	Score A	Score B	Score C	Rank A	Rank B	Rank C
1	16	31	28	1	3	2
2	6	5	10	2	1	3
3	39	45	62	1	2	3
4	0	3	2	1	3	2
5	15	16	21	1	2	3
6	13	10	17	2	1	3
			Sum	8	12	16

$$\chi_R^2 = \frac{12}{NJ(J+1)} \Sigma R_j^2 - 3N(J+1)$$

$$= \frac{12}{6(3)(4)} (8^2 + 12^2 + 16^2) - 3(6)(3+1)$$

$$= 77.333 - 72.000 = 5.333$$

We assess the significance of χ_R^2 using critical values of χ^2 presented in Table D in Appendix A. In the present example, $df = J - 1 = 3 - 1 = 2$. The critical value of χ^2 (2), $\alpha = .05$, is found to be 5.99. Since the observed value is smaller than the critical value we may not reject the null hypothesis.

In summary, a nonparametric alternative to the one-way within-subject analysis of variance is the **Friedman test**, which involves rank ordering scores for each subject across J conditions. These ranks are then summed and entered into Equation 13.4 in order to calculate χ_R^2. Since this statistic is approximately distributed as χ^2 with $(J - 1)$ df, the critical values for χ_R^2 are found in Table D of Appendix A. If the observed value exceeds the critical value, the null hypothesis of no differences among the groups may be rejected.

13.5

ALTERNATIVE TO A BETWEEN-SUBJECT ANALYSIS OF VARIANCE

The **Kruskal-Wallis test** is an extension of the Mann-Whitney test of two groups. As such, it permits the simultaneous test of the differences among J groups and provides an alternative to the one-way between-subject analysis of variance.

As in the Mann-Whitney, the first step in performing a Kruskal-Wallis test is to combine and rank order all scores. Group j has N_j subjects and, in keeping with the notation in Chapter 9, the total number of subjects, N_T, is

$$N_T = \sum_j N_j$$

As with the Friedman test, R_j is defined as the sum of ranks in group j. All such R_j are computed and entered into

$$H = \frac{12}{N_T(N_T + 1)} \sum \frac{R_j^2}{N_j} - 3(N_T + 1) \qquad (13.5)$$

Calculation of H, the test statistic for the Kruskal-Wallis test, is illustrated in Table 13.9, which shows the results of five subjects in each of three groups. In this example, there are 15 subjects, so $N_T = 15$. Notice that the scores in group 1 tend to be smaller than those in group 3, which in turn seem to be smaller than those in group 2. The question is whether these differences are reliable.

TABLE 13.9
An example of the Kruskal-Wallis test.

| | Group 1 | | | Group 2 | | | Group 3 | |
Subject	Score	Rank	Subject	Score	Rank	Subject	Score	Rank
1	22	2	6	31	8	11	24	4
2	23	3	7	34	11	12	36	13
3	21	1	8	37	14	13	30	7
4	32	9	9	35	12	14	33	10
5	26	4.5	10	40	15	15	29	6
	Sum	20		Sum	60		Sum	40

$$\sum \frac{R_j^2}{N_j} = \frac{20^2}{5} + \frac{60^2}{5} + \frac{40^2}{5} = 1120$$

$$H = \frac{12}{N_T(N_T + 1)} \sum \frac{R_j^2}{N_j} - 3(N_T + 1) = \frac{12}{15(16)}(1120) - 3(15 + 1) = 8$$

How do we evaluate an observed value of H? By taking advantage of the fact that the sampling distribution of H, when H_0 is true, is approximately χ^2 with $(J - 1)$ degrees of freedom, at least when the samples are not too small. When all N_j are less than 5, see Siegel (1956).

To illustrate, we compare the observed value of H in Table 13.9 to χ^2 with $(J - 1)$ df. Consulting Table D in Appendix A, we see that the critical value for $\chi^2(2)$, $\alpha = .05$, is 5.99. Since the observed value is larger than the critical value, we may reject the null hypothesis of no difference.

In summary, the **Kruskal-Wallis test** is a nonparametric alternative to the one-way between-subject analysis of variance on groups. Scores of N subjects are combined and rank ordered. The sums of the ranks of each condition are entered into Equation 13.5 to calculate H. When the null hypothesis is true, H is distributed approximately as χ^2 with $(J - 1)$ df. If the observed value of H exceeds the critical value, H_0 may be rejected.

13.6

ALTERNATIVE TO PEARSON'S r

Spearman's r_S is a measure of correlation based on ranks. N pairs of X and Y scores are rank ordered separately on X and Y. We then define d_i as the difference between the X rank and the Y rank for the ith pair. Suppose, for example, that subject 3 is ranked 10 on the X variable and 6 on the Y variable. Then $d_3 = 10 - 6 = 4$.

Spearman's r_S is defined as

$$r_S = 1 - \frac{6\Sigma d_i^2}{N(N^2 - 1)} \tag{13.6}$$

Equation 13.6 requires that the differences in rank be squared and summed. Notice that here d_i is defined as the difference between the *rank* of X_i and the *rank* of Y_i, not the difference between X_i and Y_i. An example is given next on the data from Table 11.4.

TABLE 13.10

An example of the calculation of Spearman's r_S.

Pair (i)	Scores X	Scores Y	Ranks X	Ranks Y	d_i	d_i^2
1	5	9	2.5	5	−2.5	6.25
2	8	7	5	3	2	4
3	1	4	1	1.5	.5	.25
4	5	4	2.5	1.5	1	1
5	7	8	4	4	0	0
6	10	10	6	6	0	0
					Sum	11.5

TABLE 13.10
(*continued*)

$$\Sigma d_i^2 = 11.5$$

$$r_S = 1 - \frac{6 \Sigma d_i^2}{N(N^2 - 1)}$$

$$r_S = 1 - \frac{6(11.5)}{(6)(36 - 1)} = 1 - .329 = .671$$

Notice that Σd_i^2 includes those pairs with zero differences. That is, in contrast to the Wilcoxon test where ties (zero differences) are placed aside, the N associated with r_S is not diminished by tied ranks. The observed value of r_S compares with .74 calculated for Pearson's r on the same data.

For small sample sizes, the significance of r_S may be evaluated by consulting Table H in Appendix A. For example, for $N = 6$, $\alpha = .05$, one-tailed test, the critical value of r_S is found to be .829. Since the observed value in Table 13.10 is smaller than the critical value, we can not reject H_0 with these data.

For Ns larger than covered by Table H, we use the fact that

$$t = r_S \sqrt{\frac{N - 2}{1 - r_S^2}} \qquad (13.7)$$

is distributed approximately as t with $(N - 2)$ degrees of freedom. For a large N, then, r_S can be inserted into Equation 13.7 and its significance evaluated by consulting Table A in Appendix A.

In summary, **Spearman's r_S** is a nonparametric measure of correlation based on ranks. For each of the N pairs, the difference between the ranks on the two measures is obtained and entered into Equation 13.6 to obtain r_S. When N is small, critical values of r_S are found in Table H. When N is large, r_S is transformed to t by Equation 13.7 and compared to critical values in Table A.

REFERENCES

Bradley, J. V. 1968. *Distribution-free statistical tests.* Englewood Cliffs, N.J.: Prentice-Hall.

Siegel, S. 1956. *Nonparametric statistics for the behavioral sciences.* New York: McGraw-Hill.

QUESTIONS

1. If data are suitable for either parametric or nonparametric analysis (i.e., not categorical or ordinal),
   (A) What are the benefits of nonparametric tests?
   (B) What is the major disadvantage of a nonparametric test?

2. Distinguish between nonparametric statistics that are based on counting and those based on ordering.

3. Most statistical tests discussed in this text have stipulated that when the calculated statistic is greater than the critical value, reject the null hypothesis. In this chapter, which tests of the null hypothesis deviate from this procedure? Describe how significance is determined in each case.

4. Complete the following table:

Parametric	Nonparametric
t test (related)	?
t test (independent)	?
?	Friedman test
between-subject analysis of variance	?
?	Spearman's r_S

5. The math clubs from two high schools participated in a contest. High school A won an award for having the highest scoring student. Another award was to be given for the team that scored better overall. Being quite knowledgeable in statistics, the students decided that neither team would win the award if there was no significant difference (.05 level) between them. Conduct a Mann-Whitney test and decide what should be done with the award.

   High school A: 128 123 118 115 110 109 98 79 72 65
   High school B: 124 120 119 114 113 105 101 99 85 80

6. Twenty rats were paired according to equal weight in order to test two types of food. One rat of each pair was fed ration A, and the other received ration B. Both groups were allowed unlimited access to food. The animals were weighed again after one month. The data represent the weight (in grams) gained by each rat:

   Ration A: 21 21 19 16 26 19 18 29 22 19
   Ration B: 30 25 25 16 29 18 18 19 24 22

   Perform a Wilcoxon test to determine if there is a significant difference between the rations.

7. The following are the weekly food expenditures for 10 families and the total number of members in each family. Using a significance level of

.05 and Spearman's r_S, what can you conclude about food expenditures and family size?

Family Size	Expenditures
1	28.19
2	36.00
3	42.45
3	45.57
4	48.29
4	52.96
1	20.19
2	34.92
5	55.26
6	65.23

8. Dr. Gregor Zilstein is interested in improving memory for prose. Five subjects are randomly assigned to three groups. One group is shown pictures that correspond to the topic of the prose, one group hears a recording of the prose, and one groups reads the prose. Based on the following scores (scores on a test of retention), evaluate the data using a Kruskal-Wallis test:

Group 1	Group 2	Group 3
41	50	43
42	59	48
45	53	52
51	54	55
40	56	49

9. Two observers were rating videotapes of seagulls for certain behaviors. They made 24 observations and compared their data to see how much they agreed. Of the 24 items, observer 1 rated higher than observer 2 five times, observer 1's rating was lower than observer 2's 17 times, and they were the same two times. Is there evidence that the ratings were significantly different?

10. A statistics professor wants to know if there is a relationship between the order in which in-class examinations are turned in and scores on the test. The data he collected from the midterm exam are as follows:

Order	Score
1	92
2	99
3	46
4	87
5	96
6	61
7	88
8	90

9	69
10	86
11	58
12	85
13	65
14	93
15	71
16	93
17	68

Is there evidence for a relationship between order and test score?

11. Six subjects are randomly chosen to participate in an experiment where each subject serves in three conditions A, B or C:

Subject	A	B	C
1	23	20	27
2	30	31	36
3	3	6	5
4	60	66	83
5	9	8	13
6	35	50	47

Use the Friedman test to test the null hypothesis that there are no differences among the groups.

12. The following are the number of hours of television viewing per week as reported by 15 men and 12 women:

Men: 3.4 6.9 3.9 6.7 5.4 6.1 5.9 5.1 7.7 5.6 3.6 5.0 4.8 3.3 5.8

Women: 9.5 8.7 9.2 7.5 11.4 7.2 10.2 9.3 10.6 9.7 10.8 8.5

Use Mann-Whitney U to test the null hypothesis that the two groups are sampled from populations with identical populations.

13. Rats were tested under three reinforcement schedules to determine whether there was a difference in the number of bar presses they made to receive food pellets. Schedule A rewarded the animal for every tenth bar press; on schedule B, the animal was rewarded after every twentieth press; and schedule C was random: the animal might sometimes be rewarded after the fifth, tenth, twentieth, or fiftieth bar press. Eight rats were tested for five minutes in all five conditions, and the number of bar presses they made in each are as follows. Is there a significant effect of reinforcement schedule on the rat's behavior?

Rat	A	B	C
1	520	610	1128
2	145	156	339
3	360	342	598

Rat	A	B	C
4	641	710	699
5	225	311	436
6	492	523	704
7	347	310	298
8	156	235	472

14. In a telephone survey, a large number of people were asked to grade the current administration's handling of domestic affairs. The grades given by nine respondents from each of three income levels were selected at random. Conduct a Kruskal-Wallis test to determine whether respondents from different income levels differed in the grades they gave.

Income Level

Low	Middle	Upper
75	82	88
73	73	92
67	81	85
82	66	84
80	67	68
70	71	70
95	83	74
32	80	86
65	74	94

15. A large university is interested in buying a new computer. Research has been conducted on the performance of two brands of computers. "Down-time" (periods when the computer is inoperative due to failure) has been collected for a 22-month period. The data are for minutes of down-time per month.

CED-01: 39 63 66 76 61 66 65 64 88 79 99 63 49 70
         58 91 66 64 67 53 67 72
Xorex: 79 88 55 78 74 78 75 67 69 97 73 64 72 49
         66 87 58 103 69 65 76 71

Using the sign test, test the null hypothesis that the average down-time is the same for both computers.

16. Suppose that the university described in Question 15 has acquired information on down-time from 11 months of operation for the MBI computer. Use the Wilcoxon test to test the following data:

MBI	XOREX
99	79
108	83
75	56
98	78
92	74

98	83
82	75
52	67
54	70
82	93
58	71

What conclusions can you draw?

17. Enlisted soldiers entering an officer training program were given a test on general army procedures. Two years later the same enlistees were rated by their superiors for their competencies as officers. Was there any relationship between the test scores and success as officers?

Test	Compentency Rating
68	93
75	82
59	47
60	61
47	57
95	40
36	37
52	74
73	68
64	90
55	88
71	75

18. A professor wondered whether there was any difference between men and women students in the number of questions they asked during class lectures. Over the course of a semester, the number of questions asked by each student was recorded. The 18 students were ranked by gender in order from most to least number of questions: F F F M F M M M F M F M F F M M F F. Is there any difference between the number of questions asked by men and women students?

19. A class of 28 EMR (educable mentally retarded) students was given an experimental thinking skills training program for one school year. A test of problem-solving ability was administered at the beginning and end of the year. Does a Wilcoxon test give evidence that the thinking skills program had any effect on the students' problem-solving skills?

May:	52	41	37	45	36	49	36	23	31	46	25	41	35
	47	49	41	48	30	24	62	24	62	34	39	35	23
	33	40											
September:	45	39	28	53	36	47	28	19	30	55	21	32	20
	43	53	38	47	36	29	52	31	50	37	28	15	17
	34	42											

14 Computer Statistics Packages

In this chapter, three statistical program packages are described that are widely available on university computer systems. Each package has programs available for performing almost all of the statistics described in this book, and much more. These three packages are MINITAB, BMDP, and SPSS. The easiest to use is MINITAB, which requires virtually no prior computer experience yet can be used in a large variety of applications by students and researchers. BMDP and SPSS are more complex, but they can handle enormous statistical problems.

This chapter is designed to relate the statistics described in earlier chapters to specific commands and programs offered by these packages. This chapter should aid in determining which command or program will perform those computations described in earlier chapters. It is assumed here that the reader is familiar with at least one of the packages and knows how to gain access to it (i.e., run or execute it) on the computer. Excellent manuals for each package have been published, including *MINITAB student handbook* by Ryan, Joiner, and Ryan (1976); *BMDP biomedical computer programs* (1977); and *SPSS primer* by Klecka, Nie, and Hull (1975). In addition, documentation available from the computer should be reviewed along with the manuals to see whether the package has been modified for local use.

All three packages may be used in either the *batch* mode or the *interactive* mode. In the batch mode, each job is submitted on cards along with other jobs and executed in turn. In the interactive mode, the user is connected to the computer via a remote terminal and commands are acted on immediately. The examples given here are useful for either mode. If the mode is batch, the series of lines in the examples may be thought of as a sequence of submitted cards. If the mode is interactive, the series of lines may be thought of as a series of commands, each acted on immediately (by MINITAB), or as a series of lines in a program file executed with BMDP or SPSS.

In the next sections, each package is treated separately. Within each section the computational examples described in earlier chapters are computed anew using the statistical package. The original order of the examples is retained.

We begin with a summary, in Table 14.1, which reports the computer command or program used to compute the same statistics calculated in the original example.

The rows of Table 14.1 correspond to the sixteen statistics defined in numbered equations and illustrated in tables throughout this text. For example, "one *M t*" refers to the *t* test on a single mean, which is used on related pairs in Chapter 5 and matched pairs in Chapter 7. The next column (Equation) reports the number of the equation used in the text for the specified statistic. In those cases (chapters 9 and 10) in

TABLE 14.1
A summary of the programs or commands used to calculate the statistics examples.

Statistic	Equation	Table	MINITAB	BMDP	SPSS
Median		2.1	MEDI	BMDP2D	FREQUENCIES
M	2.1	2.6	AVER	BMDP2D	FREQUENCIES
s	2.7–2.8		STAN	BMDP2D	FREQUENCIES
s^2	2.4–2.6	2.6	—	BMDP2D	FREQUENCIES
One M t	5.3	5.1	TTES	BMDP3D	T-TEST PAIRS =
Two M t	7.6	7.1	POOL	BMDP3D	T-TEST GROUPS =
One-way analysis of variance between S	Table 9.2	9.1	AOVO	BMDP1V	ANOVA
One-way analysis of variance within S	Table 9.6	9.5	TWOW	BMDP2V	RELIABILITY
Two-way analysis of variance	Table 10.7	10.8	TWOW	BMDP2V	ANOVA
r	11.6B	11.4	CORR	BMDP3D CORR.	PEARSON CORR
χ^2	12.1	12.1	CHIS	BMDP1F	CROSSTABS
Wilcoxon		13.3	WILC	BMDP3S WILCOXON.	NPAR TESTS WILCOXON =
Mann-Whitney	13.2	13.6	MANN	BMDP3S KRUSKAL.	NPAR TESTS M − W =
Friedman	13.4	13.8	—	BMDP3S FRIED.	NPAR TESTS FRIEDMAN =
Kruskal-Wallis	13.5	13.9	—	BMDP3S KRUSKAL.	NPAR TESTS K − W =
Spearman r_S	13.6	13.10	CORR	BMDP3S SPEARMAN.	NONPAR CORR

Note: A dash, —, in the cell indicates that the package does not compute the statistic directly. See text for explanation.

which the statistic is based on a number of equations, the reference is to a table number where the equations are given. In other cases (Median and Wilcoxon), the statistic has been defined in text and no equation number is reported. The next column (Table) reports the number of the table in which data are given in the text. These data are used to illustrate each statistic and they are used again in this chapter for each of the three statistical packages. The last three columns of Table 14.1 correspond to the three statistical packages, and the cell entries are the commands or programs that are used to compute the specified statistic. For example, the command MEDI is used in MINITAB to calculate the median of a set of previously entered scores. The table should make clear, therefore, exactly which package command corresponds to the statistics described in chapters 2–13. Some statistics may be computed by more than one command or program. Only those illustrated here are entered in the table.

Applying the packages to the examples used here is often a case of overkill. For example, the mean of six scores could almost certainly be determined with a hand calculator faster than with SPSS. The advantages of the packages are not always apparent with small samples or simple problems. These examples also fail to show the easy flexibility of the packages. For example, the command used to perform a correlation between two variables typically needs but a slight modification in order to perform the correlations between all possible pairs of thirty variables.

14.1

MINITAB

MINITAB works on columns or tables entered into memory by READ or SET commands. After the data have been entered in this manner, most of the statistics may be performed by entering a single command. For example, the numbers 2, 7, and 9 could be entered into column 1 (C1) by

$$- - \text{SET C1}$$

$$- - 2 \ 7 \ 9$$

The average, or arithmetic mean, M, could be computed by

$$- - \text{AVERAGE C1}$$

Notice that these lines are preceded by "– –". When used in an interactive mode, MINITAB prompts the user with these characters. They indicate the program is ready to accept a command or data. In the examples that follow, the "– –" before a line will therefore indicate input to the computer provided by the user, either by typing at a terminal in the interactive mode, or by a series of cards in the batch mode. The output to the user will be shown in the following examples as lines without the "– –" characters. All commands are straightforward. The mean is computed from an AVERAGE command, Pearson's *r* is computed from a CORRELATE command, and so forth. You may spell out the entire word, but the computer attends only to the first four letters of a command.

The version of MINITAB used here is the one documented by Ryan (1980).

14.1.1   *CH. 2*: DESCRIPTIVE STATISTICS

The use of MINITAB to compute the *median* of the data in Table 2.1 is shown in Fig. 14.1.

FIGURE 14.1

```
    -- set c1

    -- 1  1  2  2  4

    -- median c1

       MEDIAN =        2.0000
```

The first line instructs the computer to accept the next line as data to be entered in column 1. The third line is the command to compute the median, which is computed, then output as line four.

The *mean* and *standard deviation* of the data in Table 2.6 are computed in Fig. 14.2.

```
   -- set c2

   -- 2  5  3  3  2  3

   -- aver c2

      AVERAGE =        3.0000
   -- stan c2

      ST.DEV.  =       1.0954
   -- describe c2

      C2        N =   6     MEAN =      3.0000     ST.DEV. =      1.10
```

FIGURE 14.2

This example shows the AVER command which results in the computer printing AVERAGE = 3.0000. The figure also demonstrates the useful DESCRIBE command, which results in the determination of *N*, *M,* and *s*.

14.1.2   *CH. 5*: INFERENCES ABOUT ONE MEAN

A one-sample t test on the Katahn data is performed in Fig. 14.3.

```
-- set c1

-- 261.5 212.75 218.25 185.75 204.25 152 207 293.75

-- set c2

-- 256.5 214 215 184.75 203.75 151.75 204.5 290.5

-- subtract c2 from c1, put into c3

-- ttest 0,c3

   C3          N =    8      MEAN =        1.8125      ST.DEV. =        2.03

   TEST OF MU =      0.0000 VS. MU N.E.      0.0000
   T =   2.520
   THE TEST IS SIGNIFICANT AT    0.0398
```

FIGURE 14.3

First, the pretest scores are placed in column 1, then the posttest scores in column 2. The SUBTRACT command causes the difference scores to be placed in column 3, and finally the TTEST command performs a one-sample t test on the difference scores. The "0" after TTEST asks for a two-tailed test. In some versions of MINITAB, apparently, alternative hypotheses may be specified by using $+1$ or -1 for greater than zero or less than zero, respectively. This option fails to work properly on the version being demonstrated here. Notice that the significance level is reported in such a way that we can compare it against α (instead of comparing t against the critical value of t, as was done in Chapter 7). Since the significance is smaller than .05, we may reject H_0 at the .05 level.

14.1.3   *CH. 7*: INFERENCES ABOUT TWO MEANS

The two-sample t test comparing the experimental and control groups in Table 7.1 is shown in Fig. 14.4.

The command POOL performs the t test and conveniently DE-SCRIBES the data as well. MINITAB also has a TWOS command which performs a two-sample t test *without* pooling the data from the two groups in a combined estimate of σ^2. This test was not treated in Chapter 7.

```
--  set c1

--  12 21 30 20 26 18 16 21 19

--  set c2

--  14 16 14 6 24 18 19 10 9 13

--  pool 0,c1 c2

   C1          N =    9      MEAN =       20.333      ST.DEV. =       5.27
   C2          N =   10      MEAN =       14.300      ST.DEV. =       5.27

   DEGREES OF FREEDOM =    17

   A 95.00  PERCENT C.I. FOR MU1-MU2 IS (      0.9236,      11.1431)

   TEST OF MU1 = MU2 VS. MU1 N.E. MU2
   T =   2.492
   THE TEST IS SIGNIFICANT AT   0.0233
```

FIGURE 14.4

14.1.4 *CH. 9*: ONE-WAY ANALYSIS OF VARIANCE

A *between-subject one-way analysis of variance* is performed in Fig.
14.5 on the next page for the data of the three groups in Table 9.1.

After entering the data in three separate columns, the command
AOVO causes the appropriate analysis. Notice that what is referred to
as "Within Group" in Chapter 9 is referred to as "Error" here and that
"Between Group" is labelled "Factor" here. The critical part of this
report is under "F-ratio" where the value of 3.88 is reported (Compare
this with Table 9.4. The difference is due to rounding only). This time
the user is provided with no report of significance and has to consult a
table of F.

Using MINITAB to perform a *within-subject one-way analysis of
variance* requires some explanation. Recall that the within-subject
analysis of variance first partitions the total sums of squares into
between-subject and within-subject sources. This may be accomplished
in MINITAB by using the command for a two-way analysis of variance
and treating subjects as a main effect. In this way, the condition main
effect is tested against the subject condition interaction. (We labelled
this interaction "Residual" in Chapter 9. MINITAB labels it "Error" in
the following example.) The analysis is performed in Fig. 14.6 on page
311 for the data from Table 9.5.

```
-- set c1

-- 0  1  1  3  3

-- set c2

-- 1  2  2  3  4

-- set c3

-- 2  3  4  5  5

-- aovo c1-c3
ANALYSIS OF VARIANCE

DUE TO            DF            SS         MS=SS/DF        F-RATIO
FACTOR            2          12.40          6.20           3.88
ERROR            12          19.20          1.60
TOTAL            14          31.60


LEVEL            N            MEAN         ST. DEV.
C1               5            1.60           1.34
C2               5            2.40           1.14
C3               5            3.80           1.30

POOLED ST. DEV. =            1.26

INDIVIDUAL 95 PERCENT C. I. FOR LEVEL MEANS
(BASED ON POOLED STANDARD DEVIATION)
        +---------+---------+---------+---------+---------+---------+
C1          I***********I***********I
C2               I***********I***********I
C3                    I***********I***********I
        +---------+---------+---------+---------+---------+---------+
       0.0       1.0       2.0       3.0       4.0       5.0       6.0
```

FIGURE 14.5

The data are placed in column 1, each score coded by the "Level", or number, of subject (1–5) in column 2 and the level of condition (1–3) in column 3. For example,

2 1 3

indicates that the first subject had a score of 2 in condition 3 (see Table 9.5). After the TWOW command, the summary table of the analysis of variance is reported. The main effect of condition is labelled C3. Notice that the F is not reported. You must compute it yourself, using the

```
-- read data in c1, level of subject in c2, level of condition in c3
-- 0 1 1
-- 1 2 1
-- 1 3 1
-- 3 4 1
-- 3 5 1
-- 1 1 2
-- 2 2 2
-- 2 3 2
-- 3 4 2
-- 4 5 2
-- 2 1 3
-- 3 2 3
-- 4 3 3
-- 5 4 3
-- 5 5 3
-- twoway aovc1, c2 c3
```

ANALYSIS OF VARIANCE

DUE TO	DF	SS	MS=SS/DF
C2	4	18.267	4.567
C3	2	12.400	6.200
ERROR	8	0.933	0.117
TOTAL	14	31.600	

OBSERVATIONS
ROWS ARE LEVELS OF C2   COLS ARE LEVELS OF C3

	1	2	3	ROW MEANS
1	0.000	1.000	2.000	1.000
2	1.000	2.000	3.000	2.000
3	1.000	2.000	4.000	2.333
4	3.000	3.000	5.000	3.667
5	3.000	4.000	5.000	4.000
COL. MEANS	1.600	2.400	3.800	2.600

POOLED ST. DEV. =   0.342

FIGURE 14.6

"Error" mean sum of squares (*MS*) as the denominator. Thus, the *F* for the condition effect is

$$F = \frac{MS(\text{C3})}{MS(\text{Error})} = \frac{6.200}{.117} = 53.0$$

The results are identical to those reported in Table 9.8.

14.1.5 *CH. 10*: TWO-WAY ANALYSIS OF VARIANCE

We now apply the TWOW command to perform a *two-way analysis of variance* on the data in Table 10.8 (Fig. 14.7).

This time we do not show all the data lines. For each subject, the data is in column 1, the level of *A* is coded in column 2, and the level of

FIGURE 14.7

```
-- read data in c1, level of A in c2, level of B in c3

-- 24 1 1

-- 14 1 1

-- 23 1 1

-- 28 1 1
      .
      .
      .
      .
-- 46 2 3

-- 50 2 3

-- twoway aov c1, c2 c3

ANALYSIS OF VARIANCE

DUE TO          DF          SS          MS=SS/DF
C2              1          38.5          38.5
C3              2        2474.0        1237.0
C2    * C3      2         418.2         209.1
ERROR          42        1212.6          28.9
TOTAL          47        4143.3

CELL MEANS
ROWS ARE LEVELS OF C2      COLS ARE LEVELS OF C3
                                                    ROW
                    1          2          3       MEANS
        1         23.63      29.38      34.25      29.08
        2         16.25      25.00      40.63      27.29
COL.
MEANS             19.94      27.19      37.44      28.19
```

B is coded in column 3. In the summary table, presented after the TWOW command, A is labelled C2, B is labelled C3, and the $A \times B$ interaction is labelled C2*C3. Again, the F is not computed by MINITAB and must be determined by the user. The denominator *MSS* (labelled "Within Group *MSS*" in Chapter 10) is the "Error" *MSS* (See Table 10.10).

14.1.6 *CH. 11*: CORRELATION AND REGRESSION

Fig. 14.8 shows the MINITAB computation of Pearson's r for the data in Table 11.4.

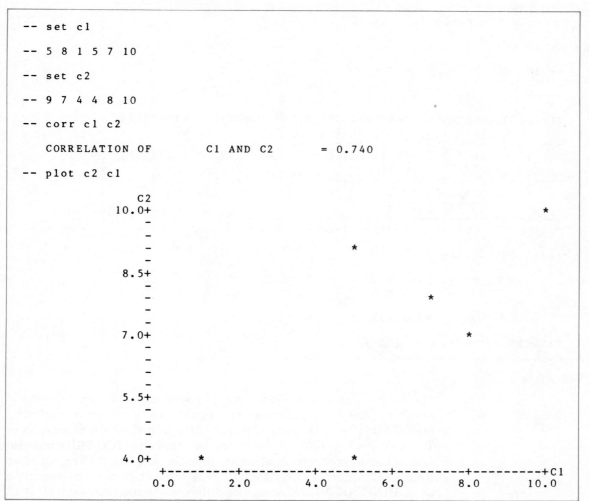

FIGURE 14.8

Remember, when entering pairs of scores in two columns in the manner shown in Fig. 14.8, the integrity of the pairs must be preserved. That is, if a subject's score is the fifth reported in one condition, it must be the fifth score reported in the other condition. Fig. 14.8 also demonstrates the useful PLOT command, which gives us a scatterplot of the data.

14.1.7   *CH. 12*: CATEGORICAL DATA

The use of the CHIS command is shown in Fig. 14.9, where a *chi-square test* is performed on the data in Table 12.1.

```
-- read the table into c1,c2

-- 60 40

-- 20 40

-- chis c1,c2

EXPECTED FREQUENCIES ARE PRINTED BELOW OBSERVED FREQUENCIES
          I  C1   I   C2   I   TOTALS
-------I-------I-------I-------
     1  I   60  I   40  I    100
        I   50.0I   50.0I
-------I-------I-------I-------
     2  I   20  I   40  I     60
        I   30.0I   30.0I
-------I-------I-------I-------
TOTALS I   80  I   80  I    160

TOTAL CHI SQUARE =

          2.00 +   2.00 +
          3.33 +   3.33 +

          =   10.67

DEGREES OF FREEDOM = ( 2-1) X ( 2-1) =    1
```

FIGURE 14.9

The figure also demonstrates another way to enter two columns of numbers. The slight difference between the values of χ^2 calculated by MINITAB ($\chi^2 = 10.67$) and the value determined for the same data in Table 12.8 ($\chi^2 = 9.626$) is due to the fact that the MINITAB computation is not corrected for continuity. (See Section 12.2.2.) The equation used in Chapter 12 (Equation 12.1) is somewhat more conservative than MINITAB. In this case, the difference is inconsequential. Both calculations lead to the same decision (reject H_0), since the critical value for $\chi^2(1\ df)$, $\alpha = .05$, is 3.84.

14.1.8 *CH. 13*: NONPARAMETRIC STATISTICS

Some versions of MINITAB perform a *Wilcoxon test* with a WILC command. However, the WILC command does not work on the version used to produce these examples.

The *Mann-Whitney test* on the data in Table 13.6 is shown in Fig. 14.10.

```
-- set c1

-- 12 21 30 20 26 18 16 21 19

-- set c2

-- 14 16 14 6 24 18 19 10 9 13

-- mann c1 c2

   C1           N =   9     MEDIAN =      20.000
   C2           N =  10     MEDIAN =      14.000

   A POINT ESTIMATE FOR ETA1-ETA2 IS        6.0
   A 95.5 PERCENT C.I. FOR ETA1-ETA2 IS (        1.0,        12.0)

   W =      117.5
   TEST OF ETA1 = ETA2 VS. ETA1 N.E. ETA2
   THE TEST IS SIGNIFICANT AT 0.0275
```

FIGURE 14.10

MINITAB uses W as the critical statistic for the Mann-Whitney instead of U, used in Chapter 13. W is the sum of the ranks of one of the groups. Here we see W is computed to be 117.5, which we see back in Table 13.6 is equal to the sum of ranks for the experimental group. The Mann-Whitney may be performed on either U or W. However, they have different sampling distributions and have to be evaluated by different tables. A more important difference between Fig. 14.10 and Table 13.6 is in the decision that is reached about H_0. In Chapter 13, U is computed to be equal to 17.5 and compared to the tabled value of $N_1 = 9$ and $N_2 = 10$ with the conclusion that H_0 not be rejected. In contrast, MINITAB concludes that the test is significant at .027. The difference is not due to the calculation of U as opposed to W. Rather, the discrepancy seems to be due to the fact that MINITAB is comparing its test statistic against a normal approximation of its sampling distribution. If a normal approximation to the distribution of U is used in the present case, the same decision is reached; i.e., reject H_0. However, the normal approximation was not used in Chapter 13, because exact probability values are available for $N_1 = N_2 = 20$ or less. Recall that the quality of fit to normal depends on sample size. The larger the sample, the better the fit. Apparently in this example we see an instance where the approximation leads to a conclusion different from that used from the tables of U. Such outcomes are doubtless unlikely, but this one does suggest that when sample sizes are small (for instance, less than 20), conclusions stemming from a normal approximation by MINITAB or other computer packages should be checked by the user by referring to tabled critical values.

Spearman's r_S is computed in Fig. 14.11 on the data in Table 13.10.

FIGURE 14.11

```
-- read c1, c2

-- 5  9

-- 8  7

-- 1  4

-- 5  4

-- 7  8

-- 10 10

-- rank c1, c3

-- rank c2, c4

-- corr c3 c4

   CORRELATION OF        C3 AND C4      = 0.662
```

Notice that Spearman's r_S is computed by first ranking the scores in each group and then performing a Pearson's r on the ranks. This is not only computationally simple with MINITAB, it also has the advantage of correcting for ties among the ranks. The value of r_S in Table 13.10 is not corrected for ties (zero differences). If it were, the value would be identical to the one computed by MINITAB.

14.2

BMDP

The forerunner to **BMDP** was BMD, a package of biomedical programs first appearing in 1961. BMDP has an English-based command language and a wide range of programs. Unlike MINITAB, which performs most of its analyses with a single command, BMDP requires a sequence of instructions. The number and form of the instructions depend on the analysis desired.

A successful run of BMDP will require three sets of program lines (or cards). First, a number of system commands are needed in order to get on the system and run or execute BMDP. These commands vary from system to system and cannot be covered fully here. However, the command to run BMDP must be accompanied by a specification of which BMDP program is to be used (e.g., BMDP2D or BMDP1V). These program labels are specified here.

The second set of program lines are the BMDP control language instructions. These include, for example, the title of the problem, an indication of the number of variables, the variable names, and so forth. These lines are reported for the specific examples here. All control language instruction lines are preceded by "/" in the examples. They vary among the program examples, but all have a /PROBLEM line to title the problem and output, and an /INPUT paragraph to indicate the number of variables in the data and their format. The remaining control language instructions give details as needed by the separate statistical tests. These are not explained in detail here. It is assumed the reader is familiar with the BMDP manual. However, a detailed explanation is given for the one-way analysis of variance in Section 14.2.4. A review of this example first may be helpful.

The third set of program lines (cards) are the data. These lines are shown in the examples just after the /END control instruction. The format of the data is specified in the control language instructions. BMDP requires fixed format for the data. Examples which have many data lines are not fully copied here.

In the following figures, a copy of the input (command language instructions and data) is presented first, followed by the relevant portions of the resulting output. The version of BMDP used for these examples is BMDP P-series 1977. All programs are designated by a six-character string, the first three characters of which are always BMD. Therefore, each is referenced by its last three characters. For example, BMDP2D is referred to as P2D.

14.2.1 *CH. 2*: DESCRIPTIVE STATISTICS

Program P2D computes a number of descriptive statistics, including the M, median, mode, s, and s^2. Only a portion of the output is shown for the Table 2.1 data in Fig. 14.12 and the Table 2.6 data in Fig. 14.13.

In the example, descriptive statistics are computed on a single set of data. However, P2D is able to compute the same statistics on a large number of variables using the same set of instructions. The variables

Instructions

```
/CONTROL        COLUMN=72.        /END
/PROBLEM        TITLE IS 'TABLE 2.1 DATA'.
/INPUT          VARIABLE IS 1.
                FORMAT IS '(F1.0)'.
/END
```

Data

```
1
1
2
2
2
4
```

Output

```
    NUMBER OF CASES READ. . . . . . . . . . . . .        6
1PAGE    2       TABLE 2.1 DATA

0   ************
    *  X(1)     *
    ************
                                            MAXIMUM         4.000000
VARIABLE NUMBER . . . . . .         1        MINIMUM         1.000000
NUMBER OF DISTINCT VALUES .         3        RANGE           3.000000
NUMBER OF VALUES COUNTED. .         6        VARIANCE        1.200000
NUMBER OF VALUES NOT COUNTED        0        ST.DEV.         1.095445
                                            (Q3-Q1)/2       0.500000


LOCATION ESTIMATES                                  ST.ERROR
                        MEAN       2.0000000        0.4472136
                        MEDIAN     2.0000000        0.8660258
                        MODE       2.0000000
```

FIGURE 14.12

in the examples are labelled "X(1)" in the output because they were not named in the control instructions. P2D also plots a histogram, which is not shown in the figures.

14.2.2   *CH. 5*: INFERENCES ABOUT ONE MEAN

Program P3D performs t tests on one or two groups. If the data of only one group are given, P3D computes a t test on the group M against a null hypothesis that $\mu = 0$. In Fig. 14.14, P3D performs a t test on the difference scores in Table 5.1.

The "P Value" is based on a two-tailed test. The output also includes a histogram of the scores.

```
Instructions    /CONTROL        COLUMN=72.        /END
                /PROBLEM        TITLE IS 'TABLE 2.6 DATA'.
                /INPUT          VARIABLE IS 1.
                                FORMAT IS '(F1.0)'.
                /END
```

```
Data            2
                5
                3
                3
                2
                3
```

```
Output      NUMBER OF CASES READ. . . . . . . . . . .          6
        1 PAGE    2       TABLE 2.6 DATA


        0  ************
           *  X(1)    *
           ************
                                            MAXIMUM          5.0000000
        VARIABLE NUMBER . . . . . .     1   MINIMUM          2.0000000
        NUMBER OF DISTINCT VALUES .     3   RANGE            3.0000000
        NUMBER OF VALUES COUNTED. .     6   VARIANCE         1.2000000
        NUMBER OF VALUES NOT COUNTED   0    ST.DEV.          1.0954451
                                            (Q3-Q1)/2        0.5000000
        LOCATION ESTIMATES                          ST.ERROR
                        MEAN        3.0000000       0.4472136
                        MEDIAN      3.0000000       0.8660258
                        MODE        3.0000000
```

FIGURE 14.13

```
Instructions    /CONTROL        COLUMN=72.        /END
                /PROBLEM        TITLE IS 'TABLE 5.1 DATA'.
                /INPUT          VARIABLE IS 1.
                                FORMAT IS '(F5.)'.
                /END
```

```
Data            5.00
               -1.25
                3.25
                1.00
                 .50
                 .25
                2.50
                3.25
```

```
Output                                          MEAN          1.8125
                T STATISTIC  P VALUE  DF  STD DEV        2.0343
                                          S.E.M.         0.7192
                    2.52       0.040   7  SAMPLE SIZE         8
                                          MAXIMUM        5.0000
                                          MINIMUM       -1.2500
```

FIGURE 14.14

319

14.2.3 *CH. 7*: INFERENCES ABOUT TWO MEANS

Program P3D tests the difference between the means of the experimental and control group data in Table 7.1 (Fig. 14.15).

Instructions

```
/PROBLEM        TITLE IS 'TABLE 7.1 DATA'.
/INPUT          VARIABLES ARE 2.
                FORMAT IS '(2F3.0)'.
/VARIABLE       NAMES ARE CODE,GROUP.
                GROUPING IS CODE.
/GROUP          CODES(1) ARE 0,1.
                NAMES(1) ARE EXPERIMENTAL,CONTROL.
/END
```

Data

```
0  12
0  21
0  30
0  20
0  26
0  18
0  16
0  21
0  19
1  14
1  16
1  14
1   6
1  24
1  18
1  19
1  10
1   9
1  13
```

Output

```
************
*  GROUP   *   VARIABLE NUMBER   2      GROUP     1 EXPERIME   2 CONTROL
************                            MEAN        20.3333     14.3000
             STATISTICS   P VALUE   DF  STD DEV      5.2678      5.2715
                                        S.E.M.       1.7559      1.6670
T (SEPARATE)     2.49      0.023   16.8 SAMPLE SIZE        9          10
T (POOLED)       2.49      0.023   17   MAXIMUM     30.0000     24.0000
                                        MINIMUM     12.0000      6.0000
```

FIGURE 14.15

14.2.4 *CH. 9*: ONE-WAY ANALYSIS OF VARIANCE

A *one-way between-subject analysis of variance* can be accomplished with program P1V, as shown on the data from Table 9.1 in Fig. 14.16.

```
/PROBLEM        TITLE IS 'TABLE 9.1 DATA'.
/INPUT          VARIABLES ARE 2.
                FORMAT IS '(2F2.0)'.
/VARIABLE       NAMES ARE GROUP, SCORE.
                GROUPING=GROUP.
/GROUP          CODES (1) ARE 1,2,3.
                NAMES ARE GROUP1, GROUP2, GROUP3.
/DESIGN
/END
```

Data

```
1 0
1 1
1 1
1 3
1 3
2 1
2 2
2 2
2 3
2 4
3 2
3 3
3 4
3 5
3 5
```

Output

ESTIMATES OF MEANS

		GROUP1	GROUP2	GROUP3	TOTAL
		1	2	3	4
SCORE	2	1.6000	2.4000	3.8000	2.6000

ONE WAY ANALYSIS OF VARIANCE FOR VARIABLE SCORE
****

ANALYSIS OF VARIANCE

SOURCE OF VARIANCE	D.F.	SUM OF SQ.	MEAN SQ.	F-VALUE
EQUALITY OF CELL MEANS	2	12.4000	6.2000	3.8750
ERROR	12	19.2000	1.6000	

T-TEST MATRIX FOR GROUP MEANS ON   12 DEGREES OF FREEDOM

		GROUP1	GROUP2	GROUP3
		1	2	3
GROUP1	1	0.0000		
GROUP2	2	1.0000	0.0000	
GROUP3	3	2.7500	1.7500	0.0000

FIGURE 14.16

We describe this example in detail. First, the instruction lines. The first line, /PROBLEM, is used here simply to title the problem and the output. (Incidentally, the title of the output is not shown in the figure in order to save space). Next is an /INPUT paragraph, which is used to specify the number of variables (2) and to indicate the format (2F2.0) that the data are to be read-in. There are two variables, the dependent variable that is being analyzed and a coding variable that tells the computer which group each subject is in. (There is one data line, or card, per subject.) The /VARIABLE paragraph, then, is used to name the variables (called "Group" and "Score" here, the latter our dependent variable), and to indicate that the grouping is done according to the variable labelled "Group". Next, the /GROUP paragraph gives the CODES for variable 1 (the first named variable, also the first one found on the data line) which are to be used with the names "Group1", "Group2", and so forth. This simply instructs the computer to put scores coded 1 into group 1, those coded 2 into group 2, etc. The /DESIGN paragraph, written here without qualifiers, commands a one-way analysis of variance on all variables except the "Grouping" variable. Since in this case there is only one other variable, the analysis is done on the dependent variable. Finally, /END terminates the control language instructions.

Fifteen lines of data follow, one line per subject. Each line has two numbers, the first indicating the subject's group number, the second the subject's score.

The last part of the figure is the computer output. The three group means and the mean of all subjects combined are reported first. Next is the summary table. The summary table labels between-groups as "Equality of Cell Means" and within-groups as "Error". A probability value for the obtained F is also reported but not shown here for lack of space. P1V also reports the mean score for each group and a t test on all possible pairs of means.

A *one-way within-subject analysis of variance* is accomplished by program P2V. Fig. 14.17 shows an analysis of the Table 9.5 data.

Each subject's data (one score for each of the three conditions) are input on a single line (card). After reporting M and s for each condition, P2V outputs a summary table (p values not shown). The sources are not labelled as they are in Table 9.8. "Between S" is labelled "1 Error" by P2V, "Condition" is labelled "R", and "Residual" is labelled "2 Error". The small discrepancy in F (53.10 in Table 9.8 versus 53.14 here) is due to rounding.

14.2.5   *CH. 10*: TWO-WAY ANALYSIS OF VARIANCE

Program P2V is used to perform a *two-way between-subject analysis of variance* on the data of Table 10.8. The data input, not all of which is

```
/PROBLEM          TITLE IS 'TABLE 9.5 DATA'.
/INPUT            VARIABLES ARE 3.
                  FORMAT IS '(3F2.0)'.
/VARIABLE         NAMES ARE COND1,COND2,COND3.
/DESIGN              DEPENDENT ARE 1 TO 3.
                  LEVEL IS 3.
/END
```

Data→

```
0 1 2
1 2 3
1 2 4
3 3 5
3 4 5
```

Output→

```
            CELL MEANS   FOR   1-ST DEPENDENT VARIABLE

                         MARGINAL
              R
COND1         1     1.60000      1.60000
COND2         2     2.40000      2.40000
COND3         3     3.80000      3.80000

       MARGINAL     2.60000      2.60000

       COUNT           5             5

STANDARD DEVIATIONS   FOR   1-ST DEPENDENT VARIABLE

              R
COND1         1     1.34164
COND2         2     1.14018
COND3         3     1.30384
1PAGE    3       TABLE 9.5 DATA

ANALYSIS OF VARIANCE FOR   1-ST
DEPENDENT VARIABLE - COND1    COND2     COND3
```

	SOURCE	SUM OF SQUARES	DEGREES OF FREEDOM	MEAN SQUARE	F
	MEAN	101.40000	1	101.40000	22.20
1	ERROR	18.26667	4	4.56667	
	R	12.40000	2	6.20000	53.14
2	ERROR	0.93333	8	0.11667	

FIGURE 14.17

323

Instructions

```
/PROBLEM        TITLE IS 'TABLE 10.8 DATA'.
/INPUT          VARIABLES ARE 3.
                FORMAT IS '(2F2.0,F3.0)'.
/VARIABLE       NAMES ARE A,B,SCORE.
/DESIGN              GROUPING=A,B.
                DEPENDENT IS SCORE.
/GROUP          CODES (1) ARE 1,2.
                NAMES (1) ARE A1,A2.
                CODES (2) ARE 1,2,3.
                NAMES (2)  ARE B1,B2,B3.
/END
```

Data

```
 1 1 24
 1 1 14
     .
     .
     .
 2 3 38
 2 3 46
 2 3 50
```

Output

```
          CELL MEANS  FOR  1-ST DEPENDENT VARIABLE

A         =  A1          A1          A1          A2          A2          A2
B         =  B1          B2          B3          B1          B2          B3

SCORE        23.62500    29.37500    34.25000    16.25000    25.00000    40.62500

COUNT               8           8           8           8           8           8

STANDARD DEVIATIONS  FOR  1-ST DEPENDENT VARIABLE

A         =  A1          A1          A1          A2          A2          A2
B         =  B1          B2          B3          B1          B2          B3

SCORE         5.87823     5.70557     5.47070     4.77344     4.65986     5.62996
1PAGE   3     TABLE 10.8 DATA

ANALYSIS OF VARIANCE FOR  1-ST
DEPENDENT VARIABLE - SCORE

        SOURCE                SUM OF     DEGREES OF      MEAN              F
                              SQUARES     FREEDOM       SQUARE

     MEAN                 38137.68750         1     38137.68750        1320.92
     A                       38.52083         1        38.52083           1.33
     B                     2474.00000         2      1237.00000          42.84
     AB                     418.16667         2       209.08333           7.24
  1  ERROR                 1212.62500        42        28.87202
```

FIGURE 14.18

shown in Fig. 14.18, is coded by level of the A variable (column 1) and level of B (column 2). Thus, a "1 1   24" in the data indicates a subject in group A_1B_1 that received a score of 24.

The labels of the output are straightforward except that P2V uses "Error" to refer to "Within Group".

14.2.6 *CH. 11*: CORRELATION AND REGRESSION

Program P3D, just used for t tests, can also be used to compute Pearson's r between two variables. This is accomplished by using a TEST CORR instruction as shown on the data in Table 11.4 (Fig. 14.19).

FIGURE 14.19

```
Instructions
        /PROBLEM          TITLE IS 'TABLE 11.4 DATA'.
        /INPUT            VARIABLES ARE 2.
                          FORMAT IS '(2F3.0)'.
        /VARIABLE         NAMES ARE X,Y.
        /TEST             CORR.
        /END

Data
          5    9
          8    7
          1    4
          5    4
          7    8
         10   10

Output
        0CORRELATION MATRIX FOR GROUP      0


                              X           Y
                                1           2

        X           1       1.0000
        Y           2       0.7400      1.0000
```

14.2.7 *CH. 12*: CATEGORICAL DATA

A χ^2 test of independence in an $R \times C$ contingency table can be accomplished by program P1F. This is demonstrated in Fig. 14.20 on the data that were summarized in Table 12.1.

Note that the data which are read-in here are not the contingency table frequencies. Rather, P1F is designed to read the coded data for each subject ($N = 160$ in this case) and *construct* the contingency table. Then the χ^2 test is performed. P1F reports χ^2 both uncorrected (10.667) and corrected (9.627). The corrected form is used in Chapter 12 (Equation 12.1).

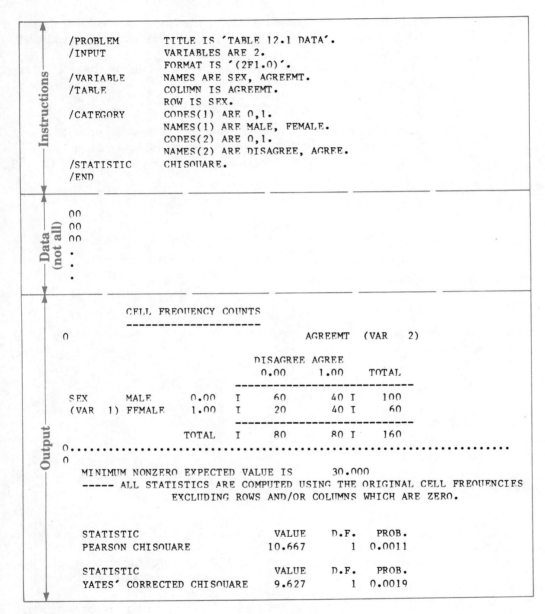

```
            /PROBLEM        TITLE IS 'TABLE 12.1 DATA'.
            /INPUT          VARIABLES ARE 2.
                            FORMAT IS '(2F1.0)'.
            /VARIABLE       NAMES ARE SEX, AGREEMT.
            /TABLE          COLUMN IS AGREEMT.
                            ROW IS SEX.
            /CATEGORY       CODES(1) ARE 0,1.
                            NAMES(1) ARE MALE, FEMALE.
                            CODES(2) ARE 0,1.
                            NAMES(2) ARE DISAGREE, AGREE.
            /STATISTIC      CHISQUARE.
            /END
```

Instructions

```
            00
            00
            00
             .
             .
             .
```

Data (not all)

Output

```
            CELL FREQUENCY COUNTS
  0         ---------------------                    AGREEMT  (VAR   2)

                                        DISAGREE AGREE
                                         0.00     1.00    TOTAL
                                        ------------------------
            SEX     MALE      0.00   I    60       40 I    100
            (VAR 1) FEMALE    1.00   I    20       40 I     60
                                        ------------------------
                              TOTAL  I    80       80 I    160
  0 ....................................................................
  0
            MINIMUM NONZERO EXPECTED VALUE IS        30.000
            ----- ALL STATISTICS ARE COMPUTED USING THE ORIGINAL CELL FREQUENCIES
                  EXCLUDING ROWS AND/OR COLUMNS WHICH ARE ZERO.


            STATISTIC                      VALUE    D.F.    PROB.
            PEARSON CHISQUARE             10.667      1    0.0011

            STATISTIC                      VALUE    D.F.    PROB.
            YATES' CORRECTED CHISQUARE     9.627      1    0.0019
```

FIGURE 14.20

14.2.8   *CH. 13*: NONPARAMETRIC STATISTICS

All nonparametric statistics are accomplished by program P3S. Different tests are selected using the TEST paragraph.

Both the *sign* and *Wilcoxon tests* are performed on the Katahn data from Table 5.1 in Fig. 14.21. Recall that these tests are performed

Instructions

```
/CONTROL         COLUMN=72.        /END
/PROBLEM         TITLE IS 'TABLE 5.1 DATA'.
/INPUT           VARIABLE ARE 2.
                 FORMAT IS '(F6.2,F7.2)'.
/TEST            TITLE IS 'WILCOXON TEST'.
                 SIGN.
                 WILCOXON.
/END
```

Data

```
261.50 256.50
212.75 214.00
218.25 215.00
185.75 184.75
204.25 203.75
152.00 151.75
207.00 204.50
293.75 290.50
```

Output

```
0 LEVEL OF SIGNIFICANCE OF SIGN TEST USING NORMAL APPROXIMATION (TWO-TAIL)

                 X(1)         X(2)
                       1            2

 X(1)       1      1.0000
 X(2)       2      0.0703       1.0000
1PAGE    3      WILCOXON TEST

0 WILCOXON SIGNED RANKS TEST RESULTS

0 NUMBER OF NON-ZERO DIFFERENCES

                 X(1)         X(2)
                       1            2

 X(1)       1            0
 X(2)       2            8            0


0 SMALLER SUM OF LIKE-SIGNED RANKS

                 X(1)         X(2)
                       1            2

 X(1)       1      0.0000
 X(2)       2      4.0000       0.0000


0 LEVEL OF SIGNIFICANCE OF WILCOXON SIGNED RANKS TEST USING NORMAL

                 X(1)         X(2)
                       1            2

 X(1)       1      1.0000
 X(2)       2      0.0500       1.0000
```

FIGURE 14.21

on *difference* scores. However, the differences are computed by P3S, so the pretest and posttest scores are entered for each subject.

The variables (pretest and posttest) are not labelled in the output because they were not named in the instructions. The output may be difficult to read since the values are reported in matrices. P3S is written to provide a test on all possible pairs of a number of variables. When two variables are involved, as here, the values are to be found in row 2 column 1 of the tables. For example, the relevant report of "Smaller Sum of Like-signed Ranks" is 4. This is the value of the Wilcoxon statistic found in Table 13.3. For both the sign test and the Wilcoxon test, the level of significance is evaluated by a normal approximation of the test statistic. In general, however, whenever the sample sizes fall within the range of tabled test statistics, the use of tables to check BMDP is advised.

Fig. 14.22 shows the application of P3S to the data in Table 13.6. The *Mann-Whitney test* can be performed by the KRUSKAL instruction, which computes the Kruskal-Wallis and, when two groups

FIGURE 14.22

Instructions

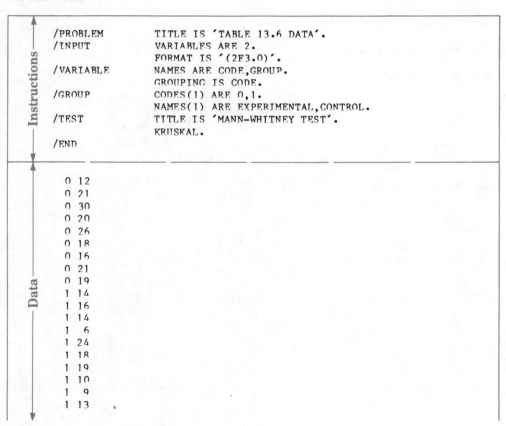

```
/PROBLEM        TITLE IS 'TABLE 13.6 DATA'.
/INPUT          VARIABLES ARE 2.
                FORMAT IS '(2F3.0)'.
/VARIABLE       NAMES ARE CODE,GROUP.
                GROUPING IS CODE.
/GROUP          CODES(1) ARE 0,1.
                NAMES(1) ARE EXPERIMENTAL,CONTROL.
/TEST           TITLE IS 'MANN-WHITNEY TEST'.
                KRUSKAL.
/END
```

Data

```
  0 12
  0 21
  0 30
  0 20
  0 26
  0 18
  0 16
  0 21
  0 19
  1 14
  1 16
  1 14
  1  6
  1 24
  1 18
  1 19
  1 10
  1  9
  1 13
```

```
Output
0 KRUSKAL-WALLIS ONE WAY ANALYSIS OF VARIANCE TEST RESULTS

0  VARIABLE        2 GROUP
   GROUP           FREQUENCY      RANK
  NO. NAME                        SUM
   1    EXPERIME        9         117.5
   2    CONTROL        10          72.5
0  KRUSKAL-WALLIS TEST STATISTIC =    5.06388
   LEVEL OF SIGNIFICANCE =  .0244 USING CHI-SQUARE DISTRIBUTION
0  MANN-WHITNEY TEST STATISTIC =       72.50
   LEVEL OF SIGNIFICANCE =  .0244 USING NORMAL TWO-TAIL APPROXIMATION
```

FIGURE 14.22
(continued)

```
Instructions
/PROBLEM          TITLE IS 'TABLE 13.8 DATA'.
/INPUT            VARIABLES ARE 3.
                  FORMAT IS '(3F3.0)'.
/VARIABLE         NAMES ARE A,B,C.
/TEST             TITLE IS 'FRIEDMAN TEST'.
                  FRIEDMAN.
/END

Data
 16 31 28
  6  5 10
 39 45 62
  0  3  2
 15 16 21
 13 10 17

Output
0 FRIEDMAN TWO WAY ANALYSIS OF VARIANCE TEST RESULTS

0  VARIABLE           RANK
   NO. NAME           SUM
    1   A              8.0
    2   B             12.0
    3   C             16.0
0  FRIEDMAN TEST STATISTIC =    5.33333
   LEVEL OF SIGNIFICANCE =  .0695 ASSUMING CHI-SQUARE DISTRIBUTION
```

FIGURE 14.23

are involved, computes the Mann-Whitney as well.

The *Friedman* test is accomplished with the FRIEDMAN instruction. The data in Fig. 14.23 on the previous page are from Table 13.8.

Next, the *Kruskal-Wallis* is performed on the data in Table 13.9 (Fig. 14.24). Finally, the *Spearman* is perfomed on the data in Table 13.10 (Fig. 14.25).

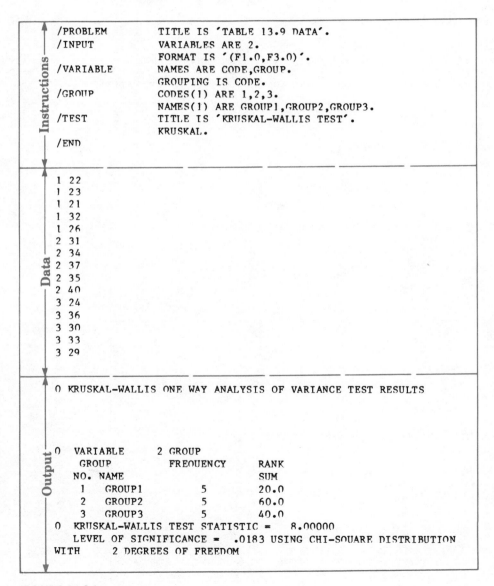

```
            /PROBLEM        TITLE IS 'TABLE 13.9 DATA'.
            /INPUT          VARIABLES ARE 2.
                            FORMAT IS '(F1.0,F3.0)'.
            /VARIABLE       NAMES ARE CODE,GROUP.
                            GROUPING IS CODE.
            /GROUP          CODES(1) ARE 1,2,3.
                            NAMES(1) ARE GROUP1,GROUP2,GROUP3.
            /TEST           TITLE IS 'KRUSKAL-WALLIS TEST'.
                            KRUSKAL.
            /END
```

Instructions

```
            1  22
            1  23
            1  21
            1  32
            1  26
            2  31
            2  34
            2  37
            2  35
            2  40
            3  24
            3  36
            3  30
            3  33
            3  29
```

Data

```
0 KRUSKAL-WALLIS ONE WAY ANALYSIS OF VARIANCE TEST RESULTS


0  VARIABLE      2 GROUP
   GROUP         FREQUENCY     RANK
   NO. NAME                    SUM
     1   GROUP1        5        20.0
     2   GROUP2        5        60.0
     3   GROUP3        5        40.0
0  KRUSKAL-WALLIS TEST STATISTIC =    8.00000
   LEVEL OF SIGNIFICANCE =   .0183 USING CHI-SQUARE DISTRIBUTION
WITH      2 DEGREES OF FREEDOM
```

Output

FIGURE 14.24

FIGURE 14.25

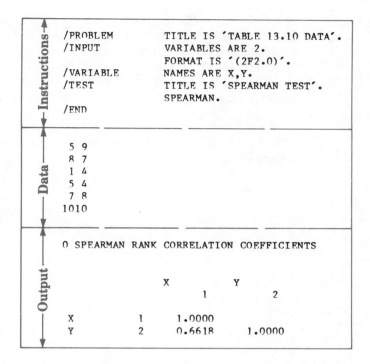

```
Instructions  /PROBLEM        TITLE IS 'TABLE 13.10 DATA'.
              /INPUT          VARIABLES ARE 2.
                              FORMAT IS '(2F2.0)'.
              /VARIABLE       NAMES ARE X,Y.
              /TEST           TITLE IS 'SPEARMAN TEST'.
                              SPEARMAN.
              /END

Data          5 9
              8 7
              1 4
              5 4
              7 8
             1010

Output        0 SPEARMAN RANK CORRELATION COEFFICIENTS


                              X              Y
                                    1              2

              X           1        1.0000
              Y           2        0.6618         1.0000
```

14.3

SPSS

Statistical Package for the Social Sciences (SPSS), has been widely available since 1970. As with BMDP, statistical analyses require a sequence of commands given on "control cards" or control lines. Different control lines are used for different tests, but all of the following examples have three lines in common. The first command line is RUN NAME, which is used simply to title the output. Near the middle of the lines you will find N OF CASES, followed by a number. In SPSS, "Case" is the basic unit of analysis, and in all our examples the basic unit is the subject. Thus, the number of cases is the number of subjects. There is one data line (card) for each subject, so N OF CASES 6, as an example, instructs the computer to expect 6 lines of data. The last control line is READ INPUT DATA, which instructs the computer to begin reading the data, which it will now expect in the next N OF CASES lines. There are other ways to read data into SPSS, but we illustrate the simplest here.

Not all SPSS commands used are explained here. It is assumed that the reader will have access to an SPSS manual. However, a detailed explanation is provided for a one-way analysis of variance in Section 14.3.4. It may be helpful to first review this example. In the following figures, a copy of the control lines and data are first presented, followed by relevant portions of the resulting output.

The version of SPSS used here is SPSS for DECsystem-10, version H, Release 8.1, August 15, 1980. It is documented in Klecka, Nie & Hull (1975); Hull & Nie (1979); and Nie, Hull, Jenkins, Steinbrenner & Bent (1975).

14.3.1 *CH. 2*: DESCRIPTIVE STATISTICS

A number of descriptive statistics are calculated by the FREQUENCIES command, as shown in Fig. 14.26 for the data in Table 2.1 and Fig. 14.27 for the data in Table 2.6.

```
RUN NAME            DESCRIPTIVE STATISTICS
DATA LIST           FIXED/ 1 SCORES 1
INPUT MEDIUM        CARDS
N OF CASES          6
FREQUENCIES         GENERAL=SCORES
STATISTICS          ALL
READ INPUT DATA
```

```
1
1
2
2
2
4
FINISH
```

	Category label		Code	Absolute freq	Relative freq (%)	Adjusted freq (%)	Cum freq (%)
0			1.	2	33.3	33.3	33.3
0			2.	3	50.0	50.0	83.3
0			4.	1	16.7	16.7	100.0
			Total	6	100.0	100.0	
0							

Mean	2.000	Std err	0.447	Median		1.833	
Mode	2.000	Std dev	1.095	Variance		1.200	
Kurtosis	2.500	Skewness	1.369	Range		3.000	
Minimum	1.000	Maximum	4.000				

FIGURE 14.26

```
RUN NAME          TABLE 2.6 DATA
DATA LIST         FIXED/ 1 SCORES 1
INPUT MEDIUM      CARDS
N OF CASES        6
FREQUENCIES       GENERAL=SCORES
STATISTICS        ALL
READ INPUT DATA
```

```
2
5
3
3
2
3
FINISH
```

```
-
                                        Relative  Adjusted    Cum
                               Absolute    freq      freq      freq
 Category label        Code      freq      ( % )     ( % )     ( % )
0                       2.         2       33.3      33.3      33.3
0                       3.         3       50.0      50.0      83.3
0                       5.         1       16.7      16.7     100.0
                                 ------    ------    ------
                     Total         6      100.0     100.0
0
 Mean         3.000   Std err   0.447    Median     2.833
 Mode         3.000   Std dev   1.095    Variance   1.200
 Kurtosis     2.500   Skewness  1.369    Range      3.000
 Minimum      2.000   Maximum   5.000
```

FIGURE 14.27

The values of the medians computed by SPSS differ slightly from those computed in Chapter 2, by MINITAB, and by BMDP. These differences result because SPSS uses a slightly different definition of the median, one that is based on the assumption that the dependent variable is continuous (see Nie, Hull, Jenkins, Steinbrenner & Bent, 1975, p. 183). The differences in median values computed from the two definitions are usually small and inconsequential.

14.3.2   CH. 5: INFERENCES ABOUT ONE MEAN

There are two options for the SPSS T-TEST command. One, PAIRS, performs a t test on related pairs. The other, GROUPS, performs a t test on the means of two independent samples. The PAIRS analysis is equivalent to the t test performed on difference scores described in chapters 5 and 7. However, in Chapter 5 the t test was performed directly on differences between paired scores, and SPSS requires that both members of the pair be entered as data. This is shown in Fig. 14.28, which reports the analysis of the pretest and posttest scores in Table 5.1 and replicates the analysis performed in Section 5.3.3.

```
RUN NAME          TABLE 5.1 DATA
DATA LIST         FIXED/1 PRETEST 1-6, POSTTEST 8-13
INPUT MEDIUM      CARDS
N OF CASES        8
T-TEST            PAIRS=PRETEST WITH POSTTEST
READ INPUT DATA
```

```
261.50 256.50
212.75 214.00
218.25 215.00
185.75 184.75
204.25 203.75
152.00 151.75
207.00 204.50
293.75 290.50
FINISH
```

0Variable	Number of cases	Mean	Standard Deviation	Standard Error	*(Difference) Mean
PRETEST					*
	8	216.9063	43.678	15.443	*
		215.0937	42.386	14.986	* 1.8125
POSTTEST					*

Standard Deviation	Standard Error	* Corr.	2-tail Prob.	*	T Value	Degrees of Freedom	2-tail Prob.
2.034	0.719	* 0.999	0.000	*	2.52	7	0.040

FIGURE 14.28

SPSS also reports the Pearson correlation between the pairs (under "Corr.") and gives probability values based on two-tailed tests.

14.3.3 *CH. 7*: INFERENCES ABOUT TWO MEANS

The GROUPS option of T-TEST is demonstrated with the data from Table 7.1 in Fig. 14.29.

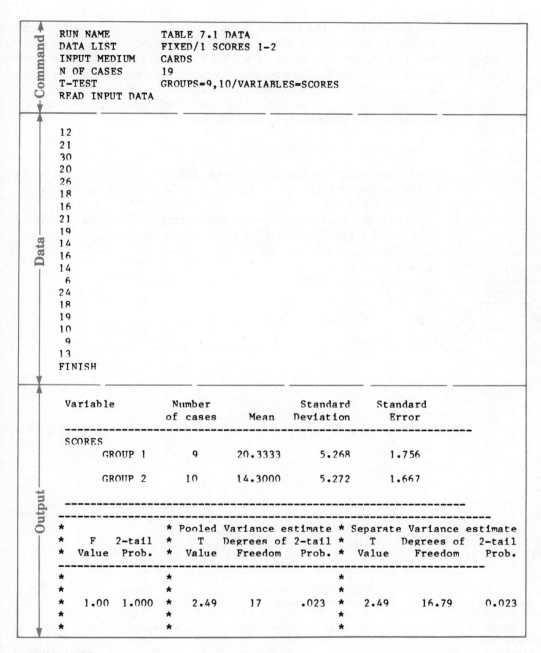

FIGURE 14.29

This figure illustrates the data format in which each subject's score is on a separate line and all the data of one group (the experimental group, called "Group 1" by SPSS) are read-in before the other group's data are entered. (The control group is labelled Group 2. Both groups could have been named in the GROUPS command). Notice that two t tests have been performed, one in which the variances of the two groups have been pooled and one in which they have not been pooled. Equation 7.6, which was used to calculate the t on these data in Chapter 7 uses a pooled estimate of variance. Again, SPSS assigns a probability to the t based on a two-tailed test.

14.3.4 *CH. 9*: ONE-WAY ANALYSIS OF VARIANCE

A *one-way between-subject analysis of variance* is accomplished by an ANOVA command. This is shown in Fig. 14.30 for the data of the three groups in Table 9.1.

There are six control lines in this example. Three (RUN NAME, N OF CASES, and READ INPUT DATA) have already been explained.

Command

```
RUN NAME        TABLE 9.1 DATA
DATA LIST       FIXED/1 GROUP 1, SCORE 3
INPUT MEDIUM    CARDS
N OF CASES      15
ANOVA           SCORE BY GROUP(1,3)
READ INPUT DATA
```

Data

```
1 0
1 1
1 1
1 3
1 3
2 1
2 2
2 2
2 3
2 4
3 2
3 3
3 4
3 5
3 5
FINISH
```

Output

Source of variation	Sum of Squares	df	Mean Square	F
0Main effects	12.400	2	6.200	3.87
GROUP	12.400	2	6.200	3.87
0Explained	12.400	2	6.200	3.87
0Residual	19.200	12	1.600	
0Total	31.600	14	2.257	

FIGURE 14.30

(However, the output shown here does not include the title "Table 9.1 Data" in order to save space.) The first new control line, DATA LIST, informs SPSS where to expect the data in the data lines. The command, FIXED/, in this case, tells the computer that each variable will always be found in the same columns of each data line. Following the FIXED/ is the variable list. GROUP 1 means the first variable is labelled "Group" and its values are to be found in column 1 of the data lines. SCORE 3 means that the second variable is labelled "Score" and its values are to be found in column 3 of the data lines.

The INPUT MEDIUM control line informs the computer as to whether the data are to be found on cards, tape, disk file, or elsewhere. The word CARDS on this line means that the data are supplied on cards (or lines) immediately after the command cards (or lines).

Finally, the ANOVA procedure line instructs the computer to perform an analysis of variance on the data. This line has a variable list. The variable named before the BY is the dependent variable (there could be more than one dependent variable). In this example, SCORE is identified as the independent variable and GROUP(1,3) means that "Group" is the independent variable, its minimum value being 1, its maximum being 3.

Command

```
RUN NAME         TABLE 9.5 DATA
VARIABLES        CND1, CND2, CND3
INPUT FORMAT     FIXED (1F1.0,2F2.0)
INPUT MEDIUM     CARDS
N OF CASES       5
RELIABILITY      VARIABLES=CND1, CND2, CND3/ SCALE(SCORE)=CND1, CND2, CND3
STATISTICS       1 10
READ INPUT DATA
```

Data

```
0 1 2
1 2 3
1 2 4
3 3 5
3 4 5
FINISH
```

Output

Analysis of Variance

	Source of Variation	SS	DF	Mean Square	F
0					
0	Between People	18.26667	4	4.56667	
	Within People	13.33333	10	1.33333	
	Between Measures	12.40000	2	6.20000	53.1428
	Residual	0.93333	8	0.11667	
	Total	31.60000	14	2.25714	
0	Grand Mean =	2.60000			

FIGURE 14.31

The data in Fig. 14.30 are listed in the next 15 lines, one line per subject, followed by some of the output from the program. Included is the summary table. The terms "Between Group" and "Within Group" in Chapter 9 become "Group" and "Residual", respectively.

A *one-way within-subject analysis of variance* is accomplished by the RELIABILITY command using a program that was not available before SPSS Update 7-9. It is illustrated in Fig. 14.31 on the previous page for the data in Table 9.5.

Each subject's data are on separate lines, one score for each of the three conditions. In the summary table for the analysis of variance, "Between Measures" corresponds to "Conditions" used in Table 9.8. The slight difference in F between there (53.0) and here (53.14) is due to rounding.

14.3.5   *CH. 10*: TWO-WAY ANALYSIS OF VARIANCE

The data in Table 10.8 are analyzed by an ANOVA command, as shown in Fig. 14.32.

```
RUN NAME          TABLE 10.8 DATA
DATA LIST         FIXED/1 A 2, B 4, SCORE 6-7
INPUT MEDIUM      CARDS
N OF CASES        48
ANOVA             SCORE BY A(1,2), B(1,3)
READ INPUT DATA
```

```
1 1 24
1 1 14
1 1 23
  .
  .
  .
2 3 38
2 3 46
2 3 50
FINISH
```

Source of variation	Sum of Squares	df	Mean Square	F
0Main effects	2512.521	3	837.507	29.008
A	38.521	1	38.521	1.334
B	2474.000	2	1237.000	42.844
02-way interactions	418.167	2	209.083	7.242
A B	418.167	2	209.083	7.242
0Explained	2930.687	5	586.137	20.301
0Residual	1212.625	42	28.872	
0Total	4143.312	47	88.156	

FIGURE 14.32

The data (not all shown) are coded by level of A in column 2 and level of B in column 4. The labels in the summary table correspond well with those used in Chapter 10.

14.3.6 *CH. 11*: CORRELATION AND REGRESSION

A Pearson's r is computed by a PEARSON CORR command, as shown in Fig. 14.33 for the data in Table 11.4.

```
        RUN NAME          TABLE 11.4 DATA
        VARIABLE LIST     X,Y
        N OF CASES        6
        INPUT FORMAT      FREEFIELD
        PEARSON CORR      X,Y
        READ INPUT DATA


        5  9
        8  7
        1  4
        5  4
        7  8
        10 10


        0- - - - - - - - - - - -P E A R S O N   C O R R E L A T I O N
        -               X            Y

        X            1.0000        0.7400
                     (    0)       (    6)
                     P=*****       P= .046


        Y            0.7400        1.0000
                     (    6)       (    0)
                     P= .046       P=*****
```

Command / Data / Output

FIGURE 14.33

This figure demonstrates the FREEFIELD format which permits the data on a line to be separated by a space. In the FIXED format, in contrast, the same columns must be used by the same variables across all lines (cards). The value of r ($r = .74$) is reported in addition to the size of N and p. The significance level is based on a one-tailed test derived from the t approximation (see Section 11.3.2).

14.3.7 *CH. 12*: CATEGORICAL DATA

CROSSTABS can be used to perform a chi-square test of independence on a contingency table. In Fig. 14.34, CROSSTABS is used on the data summarized in Table 12.1.

```
         RUN NAME        TABLE 12.1 DATA
         DATA LIST       FIXED/1 SEX 1, AGREEMT 2,
         VALUE LABELS    SEX (0) MALE (1) FEMALE/ AGREEMT (0) DISAGREE (1) AGREE
         INPUT MEDIUM    CARDS
         N OF CASES      160
         CROSSTABS       TABLES=SEX BY AGREEMT
         STATISTICS      1
         READ INPUT DATA
```

Command (bracketed label at left)

```
  00
  00
  00
   .
   .
   .

  FINISH
```

Data (bracketed label at left)

```
                      AGREEMT
               Count  :
               Row %  :DISAGREE AGREE      Row
               Col %  :                    Total
               Total %:       0.:      1.:
  SEX          --------:--------:--------:
                  0.   :    60   :    40  :   100
  MALE               :  60.0  :  40.0  :  62.5
                     :  75.0  :  50.0  :
                     :  37.5  :  25.0  :
                   -:--------:--------:
                  1.   :    20   :    40  :    60
  FEMALE             :  33.3  :  66.7  :  37.5
                     :  25.0  :  50.0  :
                     :  12.5  :  25.0  :
                   -:--------:--------:
               Column       80       80      160
               Total       50.0     50.0    100.0

  Corrected chi square =      9.62667 with 1 degree of freedom.  Significance =  0.0019
       Raw chi square =      10.66667 with 1 degree of freedom.  Significance =  0.0011
```

Output (bracketed label at left)

FIGURE 14.34

The data, not all shown, are the coded responses of all 160 subjects. CROSSTABS constructs the contingency table, then performs the chi-square test of independence. The printed table presents in each cell the obtained frequencies, percent of column, percent of row, and percent of total. Two chi squares are reported, one corrected (same as Equation 12.1) and one uncorrected.

14.3.8   *CH. 13*: NONPARAMETRIC STATISTICS

Except for the Spearman r_S test, the following nonparametric tests are based on the NPAR TESTS command, which was not available before Update 7-9.

The *Wilcoxon test* is demonstrated in Fig. 14.35 with the data from Table 5.1.

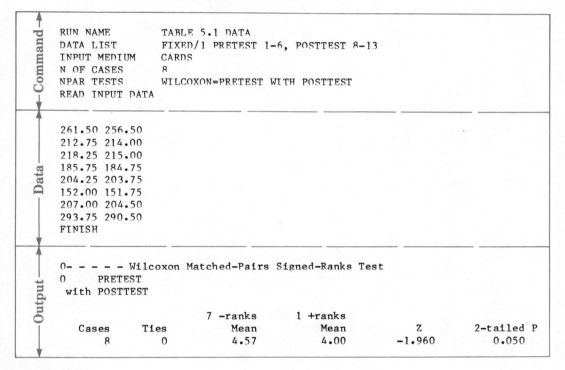

```
         RUN NAME          TABLE 5.1 DATA
         DATA LIST         FIXED/1 PRETEST 1-6, POSTTEST 8-13
         INPUT MEDIUM      CARDS
         N OF CASES        8
         NPAR TESTS        WILCOXON=PRETEST WITH POSTTEST
         READ INPUT DATA


         261.50 256.50
         212.75 214.00
         218.25 215.00
         185.75 184.75
         204.25 203.75
         152.00 151.75
         207.00 204.50
         293.75 290.50
         FINISH


         0- - - - - Wilcoxon Matched-Pairs Signed-Ranks Test
         0       PRETEST
           with POSTTEST


                                    7 -ranks      1 +ranks
            Cases       Ties          Mean          Mean           Z       2-tailed P
              8          0             4.57          4.00        -1.960       0.050
```

FIGURE 14.35

Instead of reporting the *sum* of ranks for each condition, SPSS reports the *mean* of the ranks. The significance of the Wilcoxon statistic is tested in SPSS by the normal approximation against a two-tailed test.

The *Mann-Whitney test* is demonstrated in Fig. 14.36 on the next page with the data in Table 13.6.

The data are coded by group (0 or 1) in column 1. The output includes U, as defined by Equation 13.2, and W, which is equivalent to R_1 in Equation 13.2. For samples smaller than 30, exact probability values of significance are reported based on a two-tailed test. SPSS also reports significance levels based on a normal approximation, corrected for tied ranks.

```
      RUN NAME           TABLE 13.6 DATA
      VARIABLE LIST      GROUP,SCORES
      N OF CASES         19
      INPUT FORMAT       FREEFIELD
      NPAR TESTS         M-W = SCORES BY GROUP(0,1)/
      READ INPUT DATA
```

Command (left margin label)

```
      0 12
      0 21
      0 30
      0 20
      0 26
      0 18
      0 16
      0 21
      0 19
      1 14
      1 16
      1 14
      1 6
      1 24
      1 18
      1 19
      1 10
      1 9
      1 13
```

Data (left margin label)

```
0- - - - - Mann-Whitney U - Wilcoxon Rank Sum W Test
0     SCORES
  by GROUP
```

Output (left margin label)

GROUP = 0		GROUP = 1	
Mean Rank	Number	Mean Rank	Number
13.06	9	7.25	10

		Exact	Corrected for ties	
U	W	2-tailed P	Z	2-tailed P
17.5	117.5	0.0220	-2.2503	0.0244

FIGURE 14.36

FIGURE 14.37

```
      RUN NAME           TABLE 13.8 DATA
      VARIABLE LIST      A,B,C
      N OF CASES         6
      INPUT FORMAT       FREEFIELD
      NPAR TESTS         FRIEDMAN= A B C/
      READ INPUT DATA
```

Command (left margin label)

FIGURE 14.37
(*continued*)

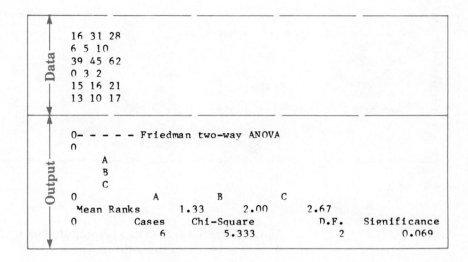

The *Friedman test* is illustrated in Fig. 14.37 for the data from Table 13.8. The Friedman statistic is called "Chi-Square" in the output.

The *Kruskal-Wallis* is illustrated in Fig. 14.38 for the data from Table 13.9.

FIGURE 14.38

```
Command

RUN NAME          TABLE 13.9 DATA
VARIABLE LIST     GROUP,SCORE
INPUT FORMAT      FREEFIELD
N OF CASES        15
NPAR TESTS        K-W = SCORE BY GROUP(1 3)
READ INPUT DATA
```

```
Data

1 22
1 23
1 21
1 32
1 26
2 31
2 34
2 37
2 35
2 40
3 24
3 36
3 30
3 33
3 29
```

```
Output

0- - - - - Kruskal-Wallis 1-Way Anova
0     SCORE
  by GROUP

0     GROUP         1        2        3
      Number        5        5        5
  Mean Ranks     4.00    12.00     8.00

                                         Corrected for ties
         Cases  Chi-Square  Significance  Chi-Square  Significance
            15       8.000         0.018       8.000         0.018
```

FIGURE 14.38
(*continued*)

Finally, the Spearman's r_S is computed using NONPAR CORR as shown in Fig. 14.39 on the data in Table 13.10.

FIGURE 14.39

```
Command

RUN NAME            TABLE 13.10 DATA
VARIABLE LIST       X,Y
N OF CASES          6
INPUT FORMAT        FREEFIELD
NONPAR CORR         X,Y
READ INPUT DATA

Data

5 9
8 7
1 4
5 4
7 8
10 10

Output

0- - - - - - - - - - - - - S P E A R M A N

    Variable            Variable
    Pair                Pair
    --------            --------


    X            0.6618
    With    N(    6)
    Y       Sig .076
```

The computed value of r_S (.6618) differs from that reported in Table 13.10, because SPSS corrects for tied ranks and Equation 13.6 does not.

REFERENCES

BMDP-77 biomedical computer program, P-series. 1977. Berkeley, California: University of California Press.

Hull, C. H., and Nie, N. H. 1979. *SPSS update 7-9: New procedures and facilities for releases 7-9.* New York: McGraw-Hill.

Klecka, W. R.; Nie, N. H.; and Hull, C. H. 1975. *SPSS primer: Statistical package for the social sciences primer.* New York: McGraw-Hill.

Nie, N. H.; Hull, C. H.; Jenkins, J. G.; Steinbrenner, K.; and Bent, D. H. 1975. *SPSS: Statistical package for the social sciences,* 2d ed. New York: McGraw-Hill.

Ryan, T. A.; Joiner, B. L.; and Ryan, B. F. 1976. *MINITAB student handbook.* North Scituate, Mass.: Duxbury Press.

QUESTIONS

1. Use MINITAB to compute descriptive statistics using the data from Question 16 in Chapter 2.

2. Compute Question 16 in Chapter 2 using BMDP BMDP2D.

3. Compute Question 16 in Chapter 2 using SPSS FREQUENCIES.

4. Compute Question 15 in Chapter 5 using MINITAB TTES.

5. Compute Question 16 in Chapter 5 using BMDP BMDP3D.

6. Compute Question 20 in Chapter 5 using SPSS T-TEST PAIRS = .

7. Compute Question 13 in Chapter 7 using MINITAB POOL.

8. Compute Question 18 in Chapter 7 using BMDP BMDP3D.

9. Compute Question 12 in Chapter 7 using SPSS T-TEST GROUPS = .

10. Compute Question 20 in Chapter 9 using MINITAB AOVO.

11. Compute Question 18 in Chapter 9 using BMDP BMDP1V.

12. Compute Question 12 in Chapter 9 using SPSS ANOVA.

13. Compute Question 19 in Chapter 10 using MINITAB TWOW.

14. Compute Question 14 in Chapter 10 using BMDP BMDP2V.

15. Compute Question 9 in Chapter 10 using SPSS RELIABILITY.

16. Compute data from Chapter 11, Question 9 using MINITAB CORR.

17. Compute data from Chapter 11, Question 16 using BMDP BMDP3D CORR.

18. Compute data from Chapter 11, Question 8 using SPSS PEARSON CORR.

19. Compute Question 13 in Chapter 12 using MINITAB CHIS.

20. Compute Question 12 in Chapter 12 using BMDP BMDP1F.

21. Compute Question 16 in Chapter 12 using SPSS CROSSTABS.

22. Compute Question 12 in Chapter 13 using MINITAB MANN.

23. Compute Question 6 in Chapter 13 using BMDP BMDP3S WILCOXON.

24. Compute Question 17 in Chapter 13 using SPSS NONPAR CORR.

APPENDIX A

Tables of Critical Values

To use Table A-1 raw scores must be transformed to z scores. Locate z in column A. The values in columns B and C represent proportions of the area under the normal distribution. Values in column B are proportions between the mean ($z = 0$) and the z given in the same row in column A. The proportions are the same for positive and negative values of z. The areas represented in column B are shown below.

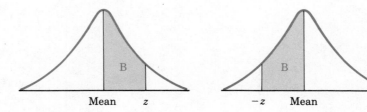

Values in column C are proportions of area beyond z. The areas represented in column C are shaded in the following graph.

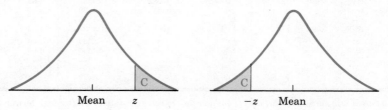

TABLE A–1
The Normal Distribution.

(A) z	(B) Area Between Mean and z	(C) Area Beyond z	(A) z	(B) Area Between Mean and z	(C) Area Beyond z	(A) z	(B) Area Between Mean and z	(C) Area Beyond z
.00	.0000	.5000	.55	.2088	.2912	1.10	.3643	.1357
.01	.0040	.4960	.56	.2123	.2877	1.11	.3665	.1335
.02	.0080	.4920	.57	.2157	.2843	1.12	.3686	.1314
.03	.0120	.4880	.58	.2190	.2810	1.13	.3708	.1292
.04	.0160	.4840	.59	.2224	.2776	1.14	.3729	.1271
.05	.0199	.4801	.60	.2257	.2743	1.15	.3749	.1251
.06	.0239	.4761	.61	.2291	.2709	1.16	.3770	.1230
.07	.0279	.4721	.62	.2324	.2676	1.17	.3790	.1210
.08	.0319	.4681	.63	.2357	.2643	1.18	.3810	.1190
.09	.0359	.4641	.64	.2389	.2611	1.19	.3830	.1170
.10	.0398	.4602	.65	.2422	.2578	1.20	.3849	.1151
.11	.0438	.4562	.66	.2454	.2546	1.21	.3869	.1131
.12	.0478	.4522	.67	.2486	.2514	1.22	.3888	.1112
.13	.0517	.4483	.68	.2517	.2483	1.23	.3907	.1093
.14	.0557	.4443	.69	.2549	.2451	1.24	.3925	.1075
.15	.0596	.4404	.70	.2580	.2420	1.25	.3944	.1056
.16	.0636	.4364	.71	.2611	.2389	1.26	.3962	.1038
.17	.0675	.4325	.72	.2642	.2358	1.27	.3980	.1020
.18	.0714	.4286	.73	.2673	.2327	1.28	.3997	.1003
.19	.0753	.4247	.74	.2704	.2296	1.29	.4015	.0985
.20	.0793	.4207	.75	.2734	.2266	1.30	.4032	.0968
.21	.0832	.4168	.76	.2764	.2236	1.31	.4049	.0951
.22	.0871	.4129	.77	.2794	.2206	1.32	.4066	.0934
.23	.0910	.4090	.78	.2823	.2177	1.33	.4082	.0918
.24	.0948	.4052	.79	.2852	.2148	1.34	.4099	.0901
.25	.0987	.4013	.80	.2881	.2119	1.35	.4115	.0885
.26	.1026	.3974	.81	.2910	.2090	1.36	.4131	.0869
.27	.1064	.3936	.82	.2939	.2061	1.37	.4147	.0853
.28	.1103	.3987	.83	.2967	.2033	1.38	.4162	.0838
.29	.1141	.3859	.84	.2995	.2005	1.39	.4177	.0823
.30	.1179	.3821	.85	.3023	.1977	1.40	.4192	.0808
.31	.1217	.3783	.86	.3051	.1949	1.41	.4207	.0793
.32	.1255	.3745	.87	.3078	.1922	1.42	.4222	.0778
.33	.1293	.3707	.88	.3106	.1894	1.43	.4236	.0764
.34	.1331	.3669	.89	.3133	.1867	1.44	.4251	.0749
.35	.1368	.3632	.90	.3159	.1841	1.45	.4265	.0735
.36	.1406	.3594	.91	.3186	.1814	1.46	.4279	.0721
.37	.1443	.3557	.92	.3212	.1788	1.47	.4292	.0708
.38	.1480	.3520	.93	.3238	.1762	1.48	.4306	.0694
.39	.1517	.3483	.94	.3264	.1736	1.49	.4319	.0681
.40	.1554	.3446	.95	.3289	.1711	1.50	.4332	.0668
.41	.1591	.3409	.96	.3315	.1685	1.51	.4345	.0655
.42	.1628	.3372	.97	.3340	.1660	1.52	.4357	.0643
.43	.1664	.3336	.98	.3365	.1635	1.53	.4370	.0630
.44	.1700	.3300	.99	.3389	.1611	1.54	.4382	.0618
.45	.1736	.3264	1.00	.3413	.1587	1.55	.4394	.0606
.46	.1772	.3228	1.01	.3438	.1562	1.56	.4406	.0594
.47	.1808	.3192	1.02	.3461	.1539	1.57	.4418	.0582
.48	.1844	.3156	1.03	.3485	.1515	1.58	.4429	.0571
.49	.1879	.3121	1.04	.3508	.1492	1.59	.4441	.0559
.50	.1915	.3085	1.05	.3531	.1469	1.60	.4452	.0548
.51	.1950	.3050	1.06	.3554	.1446	1.61	.4463	.0537
.52	.1985	.3015	1.07	.3577	.1423	1.62	.4474	.0526
.53	.2019	.2981	1.08	.3599	.1401	1.63	.4484	.0516
.54	.2054	.2946	1.09	.3621	.1379	1.64	.4495	.0505

Source: Table A of Runyon and Haber. 1971. *Fundamentals of behavioral statistics*. Reading, Mass.: Addison-Wesley Publishing Co. Reprinted with permission.

TABLE A–1
The Normal Distribution (*continued*).

(A) z	(B) Area Between Mean and z	(C) Area Beyond z	(A) z	(B) Area Between Mean and z	(C) Area Beyond z	(A) z	(B) Area Between Mean and z	(C) Area Beyond z
1.65	.4505	.0495	2.22	.4868	.0132	2.79	.4974	.0026
1.66	.4515	.0485	2.23	.4871	.0129	2.80	.4974	.0026
1.67	.4525	.0475	2.24	.4875	.0125	2.81	.4975	.0025
1.68	.4535	.0465	2.25	.4878	.0122	2.82	.4976	.0024
1.69	.4545	.0455	2.26	.4881	.0119	2.83	.4977	.0023
1.70	.4554	.0446	2.27	.4884	.0116	2.84	.4977	.0023
1.71	.4564	.0436	2.28	.4887	.0113	2.85	.4978	.0022
1.72	.4573	.0427	2.29	.4890	.0110	2.86	.4979	.0021
1.73	.4582	.0418	2.30	.4893	.0107	2.87	.4979	.0021
1.74	.4591	.0409	2.31	.4896	.0104	2.88	.4980	.0020
1.75	.4599	.0401	2.32	.4898	.0102	2.89	.4981	.0019
1.76	.4608	.0392	2.33	.4901	.0099	2.90	.4981	.0019
1.77	.4616	.0384	2.34	.4904	.0096	2.91	.4982	.0018
1.78	.4625	.0375	2.35	.4906	.0094	2.92	.4982	.0018
1.79	.4633	.0367	2.36	.4909	.0091	2.93	.4983	.0017
1.80	.4641	.0359	2.37	.4911	.0089	2.94	.4984	.0016
1.81	.4649	.0351	2.38	.4913	.0087	2.95	.4984	.0016
1.82	.4656	.0344	2.39	.4916	.0084	2.96	.4985	.0015
1.83	.4664	.0336	2.40	.4918	.0082	2.97	.4985	.0015
1.84	.4671	.0329	2.41	.4920	.0080	2.98	.4986	.0014
1.85	.4678	.0322	2.42	.4922	.0078	2.99	.4986	.0014
1.86	.4686	.0314	2.43	.4925	.0075	3.00	.4987	.0013
1.87	.4693	.0307	2.44	.4927	.0073	3.01	.4987	.0013
1.88	.4699	.0301	2.45	.4929	.0071	3.02	.4987	.0013
1.89	.4706	.0294	2.46	.4931	.0069	3.03	.4988	.0012
1.90	.4713	.0287	2.47	.4932	.0068	3.04	.4988	.0012
1.91	.4719	.0281	2.48	.4934	.0066	3.05	.4989	.0011
1.92	.4726	.0274	2.49	.4936	.0064	3.06	.4989	.0011
1.93	.4732	.0268	2.50	.4938	.0062	3.07	.4989	.0011
1.94	.4738	.0262	2.51	.4940	.0060	3.08	.4990	.0010
1.95	.4744	.0256	2.52	.4941	.0059	3.09	.4990	.0010
1.96	.4750	.0250	2.53	.4943	.0057	3.10	.4990	.0010
1.97	.4756	.0244	2.54	.4945	.0055	3.11	.4991	.0009
1.98	.4761	.0239	2.55	.4946	.0054	3.12	.4991	.0009
1.99	.4767	.0233	2.56	.4948	.0052	3.13	.4991	.0009
2.00	.4772	.0228	2.57	.4949	.0051	3.14	.4992	.0008
2.01	.4778	.0222	2.58	.4951	.0049	3.15	.4992	.0008
2.02	.4783	.0217	2.59	.4952	.0048	3.16	.4992	.0008
2.03	.4788	.0212	2.60	.4953	.0047	3.17	.4992	.0008
2.04	.4793	.0207	2.61	.4955	.0045	3.18	.4993	.0007
2.05	.4798	.0202	2.62	.4956	.0044	3.19	.4993	.0007
2.06	.4803	.0197	2.63	.4957	.0043	3.20	.4993	.0007
2.07	.4808	.0192	2.64	.4959	.0041	3.21	.4993	.0007
2.08	.4812	.0188	2.65	.4960	.0040	3.22	.4994	.0006
2.09	.4817	.0183	2.66	.4961	.0039	3.23	.4994	.0006
2.10	.4821	.0179	2.67	.4962	.0038	3.24	.4994	.0006
2.11	.4826	.0174	2.68	.4963	.0037	3.25	.4994	.0006
2.12	.4830	.0170	2.69	.4964	.0036	3.30	.4995	.0005
2.13	.4834	.0166	2.70	.4965	.0035	3.35	.4996	.0004
2.14	.4838	.0162	2.71	.4966	.0034	3.40	.4997	.0003
2.15	.4842	.0158	2.72	.4967	.0033	3.45	.4997	.0003
2.16	.4846	.0154	2.73	.4968	.0032	3.50	.4998	.0002
2.17	.4850	.0150	2.74	.4969	.0031	3.60	.4998	.0002
2.18	.4854	.0146	2.75	.4970	.0030	3.70	.4999	.0001
2.19	.4857	.0143	2.76	.4971	.0029	3.80	.4999	.0001
2.20	.4861	.0139	2.77	.4972	.0028	3.90	.49995	.00005
2.21	.4864	.0136	2.78	.4973	.0027	4.00	.49997	.00003

Locate the degrees of freedom (*df*) in the leftmost column. For a one-tailed test, use α levels reported at the top of the table. Reject H_0, if results are in predicted direction and if observed value of *t* equals or exceeds the tabled value (t_α). For a two-tailed test, use α levels reported at the bottom of the table. Reject H_0, if observed value is larger than positive value of t_α or smaller than negative value of t_α.

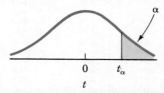

TABLE A–2
The *t* Distribution.

df	Alpha Levels for a Directional (One-tailed) Test					
	.10	.05	.025	.01	.005	.0005
1	3.078	6.314	12.706	31.821	63.657	636.619
2	1.886	2.920	4.303	6.965	9.925	31.598
3	1.638	2.353	3.182	4.541	5.841	12.941
4	1.533	2.132	2.776	3.747	4.604	8.610
5	1.476	2.015	2.571	3.365	4.032	6.859
6	1.440	1.943	2.447	3.143	3.707	5.959
7	1.415	1.895	2.365	2.998	3.499	5.405
8	1.397	1.860	2.306	2.896	3.355	5.041
9	1.383	1.833	2.262	2.821	3.250	4.781
10	1.372	1.812	2.228	2.764	3.169	4.587
11	1.363	1.796	2.201	2.718	3.106	4.437
12	1.356	1.782	2.179	2.681	3.055	4.318
13	1.350	1.771	2.160	2.650	3.012	4.221
14	1.345	1.761	2.145	2.624	2.977	4.140
15	1.341	1.753	2.131	2.602	2.947	4.073
16	1.337	1.746	2.120	2.583	2.921	4.015
17	1.333	1.740	2.110	2.567	2.898	3.965
18	1.330	1.734	2.101	2.552	2.878	3.922
19	1.328	1.729	2.093	2.539	2.861	3.883
20	1.325	1.725	2.086	2.528	2.845	3.850
21	1.323	1.721	2.080	2.518	2.831	3.819
22	1.321	1.717	2.074	2.508	2.819	3.792
23	1.319	1.714	2.069	2.500	2.807	3.767
24	1.318	1.711	2.064	2.492	2.797	3.745
25	1.316	1.708	2.060	2.485	2.787	3.725
26	1.315	1.706	2.056	2.479	2.779	3.707
27	1.314	1.703	2.052	2.473	2.771	3.690
28	1.313	1.701	2.048	2.467	2.763	3.674
29	1.311	1.699	2.045	2.462	2.756	3.659
30	1.310	1.697	2.042	2.457	2.750	3.646
40	1.303	1.684	2.021	2.423	2.704	3.551
60	1.296	1.671	2.000	2.390	2.660	3.460
120	1.289	1.658	1.980	2.358	2.617	3.373
∞	1.282	1.645	1.960	2.326	2.576	3.291
	.20	.10	.05	.02	.01	.001
	Alpha Levels for a Nondirectional (Two-tailed) Test					

Source: Table III of Fisher and Yates. 1974. *Statistical tables for biological, agricultural and medical research*, 6th ed. Published by Longman Group Ltd., London (previously published by Oliver and Boyd, Ltd., Edinburgh), and by permission of the authors and publishers.

Locate the column containing the numerator degrees of freedom and the row containing the denominator degrees of freedom. Two critical values (F_α) are reported. The value in regular type is for α = .05, and the value in boldface is for α = .01. Reject H_0, if the observed value of F equals or exceeds the tabled value.

TABLE B
F Table.

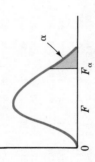

Each cell shows the regular-type value (α = .05) over the boldface value (α = .01).

Degrees of Freedom for Numerator

	1	2	3	4	5	6	7	8	9	10	11	12	14	16	20	24	30	40	50	75	100	200	500	∞
1	161 / **4052**	200 / **4999**	216 / **5403**	225 / **5625**	230 / **5764**	234 / **5859**	237 / **5928**	239 / **5981**	241 / **6022**	242 / **6056**	243 / **6082**	244 / **6106**	245 / **6142**	246 / **6169**	248 / **6208**	249 / **6234**	250 / **6261**	251 / **6286**	252 / **6302**	253 / **6323**	253 / **6334**	254 / **6352**	254 / **6361**	254 / **6366**
2	18.51 / **98.49**	19.00 / **99.00**	19.16 / **99.17**	19.25 / **99.25**	19.30 / **99.30**	19.33 / **99.33**	19.36 / **99.36**	19.37 / **99.37**	19.38 / **99.39**	19.39 / **99.40**	19.40 / **99.41**	19.41 / **99.42**	19.42 / **99.43**	19.43 / **99.44**	19.44 / **99.45**	19.45 / **99.46**	19.46 / **99.47**	19.47 / **99.48**	19.47 / **99.48**	19.48 / **99.49**	19.49 / **99.49**	19.49 / **99.49**	19.50 / **99.50**	19.50 / **99.50**
3	10.13 / **34.12**	9.55 / **30.82**	9.28 / **29.46**	9.12 / **28.71**	9.01 / **28.24**	8.94 / **27.91**	8.88 / **27.67**	8.84 / **27.49**	8.81 / **27.34**	8.78 / **27.23**	8.76 / **27.13**	8.74 / **27.05**	8.71 / **26.92**	8.69 / **26.83**	8.66 / **26.69**	8.64 / **26.60**	8.62 / **26.50**	8.60 / **26.41**	8.58 / **26.35**	8.57 / **26.27**	8.56 / **26.23**	8.54 / **26.18**	8.54 / **26.14**	8.53 / **26.12**
4	7.71 / **21.20**	6.94 / **18.00**	6.59 / **16.69**	6.39 / **15.98**	6.26 / **15.52**	6.16 / **15.21**	6.09 / **14.98**	6.04 / **14.80**	6.00 / **14.66**	5.96 / **14.54**	5.93 / **14.45**	5.91 / **14.37**	5.87 / **14.24**	5.84 / **14.15**	5.80 / **14.02**	5.77 / **13.93**	5.74 / **13.83**	5.71 / **13.74**	5.70 / **13.69**	5.68 / **13.61**	5.66 / **13.57**	5.65 / **13.52**	5.64 / **13.48**	5.63 / **13.46**
5	6.61 / **16.26**	5.79 / **13.27**	5.41 / **12.06**	5.19 / **11.39**	5.05 / **10.97**	4.95 / **10.67**	4.88 / **10.45**	4.82 / **10.29**	4.78 / **10.15**	4.74 / **10.05**	4.70 / **9.96**	4.68 / **9.89**	4.64 / **9.77**	4.60 / **9.68**	4.56 / **9.55**	4.53 / **9.47**	4.50 / **9.38**	4.46 / **9.29**	4.44 / **9.24**	4.42 / **9.17**	4.40 / **9.13**	4.38 / **9.07**	4.37 / **9.04**	4.36 / **9.02**
6	5.99 / **13.74**	5.14 / **10.92**	4.76 / **9.78**	4.53 / **9.15**	4.39 / **8.75**	4.28 / **8.47**	4.21 / **8.26**	4.15 / **8.10**	4.10 / **7.98**	4.06 / **7.87**	4.03 / **7.79**	4.00 / **7.72**	3.96 / **7.60**	3.92 / **7.52**	3.87 / **7.39**	3.84 / **7.31**	3.81 / **7.23**	3.77 / **7.14**	3.75 / **7.09**	3.72 / **7.02**	3.71 / **6.99**	3.69 / **6.94**	3.68 / **6.90**	3.67 / **6.88**
7	5.59 / **12.25**	4.74 / **9.55**	4.35 / **8.45**	4.12 / **7.85**	3.97 / **7.46**	3.87 / **7.19**	3.79 / **7.00**	3.73 / **6.84**	3.68 / **6.71**	3.63 / **6.62**	3.60 / **6.54**	3.57 / **6.47**	3.52 / **6.35**	3.49 / **6.27**	3.44 / **6.15**	3.41 / **6.07**	3.38 / **5.98**	3.34 / **5.90**	3.32 / **5.85**	3.29 / **5.78**	3.28 / **5.75**	3.25 / **5.70**	3.24 / **5.67**	3.23 / **5.65**
8	5.32 / **11.26**	4.46 / **8.65**	4.07 / **7.59**	3.84 / **7.01**	3.69 / **6.63**	3.58 / **6.37**	3.50 / **6.19**	3.44 / **6.03**	3.39 / **5.91**	3.34 / **5.82**	3.31 / **5.74**	3.28 / **5.67**	3.23 / **5.56**	3.20 / **5.48**	3.15 / **5.36**	3.12 / **5.28**	3.08 / **5.20**	3.05 / **5.11**	3.03 / **5.06**	3.00 / **5.00**	2.98 / **4.96**	2.96 / **4.91**	2.94 / **4.88**	2.93 / **4.86**
9	5.12 / **10.56**	4.26 / **8.02**	3.86 / **6.99**	3.63 / **6.42**	3.48 / **6.06**	3.37 / **5.80**	3.29 / **5.62**	3.23 / **5.47**	3.18 / **5.35**	3.13 / **5.26**	3.10 / **5.18**	3.07 / **5.11**	3.02 / **5.00**	2.98 / **4.92**	2.93 / **4.80**	2.90 / **4.73**	2.86 / **4.64**	2.82 / **4.56**	2.80 / **4.51**	2.77 / **4.45**	2.76 / **4.41**	2.73 / **4.36**	2.72 / **4.33**	2.71 / **4.31**
10	4.96 / **10.04**	4.10 / **7.56**	3.71 / **6.55**	3.48 / **5.99**	3.33 / **5.64**	3.22 / **5.39**	3.14 / **5.21**	3.07 / **5.06**	3.02 / **4.95**	2.97 / **4.85**	2.94 / **4.78**	2.91 / **4.71**	2.86 / **4.60**	2.82 / **4.52**	2.77 / **4.41**	2.74 / **4.33**	2.70 / **4.25**	2.67 / **4.17**	2.64 / **4.12**	2.61 / **4.05**	2.59 / **4.01**	2.56 / **3.96**	2.55 / **3.93**	2.54 / **3.91**

Source: P. G. Hoel. 1966. *Elementary statistics*, 2nd ed. New York: John Wiley and Sons.

TABLE B

F Table (*continued*).

Each cell shows the upper value (0.05 level) over the lower value (0.01 level).

	\ Numerator df →	1	2	3	4	5	6	7	8	9	10	11	12	14	16	20	24	30	40	50	75	100	200	500	∞
Denominator df ↓																									
11		4.84 / 9.65	3.98 / 7.20	3.59 / 6.22	3.36 / 5.67	3.20 / 5.32	3.09 / 5.07	3.01 / 4.88	2.95 / 4.74	2.90 / 4.63	2.86 / 4.54	2.82 / 4.46	2.79 / 4.40	2.74 / 4.29	2.70 / 4.21	2.65 / 4.10	2.61 / 4.02	2.57 / 3.94	2.53 / 3.86	2.50 / 3.80	2.47 / 3.74	2.45 / 3.70	2.42 / 3.66	2.41 / 3.62	2.40 / 3.60
12		4.75 / 9.33	3.88 / 6.93	3.49 / 5.95	3.26 / 5.41	3.11 / 5.06	3.00 / 4.82	2.92 / 4.65	2.85 / 4.50	2.80 / 4.39	2.76 / 4.30	2.72 / 4.22	2.69 / 4.16	2.64 / 4.05	2.60 / 3.98	2.54 / 3.86	2.50 / 3.78	2.46 / 3.70	2.42 / 3.61	2.40 / 3.56	2.36 / 3.49	2.35 / 3.46	2.32 / 3.41	2.31 / 3.38	2.30 / 3.36
13		4.67 / 9.07	3.80 / 6.70	3.41 / 5.74	3.18 / 5.20	3.02 / 4.86	2.92 / 4.62	2.84 / 4.44	2.77 / 4.30	2.72 / 4.19	2.67 / 4.10	2.63 / 4.02	2.60 / 3.96	2.55 / 3.85	2.51 / 3.78	2.46 / 3.67	2.42 / 3.59	2.38 / 3.51	2.34 / 3.42	2.32 / 3.37	2.28 / 3.30	2.26 / 3.27	2.24 / 3.21	2.22 / 3.18	2.21 / 3.16
14		4.60 / 8.86	3.74 / 6.51	3.34 / 5.56	3.11 / 5.03	2.96 / 4.69	2.85 / 4.46	2.77 / 4.28	2.70 / 4.14	2.65 / 4.03	2.60 / 3.94	2.56 / 3.86	2.53 / 3.80	2.48 / 3.70	2.44 / 3.62	2.39 / 3.51	2.35 / 3.43	2.31 / 3.34	2.27 / 3.26	2.24 / 3.21	2.21 / 3.14	2.19 / 3.11	2.16 / 3.06	2.14 / 3.02	2.13 / 3.00
15		4.54 / 8.68	3.68 / 6.36	3.29 / 5.42	3.06 / 4.89	2.90 / 4.56	2.79 / 4.32	2.70 / 4.14	2.64 / 4.00	2.59 / 3.89	2.55 / 3.80	2.51 / 3.73	2.48 / 3.67	2.43 / 3.56	2.39 / 3.48	2.33 / 3.36	2.29 / 3.29	2.25 / 3.20	2.21 / 3.12	2.18 / 3.07	2.15 / 3.00	2.12 / 2.97	2.10 / 2.92	2.08 / 2.89	2.07 / 2.87
16		4.49 / 8.53	3.63 / 6.23	3.24 / 5.29	3.01 / 4.77	2.85 / 4.44	2.74 / 4.20	2.66 / 4.03	2.59 / 3.89	2.54 / 3.78	2.49 / 3.69	2.45 / 3.61	2.42 / 3.55	2.37 / 3.45	2.33 / 3.37	2.28 / 3.25	2.24 / 3.18	2.20 / 3.10	2.16 / 3.01	2.13 / 2.96	2.09 / 2.98	2.07 / 2.86	2.04 / 2.80	2.02 / 2.77	2.01 / 2.75
17		4.45 / 8.40	3.59 / 6.11	3.20 / 5.18	2.96 / 4.67	2.81 / 4.34	2.70 / 4.10	2.62 / 3.93	2.55 / 3.79	2.50 / 3.68	2.45 / 3.59	2.41 / 3.52	2.38 / 3.45	2.33 / 3.35	2.29 / 3.27	2.23 / 3.16	2.19 / 3.08	2.15 / 3.00	2.11 / 2.92	2.08 / 2.86	2.04 / 2.79	2.02 / 2.76	1.99 / 2.70	1.97 / 2.67	1.96 / 2.65
18		4.41 / 8.28	3.55 / 6.01	3.16 / 5.09	2.93 / 4.58	2.77 / 4.25	2.66 / 4.01	2.58 / 3.85	2.51 / 3.71	2.46 / 3.60	2.41 / 3.51	2.37 / 3.44	2.34 / 3.37	2.29 / 3.27	2.25 / 3.19	2.19 / 3.07	2.15 / 3.00	2.11 / 2.91	2.07 / 2.83	2.04 / 2.78	2.00 / 2.71	1.98 / 2.68	1.95 / 2.62	1.93 / 2.59	1.92 / 2.57
19		4.38 / 8.18	3.52 / 5.93	3.13 / 5.01	2.90 / 4.50	2.74 / 4.17	2.63 / 3.94	2.55 / 3.77	2.48 / 3.63	2.43 / 3.52	2.38 / 3.43	2.34 / 3.36	2.31 / 3.30	2.26 / 3.19	2.21 / 3.12	2.15 / 3.00	2.11 / 2.92	2.07 / 2.84	2.02 / 2.76	2.00 / 2.70	1.96 / 2.63	1.94 / 2.60	1.91 / 2.54	1.90 / 2.51	1.88 / 2.49
20		4.35 / 8.10	3.49 / 5.85	3.10 / 4.94	2.87 / 4.43	2.71 / 4.10	2.60 / 3.87	2.52 / 3.71	2.45 / 3.56	2.40 / 3.45	2.35 / 3.37	2.31 / 3.30	2.28 / 3.23	2.23 / 3.13	2.18 / 3.05	2.12 / 2.94	2.08 / 2.86	2.04 / 2.77	1.99 / 2.69	1.96 / 2.63	1.92 / 2.56	1.90 / 2.53	1.87 / 2.47	1.85 / 2.44	1.84 / 2.42
21		4.32 / 8.02	3.47 / 5.78	3.07 / 4.87	2.84 / 4.37	2.68 / 4.04	2.57 / 3.81	2.49 / 3.65	2.42 / 3.51	2.37 / 3.40	2.32 / 3.31	2.28 / 3.24	2.25 / 3.17	2.20 / 3.07	2.15 / 2.99	2.09 / 2.88	2.05 / 2.80	2.00 / 2.72	1.96 / 2.63	1.93 / 2.58	1.89 / 2.51	1.87 / 2.47	1.84 / 2.42	1.82 / 2.38	1.81 / 2.36
22		4.30 / 7.94	3.44 / 5.72	3.05 / 4.82	2.82 / 4.31	2.66 / 3.99	2.55 / 3.76	2.47 / 3.59	2.40 / 3.45	2.35 / 3.35	2.30 / 3.26	2.26 / 3.18	2.23 / 3.12	2.18 / 3.02	2.13 / 2.94	2.07 / 2.83	2.03 / 2.75	1.98 / 2.67	1.93 / 2.58	1.91 / 2.53	1.87 / 2.46	1.84 / 2.42	1.81 / 2.37	1.80 / 2.33	1.78 / 2.31
23		4.28 / 7.88	3.42 / 5.66	3.03 / 4.76	2.80 / 4.26	2.64 / 3.94	2.53 / 3.71	2.45 / 3.54	2.38 / 3.41	2.32 / 3.30	2.28 / 3.21	2.24 / 3.14	2.20 / 3.07	2.14 / 2.97	2.10 / 2.89	2.04 / 2.78	2.00 / 2.70	1.96 / 2.62	1.91 / 2.53	1.88 / 2.48	1.84 / 2.41	1.82 / 2.37	1.79 / 2.32	1.77 / 2.28	1.76 / 2.26
24		4.26 / 7.82	3.40 / 5.61	3.01 / 4.72	2.78 / 4.22	2.62 / 3.90	2.51 / 3.67	2.43 / 3.50	2.36 / 3.36	2.30 / 3.25	2.26 / 3.17	2.22 / 3.09	2.18 / 3.03	2.13 / 2.93	2.09 / 2.85	2.02 / 2.74	1.98 / 2.66	1.94 / 2.58	1.89 / 2.49	1.86 / 2.44	1.82 / 2.36	1.80 / 2.33	1.76 / 2.27	1.74 / 2.23	1.73 / 2.21
25		4.24 / 7.77	3.38 / 5.57	2.99 / 4.68	2.76 / 4.18	2.60 / 3.86	2.49 / 3.63	2.41 / 3.46	2.34 / 3.32	2.28 / 3.21	2.24 / 3.13	2.20 / 3.05	2.16 / 2.99	2.11 / 2.89	2.06 / 2.81	2.00 / 2.70	1.96 / 2.62	1.92 / 2.54	1.87 / 2.45	1.84 / 2.40	1.80 / 2.32	1.77 / 2.29	1.74 / 2.23	1.72 / 2.19	1.71 / 2.17

Degrees of Freedom for Numerator — Degrees of Freedom for Denominator

TABLE B

F Table (*continued*).

Degrees of Freedom for Denominator

Degrees of Freedom for Numerator

df	1	2	3	4	5	6	7	8	9	10	11	12	14	16	20	24	30	40	50	75	100	200	500	∞
26	4.22 / 7.72	3.37 / 5.53	2.98 / 4.64	2.74 / 4.14	2.59 / 3.82	2.47 / 3.59	2.39 / 3.42	2.32 / 3.29	2.27 / 3.17	2.22 / 3.09	2.18 / 3.02	2.15 / 2.96	2.10 / 2.86	2.05 / 2.77	1.99 / 2.66	1.95 / 2.58	1.90 / 2.50	1.85 / 2.41	1.82 / 2.36	1.78 / 2.28	1.76 / 2.25	1.72 / 2.19	1.70 / 2.15	1.69 / 2.13
27	4.21 / 7.68	3.35 / 5.49	2.96 / 4.60	2.73 / 4.11	2.57 / 3.79	2.46 / 3.56	2.37 / 3.39	2.30 / 3.26	2.25 / 3.14	2.20 / 3.06	2.16 / 2.98	2.13 / 2.93	2.08 / 2.83	2.03 / 2.74	1.97 / 2.63	1.93 / 2.55	1.88 / 2.47	1.84 / 2.38	1.80 / 2.33	1.76 / 2.25	1.74 / 2.21	1.71 / 2.16	1.68 / 2.12	1.67 / 2.10
28	4.20 / 7.64	3.34 / 5.45	2.95 / 4.57	2.71 / 4.07	2.56 / 3.76	2.44 / 3.53	2.36 / 3.36	2.29 / 3.23	2.24 / 3.11	2.19 / 3.03	2.15 / 2.95	2.12 / 2.90	2.06 / 2.80	2.02 / 2.71	1.96 / 2.60	1.91 / 2.52	1.87 / 2.44	1.81 / 2.35	1.78 / 2.30	1.75 / 2.22	1.72 / 2.18	1.69 / 2.13	1.67 / 2.09	1.65 / 2.00
29	4.18 / 7.60	3.33 / 5.42	2.93 / 4.54	2.70 / 4.04	2.54 / 3.73	2.43 / 3.50	2.35 / 3.33	2.28 / 3.20	2.22 / 3.06	2.18 / 3.00	2.14 / 2.92	2.10 / 2.87	2.05 / 2.77	2.00 / 2.68	1.94 / 2.57	1.90 / 2.49	1.85 / 2.41	1.80 / 2.32	1.77 / 2.27	1.73 / 2.19	1.71 / 2.15	1.68 / 2.10	1.65 / 2.06	1.64 / 2.03
30	4.17 / 7.56	3.32 / 5.39	2.92 / 4.51	2.69 / 4.02	2.53 / 3.70	2.42 / 3.47	2.34 / 3.30	2.27 / 3.17	2.21 / 3.06	2.16 / 2.98	2.12 / 2.90	2.09 / 2.84	2.04 / 2.74	1.99 / 2.66	1.93 / 2.55	1.89 / 2.47	1.84 / 2.38	1.79 / 2.29	1.76 / 2.24	1.72 / 2.16	1.69 / 2.13	1.65 / 2.07	1.64 / 2.03	1.62 / 2.01
32	4.15 / 7.50	3.30 / 5.34	2.90 / 4.46	2.67 / 3.97	2.51 / 3.66	2.40 / 3.42	2.32 / 3.25	2.25 / 3.12	2.19 / 3.01	2.14 / 2.94	2.10 / 2.86	2.07 / 2.80	2.02 / 2.70	1.97 / 2.62	1.91 / 2.51	1.86 / 2.42	1.82 / 2.34	1.76 / 2.25	1.74 / 2.20	1.69 / 2.12	1.67 / 2.08	1.64 / 2.02	1.61 / 1.98	1.59 / 1.96
34	4.13 / 7.44	3.28 / 5.29	2.88 / 4.42	2.65 / 3.93	2.49 / 3.61	2.38 / 3.38	2.30 / 3.21	2.23 / 3.08	2.17 / 2.97	2.12 / 2.89	2.08 / 2.82	2.05 / 2.76	2.00 / 2.66	1.95 / 2.58	1.89 / 2.47	1.84 / 2.38	1.80 / 2.30	1.74 / 2.21	1.71 / 2.15	1.67 / 2.08	1.64 / 2.04	1.61 / 1.98	1.59 / 1.94	1.57 / 1.91
36	4.11 / 7.39	3.26 / 5.25	2.86 / 4.38	2.63 / 3.89	2.48 / 3.58	2.36 / 3.35	2.28 / 3.18	2.21 / 3.04	2.15 / 2.94	2.10 / 2.86	2.06 / 2.78	2.03 / 2.72	1.98 / 2.62	1.93 / 2.54	1.87 / 2.43	1.82 / 2.35	1.78 / 2.26	1.72 / 2.17	1.69 / 2.12	1.65 / 2.04	1.62 / 2.00	1.59 / 1.94	1.56 / 1.90	1.55 / 1.87
38	4.10 / 7.35	3.25 / 5.21	2.85 / 4.34	2.62 / 3.86	2.46 / 3.54	2.35 / 3.32	2.26 / 3.15	2.19 / 3.02	2.14 / 2.91	2.09 / 2.82	2.05 / 2.75	2.02 / 2.69	1.96 / 2.59	1.92 / 2.51	1.85 / 2.40	1.80 / 2.32	1.76 / 2.22	1.71 / 2.14	1.67 / 2.08	1.63 / 2.00	1.60 / 1.97	1.57 / 1.90	1.54 / 1.86	1.53 / 1.84
40	4.08 / 7.31	3.23 / 5.18	2.84 / 4.31	2.61 / 3.83	2.45 / 3.51	2.34 / 3.29	2.25 / 3.12	2.18 / 2.99	2.12 / 2.88	2.07 / 2.80	2.04 / 2.73	2.00 / 2.66	1.95 / 2.56	1.90 / 2.49	1.84 / 2.37	1.79 / 2.29	1.74 / 2.20	1.69 / 2.11	1.66 / 2.05	1.61 / 1.97	1.59 / 1.94	1.55 / 1.88	1.53 / 1.84	1.51 / 1.81
42	4.07 / 7.27	3.22 / 5.15	2.83 / 4.29	2.59 / 3.80	2.44 / 3.49	2.32 / 3.26	2.24 / 3.10	2.17 / 2.96	2.11 / 2.86	2.06 / 2.77	2.02 / 2.70	1.99 / 2.64	1.94 / 2.54	1.89 / 2.46	1.82 / 2.35	1.78 / 2.26	1.73 / 2.17	1.68 / 2.08	1.64 / 2.02	1.60 / 1.94	1.57 / 1.91	1.54 / 1.85	1.51 / 1.80	1.49 / 1.78
44	4.06 / 7.24	3.21 / 5.12	2.82 / 4.26	2.58 / 3.78	2.43 / 3.46	2.31 / 3.24	2.23 / 3.07	2.16 / 2.94	2.10 / 2.84	2.05 / 2.75	2.01 / 2.68	1.98 / 2.62	1.92 / 2.52	1.88 / 2.44	1.81 / 2.32	1.76 / 2.24	1.72 / 2.15	1.66 / 2.06	1.63 / 2.00	1.58 / 1.92	1.56 / 1.88	1.52 / 1.82	1.50 / 1.78	1.48 / 1.75
46	4.05 / 7.21	3.20 / 5.10	2.81 / 4.24	2.57 / 3.76	2.42 / 3.44	2.30 / 3.22	2.22 / 3.05	2.14 / 2.92	2.09 / 2.82	2.04 / 2.73	2.00 / 2.66	1.97 / 2.60	1.91 / 2.50	1.87 / 2.42	1.80 / 2.30	1.75 / 2.22	1.71 / 2.13	1.65 / 2.04	1.62 / 1.98	1.57 / 1.90	1.54 / 1.86	1.51 / 1.80	1.48 / 1.76	1.46 / 1.72
48	4.04 / 7.19	3.19 / 5.08	2.80 / 4.22	2.56 / 3.74	2.41 / 3.42	2.30 / 3.20	2.21 / 3.04	2.14 / 2.90	2.08 / 2.80	2.03 / 2.71	1.99 / 2.64	1.96 / 2.58	1.90 / 2.48	1.86 / 2.40	1.79 / 2.28	1.74 / 2.20	1.70 / 2.11	1.64 / 2.02	1.61 / 1.90	1.56 / 1.88	1.53 / 1.84	1.50 / 1.78	1.47 / 1.73	1.45 / 1.70

TABLE B
F Table (continued).

Degrees of Freedom for Numerator

df denom.	1	2	3	4	5	6	7	8	9	10	11	12	14	16	20	24	30	40	50	75	100	200	500	∞
50	4.03 / 7.17	3.18 / 5.06	2.79 / 4.20	2.56 / 3.72	2.40 / 3.41	2.29 / 3.18	2.20 / 3.02	2.13 / 2.88	2.07 / 2.78	2.02 / 2.70	1.98 / 2.62	1.95 / 2.56	1.90 / 2.46	1.85 / 2.39	1.78 / 2.26	1.74 / 2.18	1.69 / 2.10	1.63 / 2.00	1.60 / 1.94	1.55 / 1.86	1.52 / 1.82	1.48 / 1.76	1.46 / 1.71	1.44 / 1.68
55	4.02 / 7.12	3.17 / 5.01	2.78 / 4.16	2.54 / 3.68	2.38 / 3.37	2.27 / 3.15	2.18 / 2.98	2.11 / 2.85	2.05 / 2.75	2.00 / 2.66	1.97 / 2.59	1.93 / 2.53	1.88 / 2.43	1.83 / 2.35	1.76 / 2.23	1.72 / 2.15	1.67 / 2.06	1.61 / 1.96	1.58 / 1.90	1.52 / 1.82	1.50 / 1.78	1.46 / 1.71	1.43 / 1.66	1.41 / 1.64
60	4.00 / 7.08	3.15 / 4.98	2.76 / 4.13	2.52 / 3.65	2.37 / 3.34	2.25 / 3.12	2.17 / 2.95	2.10 / 2.82	2.04 / 2.72	1.99 / 2.63	1.95 / 2.56	1.92 / 2.50	1.86 / 2.40	1.81 / 2.32	1.75 / 2.20	1.70 / 2.12	1.65 / 2.03	1.59 / 1.93	1.56 / 1.87	1.50 / 1.79	1.48 / 1.74	1.44 / 1.68	1.41 / 1.63	1.39 / 1.60
65	3.99 / 7.04	3.14 / 4.95	2.75 / 4.10	2.51 / 3.62	2.36 / 3.31	2.24 / 3.09	2.15 / 2.93	2.08 / 2.79	2.02 / 2.70	1.98 / 2.61	1.94 / 2.54	1.90 / 2.47	1.85 / 2.37	1.80 / 2.30	1.73 / 2.18	1.68 / 2.09	1.63 / 2.00	1.57 / 1.90	1.54 / 1.84	1.49 / 1.76	1.46 / 1.71	1.42 / 1.64	1.39 / 1.60	1.37 / 1.56
70	3.98 / 7.01	3.13 / 4.92	2.74 / 4.08	2.50 / 3.60	2.35 / 3.29	2.23 / 3.07	2.14 / 2.91	2.07 / 2.77	2.01 / 2.67	1.97 / 2.59	1.93 / 2.51	1.89 / 2.45	1.84 / 2.35	1.79 / 2.28	1.72 / 2.15	1.67 / 2.07	1.62 / 1.98	1.56 / 1.88	1.53 / 1.82	1.47 / 1.74	1.45 / 1.69	1.40 / 1.62	1.37 / 1.56	1.35 / 1.53
80	3.96 / 6.96	3.11 / 4.88	2.72 / 4.04	2.48 / 3.56	2.33 / 3.25	2.21 / 3.04	2.12 / 2.87	2.05 / 2.74	1.99 / 2.64	1.95 / 2.55	1.91 / 2.48	1.88 / 2.41	1.82 / 2.32	1.77 / 2.24	1.70 / 2.11	1.65 / 2.03	1.60 / 1.94	1.54 / 1.84	1.51 / 1.78	1.45 / 1.70	1.42 / 1.65	1.38 / 1.57	1.35 / 1.52	1.32 / 1.49
100	3.94 / 6.90	3.09 / 4.82	2.70 / 3.98	2.46 / 3.51	2.30 / 3.20	2.19 / 2.99	2.10 / 2.82	2.03 / 2.69	1.97 / 2.59	1.92 / 2.51	1.88 / 2.43	1.85 / 2.36	1.79 / 2.26	1.75 / 2.19	1.68 / 2.06	1.63 / 1.98	1.57 / 1.89	1.51 / 1.79	1.48 / 1.73	1.42 / 1.64	1.39 / 1.59	1.34 / 1.51	1.30 / 1.46	1.28 / 1.43
125	3.92 / 6.84	3.07 / 4.78	2.68 / 3.94	2.44 / 3.47	2.29 / 3.17	2.17 / 2.95	2.08 / 2.79	2.01 / 2.65	1.95 / 2.56	1.90 / 2.47	1.86 / 2.40	1.83 / 2.33	1.77 / 2.23	1.72 / 2.15	1.65 / 2.03	1.60 / 1.94	1.55 / 1.85	1.49 / 1.75	1.45 / 1.68	1.39 / 1.59	1.36 / 1.54	1.31 / 1.46	1.27 / 1.40	1.25 / 1.37
150	3.91 / 6.81	3.06 / 4.75	2.67 / 3.91	2.43 / 3.44	2.27 / 3.14	2.16 / 2.92	2.07 / 2.76	2.00 / 2.62	1.94 / 2.53	1.89 / 2.44	1.85 / 2.37	1.82 / 2.30	1.76 / 2.20	1.71 / 2.12	1.64 / 2.00	1.59 / 1.91	1.54 / 1.83	1.47 / 1.72	1.44 / 1.66	1.37 / 1.56	1.34 / 1.51	1.29 / 1.43	1.25 / 1.37	1.22 / 1.33
200	3.89 / 6.76	3.04 / 4.71	2.65 / 3.88	2.41 / 3.41	2.26 / 3.11	2.14 / 2.90	2.05 / 2.73	1.98 / 2.60	1.92 / 2.50	1.87 / 2.41	1.83 / 2.34	1.80 / 2.28	1.74 / 2.17	1.69 / 2.09	1.62 / 1.97	1.57 / 1.88	1.52 / 1.79	1.45 / 1.69	1.42 / 1.62	1.35 / 1.53	1.32 / 1.48	1.26 / 1.39	1.22 / 1.33	1.19 / 1.28
400	3.86 / 6.70	3.02 / 4.66	2.62 / 3.83	2.39 / 3.36	2.23 / 3.06	2.12 / 2.85	2.03 / 2.69	1.96 / 2.55	1.90 / 2.46	1.85 / 2.37	1.81 / 2.29	1.78 / 2.23	1.72 / 2.12	1.67 / 2.04	1.60 / 1.92	1.54 / 1.84	1.49 / 1.74	1.42 / 1.64	1.38 / 1.57	1.32 / 1.47	1.28 / 1.42	1.22 / 1.32	1.16 / 1.24	1.13 / 1.19
1000	3.85 / 6.66	3.00 / 4.62	2.61 / 3.80	2.38 / 3.34	2.22 / 3.04	2.10 / 2.82	2.02 / 2.66	1.95 / 2.53	1.89 / 2.43	1.84 / 2.34	1.80 / 2.26	1.76 / 2.20	1.70 / 2.09	1.65 / 2.01	1.58 / 1.89	1.53 / 1.81	1.47 / 1.71	1.41 / 1.61	1.36 / 1.54	1.30 / 1.44	1.26 / 1.38	1.19 / 1.28	1.13 / 1.19	1.08 / 1.11
∞	3.84 / 6.64	2.99 / 4.60	2.60 / 3.78	2.37 / 3.32	2.21 / 3.02	2.09 / 2.80	2.01 / 2.64	1.94 / 2.51	1.88 / 2.41	1.83 / 2.32	1.79 / 2.24	1.75 / 2.18	1.69 / 2.07	1.64 / 1.99	1.57 / 1.87	1.52 / 1.79	1.46 / 1.69	1.40 / 1.59	1.35 / 1.52	1.28 / 1.41	1.24 / 1.36	1.17 / 1.25	1.11 / 1.15	1.00 / 1.00

Degrees of Freedom for Denominator

Locate the *df* in the leftmost column. Reject H_0, if the observed value equals or exceeds the tabled value. For example, for $df = 10$, *r* must equal or exceed .497 to be significant at the .05 level of significance.

TABLE C
Critical Values for
Pearson's *r*.

df (= N − 2; N = Number of Pairs)	.05	.025	.01	.005
	Level of Significance for One-tailed Test			
	.10	.05	.02	.01
	Level of Significance for Two-tailed Test			
1	.988	.997	.9995	.9999
2	.900	.950	.980	.990
3	.805	.878	.934	.959
4	.729	.811	.882	.917
5	.669	.754	.833	.874
6	.622	.707	.789	.834
7	.582	.666	.750	.798
8	.549	.632	.716	.765
9	.521	.602	.685	.735
10	.497	.576	.658	.708
11	.476	.553	.634	.684
12	.458	.532	.612	.661
13	.441	.514	.592	.641
14	.426	.497	.574	.623
15	.412	.482	.558	.606
16	.400	.468	.542	.590
17	.389	.456	.528	.575
18	.378	.444	.516	.561
19	.369	.433	.503	.549
20	.360	.423	.492	.537
21	.352	.413	.482	.526
22	.344	.404	.472	.515
23	.337	.396	.462	.505
24	.330	.388	.453	.496
25	.323	.381	.445	.487
26	.317	.374	.437	.479
27	.311	.367	.430	.471
28	.306	.361	.423	.463
29	.301	.355	.416	.456
30	.296	.349	.409	.449
35	.275	.325	.381	.418
40	.257	.304	.358	.393
45	.243	.288	.338	.372
50	.231	.273	.322	.354
60	.211	.250	.295	.325
70	.195	.232	.274	.302
80	.183	.217	.256	.283
90	.173	.205	.242	.267
100	.164	.195	.230	.254

Source: Table VII of Fisher and Yates. 1974. *Statistical tables for biological, agricultural and medical research*, 6th ed. Published by Longman Group Ltd., London (previously published by Oliver and Boyd, Ltd., Edinburgh), and by permission of the authors and publishers.

Observed value of χ^2 must equal or exceed tabled value in order to be significant. For example, for $df = 15$, χ^2 must equal or exceed 30.58 to be significant at the .01 level.

TABLE D
Critical Values for Chi Square.

Degrees of Freedom (*df*)	.05	.01
1	3.84	6.64
2	5.99	9.21
3	7.82	11.34
4	9.49	13.28
5	11.07	15.09
6	12.59	16.81
7	14.07	18.48
8	15.51	20.09
9	16.92	21.67
10	18.31	23.21
11	19.68	24.72
12	21.03	26.22
13	22.36	27.69
14	23.68	29.14
15	25.00	30.58
16	26.30	32.00
17	27.59	33.41
18	28.87	34.80
19	30.14	36.19
20	31.41	37.57
21	32.67	38.93
22	33.92	40.29
23	35.17	41.64
24	36.42	42.98
25	37.65	44.31
26	38.88	45.64
27	40.11	46.96
28	41.34	48.28
29	42.56	49.59
30	43.77	50.89

Source: Table IV of Fisher and Yates. 1974. *Statistical tables for biological, agricultural and medical research*, 6th ed. Published by Longman Group Ltd., London (previously published by Oliver and Boyd, Ltd., Edinburgh), and by permission of the authors and publishers.

TABLE E
Critical Values for the
Sign Test.

N	One-tailed Test .01	One-tailed Test .05	Two-tailed Test .01	Two-tailed Test .05
7	7	7	—	7
8	8	7	8	8
9	9	8	9	8
10	10	9	10	9
11	10	9	11	10
12	11	10	11	10
13	12	10	12	11
14	12	11	13	12
15	13	12	13	12
16	14	12	14	13
17	14	13	15	13
18	15	13	15	14
19	15	14	16	15
20	16	15	17	15
21	17	15	17	16
22	17	16	18	17
23	18	16	19	17
24	19	17	19	18
25	19	18	20	18
26	20	18	21	19
27	21	19	21	20
28	21	19	22	20
29	22	20	22	21
30	22	21	23	21
31	23	21	24	22
32	24	22	24	23
33	24	22	25	23
34	25	23	26	24
35	25	23	26	24
36	26	24	27	25
37	27	25	27	25
38	27	25	28	26
39	28	26	29	27
40	28	26	29	27

Tables entries are the number of positive (or negative) difference scores that are needed out of N scores to be significant at $\alpha = .05$ or $\alpha = .01$, for one- or two-tailed tests.

Note: For $N < 26$, the critical values are adapted from Table IV in H. M. Walker and J. Lev. 1953. *Statistical inference*. New York: Henry Holt. For $N > 25$, the critical values have been computed using the normal approximation to the binomial.

In the column labelled N, find the number of pairs of scores (minus the pairs with zero differences). Reject H_0 at the appropriate level, if the observed value of W is less than or equal to the tabled value.

TABLE F

Critical Values for the Wilcoxon Test.

	Level of Significance for a Directional Test					Level of Significance for a Directional Test			
	.05	.025	.01	.005		.05	.025	.01	.005
	Level of Significance for a Nondirectional Test					Level of Significance for a Nondirectional Test			
N	.10	.05	.02	.01	N	.10	.05	.02	.01
5	0	—	—	—	28	130	116	101	91
6	2	0	—	—	29	140	126	110	100
7	3	2	0	—	30	151	137	120	109
8	5	3	1	0	31	163	147	130	118
9	8	5	3	1	32	175	159	140	128
10	10	8	5	3	33	187	170	151	138
11	13	10	7	5	34	200	182	162	148
12	17	13	9	7	35	213	195	173	159
13	21	17	12	9	36	227	208	185	171
14	25	21	15	12	37	241	221	198	182
15	30	25	19	15	38	256	235	211	194
16	35	29	23	19	39	271	249	224	207
17	41	34	27	23	40	286	264	238	220
18	47	40	32	27	41	302	279	252	233
19	53	46	37	32	42	319	294	266	247
20	60	52	43	37	43	336	310	281	261
21	67	58	49	42	44	353	327	296	276
22	75	65	55	48	45	371	343	312	291
23	83	73	62	54	46	389	361	328	307
24	91	81	69	61	47	407	378	345	322
25	100	89	76	68	48	426	396	362	339
26	110	98	84	75	49	446	415	379	355
27	119	107	92	83	50	466	434	397	373

Source: From F. Wilcoxon, S. Katte, and R. A. Wilcox. 1963. *Critical values and probability levels for the Wilcoxon rank sum test and the Wilcoxon signed rank test*. New York: American Cyanamid Co., and F. Wilcoxon and R. A. Wilcox. 1964. *Some rapid approximate statistical procedures*. New York: Lederle Laboratories, as used in Runyon and Haber. 1976. *Fundamentals of behavioral statistics*, 3rd ed., Reading, Mass.: Addison-Wesley. Reprinted by permission of the American Cyanamid Company.

Four tables follow, each for different values of α. Find the page appropriate to the α and direction of your test. Find the column for N_1 and the row for N_2. Reject H_0, if the observed value of U does not fall between the two tabled values.

Source: From Mann, H. B., and Whitney, D. R., "On a Test of Whether One of Two Random Variables Is Stochastically Larger Than the Other," *Annals of mathematical statistics* 18 (1947): 50–60, and Auble, D., "Extended Tables for the Mann-Whitney Statistic," *Bulletin of the institute of educational research at Indiana University,* vol. 1, no. 2 (1953), as used in Runyon and Haber. 1976. *Fundamentals of behavioral statistics*, 3rd ed., Reading, Mass.: Addison-Wesley. Reprint permission granted by the Institute of Mathematical Statistics.

TABLE G
Critical Value for the Mann-Whitney Test.

Critical Values of the Mann-Whitney U for a Directional Test at .05 or a Nondirectional Test at .10

N_2 \ N_1	1	2	3	4	5	6	7	8	9	10	11	12	13	14	15	16	17	18	19	20
1	—	—	—	—	—	—	—	—	—	—	—	—	—	—	—	—	—	—	—	—
2	—	—	—	—	—	—	—	—	—	—	—	—	—	—	—	—	—	—	0/38	0/40
3	—	—	—	—	—	—	—	—	0/27	0/30	0/33	1/35	1/38	1/41	2/43	2/46	2/49	2/52	3/54	3/57
4	—	—	—	—	—	0/24	0/28	1/31	1/35	2/38	2/42	3/45	3/49	4/52	5/55	5/59	6/62	6/66	7/69	8/72
5	—	—	—	—	0/25	1/29	1/34	2/38	3/42	4/46	5/50	6/54	7/58	7/63	8/67	9/71	10/75	11/79	12/83	13/87
6	—	—	—	0/24	1/29	2/34	3/39	4/44	5/49	6/54	7/59	9/63	10/68	11/73	12/78	13/83	15/87	16/92	17/97	18/102
7	—	—	—	0/28	1/34	3/39	4/45	6/50	7/56	9/61	10/67	12/72	13/78	15/83	16/89	18/94	19/100	21/105	22/111	24/116
8	—	—	—	1/31	2/38	4/44	6/50	7/57	9/63	11/69	13/75	15/81	17/87	18/94	20/100	22/106	24/112	26/118	28/124	30/130
9	—	—	0/27	1/35	3/42	5/49	7/56	9/63	11/70	13/77	16/83	18/90	20/97	22/104	24/111	27/117	29/124	31/131	33/138	36/144
10	—	—	0/30	2/38	4/46	6/54	9/61	11/69	13/77	16/84	18/92	21/99	24/106	26/114	29/121	31/129	34/136	37/143	39/151	42/158
11	—	—	0/33	2/42	5/50	7/59	10/67	13/75	16/83	18/92	21/100	24/108	27/116	30/124	33/132	36/140	39/148	42/156	45/164	48/172
12	—	—	1/35	3/45	6/54	9/63	12/72	15/81	18/90	21/99	24/108	27/117	31/125	34/134	37/143	41/151	44/160	47/169	51/177	54/186
13	—	—	1/38	3/49	7/58	10/68	13/78	17/87	20/97	24/106	27/116	31/125	34/125	38/144	42/153	45/163	49/172	53/181	56/191	60/200
14	—	—	1/41	4/52	7/63	11/73	15/83	18/94	22/104	26/114	30/124	34/134	38/144	42/154	46/164	50/174	54/184	58/194	63/203	67/213
15	—	—	2/43	5/55	8/67	12/78	16/89	20/100	24/111	29/121	33/132	37/143	42/153	46/164	51/174	55/185	60/195	64/206	69/216	73/227
16	—	—	2/46	5/59	9/71	13/83	18/94	22/106	27/117	31/129	36/140	41/151	45/163	50/174	55/185	60/196	65/207	70/218	74/230	79/241
17	—	—	2/49	6/62	10/75	15/87	19/100	24/112	29/124	34/148	39/148	44/160	49/172	54/184	60/195	65/207	70/219	75/231	81/242	86/254
18	—	—	2/52	6/66	11/79	16/92	21/105	26/118	31/131	37/143	42/156	47/169	53/181	58/194	64/206	70/218	75/231	81/243	87/255	92/268
19	—	0/38	3/54	7/69	12/83	17/97	22/111	28/124	33/138	39/151	45/164	51/177	56/191	63/203	69/216	74/230	81/242	87/255	93/268	99/281
20	—	0/40	3/57	8/72	13/87	18/102	24/116	30/130	36/144	42/158	48/172	54/186	60/200	67/213	73/227	79/241	86/254	92/268	99/281	105/295

Note: Dashes in the body of the table indicate that no decision is possible at the stated level of significance.

TABLE G
Critical Values for the Mann-Whitney Test *(continued)*.

Critical Values of the Mann-Whitney U for a Directional Test at .025 or a Nondirectional Test at .05

Each cell shows the upper critical value over the (underlined) lower critical value.

N_2＼N_1	1	2	3	4	5	6	7	8	9	10	11	12	13	14	15	16	17	18	19	20
1	—	—	—	—	—	—	—	—	—	—	—	—	—	—	—	—	—	—	—	—
2	—	—	—	—	—	—	—	—	—	—	—	—	0/26	0/28	0/30	0/32	0/34	0/36	1/37	1/39
3	—	—	—	—	—	—	0/21	0/24	1/26	1/29	1/32	2/34	2/37	2/40	3/42	3/45	4/47	4/50	4/52	5/55
4	—	—	—	—	0/20	1/23	1/27	2/30	3/33	3/37	4/40	5/43	5/47	6/50	7/53	7/57	8/60	9/63	9/67	10/70
5	—	—	—	0/20	1/24	2/28	3/32	4/36	5/40	6/44	7/48	8/52	9/56	10/60	11/64	12/68	13/72	14/76	15/80	16/84
6	—	—	—	1/23	2/28	3/33	4/38	6/42	7/47	8/52	9/57	11/61	12/66	13/71	15/75	16/80	18/84	19/89	20/94	22/98
7	—	—	0/21	1/27	3/32	4/38	6/43	7/49	9/54	11/59	12/65	14/70	16/75	17/81	19/86	21/91	23/96	24/102	26/107	28/112
8	—	—	0/24	2/30	4/36	6/42	7/49	9/55	11/61	13/67	15/73	17/79	20/84	22/90	24/96	26/102	28/108	30/114	32/120	34/126
9	—	—	1/26	3/33	5/40	7/47	9/54	11/61	14/67	16/74	18/81	21/87	23/94	26/100	28/107	31/113	33/120	36/126	38/133	40/140
10	—	—	1/29	3/37	6/44	8/52	11/59	13/67	16/74	19/81	22/88	24/96	27/103	30/110	33/117	36/124	38/132	41/139	44/146	47/153
11	—	—	1/32	4/40	7/48	9/57	12/65	15/73	18/81	22/88	25/96	28/104	31/112	34/120	37/128	41/135	44/143	47/151	50/159	53/167
12	—	—	2/34	5/43	8/52	11/61	14/70	17/79	21/87	24/96	28/104	31/113	35/121	38/130	42/138	46/146	49/155	53/163	56/172	60/180
13	—	0/26	2/37	5/47	9/56	12/66	16/75	20/84	23/94	27/103	31/112	35/121	39/130	43/139	47/148	51/157	55/166	59/175	63/184	67/193
14	—	0/28	2/40	6/50	10/60	13/71	17/81	22/90	26/100	30/110	34/120	38/130	43/139	47/149	51/159	56/168	60/178	65/187	69/197	73/207
15	—	0/30	3/42	7/53	11/64	15/75	19/86	24/96	28/107	33/117	37/128	42/138	47/148	51/159	56/169	61/179	66/189	70/200	75/210	80/220
16	—	0/32	3/45	7/57	12/68	16/80	21/91	26/102	31/113	36/124	41/135	46/146	51/157	56/168	61/179	66/190	71/201	76/212	82/222	87/233
17	—	0/34	4/47	8/60	13/72	18/84	23/96	28/108	33/120	38/132	44/143	49/155	55/166	60/178	66/189	71/201	77/212	82/224	88/234	93/247
18	—	0/36	4/50	9/63	14/76	19/89	24/102	30/114	36/126	41/139	47/151	53/163	59/175	65/187	70/200	76/212	82/224	88/236	94/248	100/260
19	—	1/37	4/53	9/67	15/80	20/94	26/107	32/120	38/133	44/146	50/159	56/172	63/184	69/197	75/210	82/222	88/235	94/248	101/260	107/273
20	—	1/39	5/55	10/70	16/84	22/98	28/112	34/126	40/140	47/153	53/167	60/180	67/193	73/207	80/220	87/233	93/247	100/260	107/273	114/286

TABLE G

Critical Values for the Mann-Whitney Test *(continued)*.

Critical Values of the Mann-Whitney U for a Directional Test at .01 or a Nondirectional Test at .02

Each cell lists the critical value over its complement (upper value / lower value).

N_2 ＼ N_1	1	2	3	4	5	6	7	8	9	10	11	12	13	14	15	16	17	18	19	20
1	—	—	—	—	—	—	—	—	—	—	—	—	—	—	—	—	—	—	—	—
2	—	—	—	—	—	—	—	0/16	0/18	0/20	0/22	1/23	1/25	1/27	1/29	1/31	2/32	2/34	2/36	2/38
3	—	—	—	—	0/15	1/17	1/20	2/22	2/25	3/27	3/30	4/32	4/35	5/37	5/40	6/42	6/45	7/47	7/50	8/52
4	—	—	—	0/16	1/19	2/22	3/25	4/28	4/32	5/35	6/38	7/41	8/44	9/47	10/50	11/53	11/57	12/60	13/63	13/67
5	—	—	0/15	1/19	2/23	3/27	5/30	6/34	7/38	8/42	9/46	11/49	12/53	13/57	14/61	15/65	17/68	18/72	19/76	20/80
6	—	—	1/17	2/22	3/27	5/31	6/36	8/40	10/44	11/49	13/53	14/58	16/62	17/67	19/71	21/75	22/80	24/84	25/89	27/93
7	—	—	1/20	3/25	5/30	6/36	8/41	10/46	12/51	14/56	16/61	18/66	20/71	22/76	24/81	26/86	28/91	30/96	32/101	34/106
8	—	0/16	2/22	4/28	6/34	8/40	10/46	13/51	15/57	17/63	19/69	22/74	24/80	26/86	29/91	31/97	34/102	36/108	38/114	41/119
9	—	0/18	2/25	4/32	7/38	10/44	12/51	15/57	17/64	20/70	23/76	26/82	28/89	31/95	34/101	37/107	39/114	42/120	45/126	48/132
10	—	0/20	3/27	5/35	8/42	11/49	14/56	17/63	20/70	23/77	26/84	29/91	33/97	36/104	39/111	42/118	45/125	48/132	52/138	55/145
11	—	0/22	3/30	6/38	9/46	13/53	16/61	19/69	23/76	26/84	30/91	33/99	37/106	40/114	44/121	47/129	51/136	55/143	58/151	62/158
12	—	1/23	4/32	7/41	11/49	14/58	18/66	22/74	26/82	29/91	33/99	37/107	41/115	45/123	49/131	53/139	57/147	61/155	65/163	69/171
13	—	1/25	4/35	8/44	12/53	16/62	20/71	24/80	28/89	33/97	37/106	41/115	45/124	50/132	54/141	59/149	63/158	67/167	72/175	76/184
14	—	1/27	5/37	9/47	13/57	17/67	22/76	26/86	31/95	36/104	40/114	45/123	50/132	55/141	59/151	64/160	67/171	74/178	78/188	83/197
15	—	1/29	5/40	10/50	14/61	19/71	24/81	29/91	34/101	39/111	44/121	49/131	54/141	59/151	64/161	70/170	75/180	80/190	85/200	90/210
16	—	1/31	6/42	11/53	15/65	21/75	26/86	31/97	37/107	42/118	47/129	53/139	59/149	64/160	70/170	75/181	81/191	86/202	92/212	98/222
17	—	2/32	6/45	11/57	17/68	22/80	28/91	34/102	39/114	45/125	51/136	57/147	63/158	67/171	75/180	81/191	87/202	93/213	99/224	105/235
18	—	2/34	7/47	12/60	18/72	24/84	30/96	36/108	42/120	48/132	55/143	61/155	67/167	74/178	80/190	86/202	93/213	99/225	106/236	112/248
19	—	2/36	7/50	13/63	19/76	25/89	32/101	38/114	45/126	52/138	58/151	65/163	72/175	78/188	85/200	92/212	99/224	106/236	113/248	119/261
20	—	2/38	8/52	13/67	20/80	27/93	34/106	41/119	48/132	55/145	62/158	69/171	76/184	83/197	90/210	98/222	105/235	112/248	119/261	127/273

TABLE G
Critical Values for the Mann-Whitney Test (continued).

Critical Values of the Mann-Whitney U for a Directional Test at .005 or a Nondirectional Test at .01

N_2 \ N_1	1	2	3	4	5	6	7	8	9	10	11	12	13	14	15	16	17	18	19	20
1	—	—	—	—	—	—	—	—	—	—	—	—	—	—	—	—	—	—	0/19	0/20
2	—	—	—	—	0/10	0/12	0/14	1/15	1/17	1/19	1/21	2/22	2/24	2/26	3/27	3/29	3/31	4/32	4/34	4/36
3	—	—	0/9	0/12	1/14	2/16	2/19	3/21	3/24	4/26	5/28	5/31	6/33	7/35	7/38	8/40	9/42	9/45	10/47	11/49
4	—	—	0/12	1/15	2/18	3/21	4/24	5/27	6/30	7/33	8/36	9/39	10/42	11/45	12/48	14/50	15/53	16/56	17/59	18/62
5	—	0/10	1/14	2/18	4/21	5/25	6/29	8/32	9/36	11/39	12/43	13/47	15/50	16/54	18/57	19/61	20/65	22/68	23/72	25/75
6	—	0/12	2/16	3/21	5/25	7/29	8/34	10/38	12/42	14/46	16/50	17/55	19/59	21/63	23/67	25/71	26/76	28/80	30/84	32/88
7	—	0/14	2/19	4/24	6/29	8/34	11/38	13/43	15/48	17/53	19/58	21/63	24/67	26/72	28/77	30/82	33/86	35/91	37/96	39/101
8	—	1/15	3/21	5/27	8/32	10/38	13/43	15/49	18/54	20/60	23/65	26/70	28/76	31/81	33/87	36/92	39/97	41/103	44/108	47/113
9	—	1/17	3/24	6/30	9/36	12/42	15/48	18/54	21/60	24/66	27/72	30/78	33/84	36/90	39/96	42/102	45/108	48/114	51/120	54/126
10	—	1/19	4/26	7/33	11/39	14/46	17/53	20/60	24/66	27/73	31/79	34/86	37/93	41/99	44/106	48/112	51/119	55/125	58/132	62/138
11	—	1/21	5/28	8/36	12/43	16/50	19/58	23/65	27/72	31/79	34/87	38/94	42/101	46/108	50/115	54/122	57/130	61/137	65/144	69/151
12	—	2/22	5/31	9/39	13/47	17/55	21/63	26/70	30/78	34/86	38/94	42/102	47/109	51/117	55/125	60/132	64/140	68/148	72/156	77/163
13	—	2/24	6/33	10/42	15/50	19/59	24/67	28/76	33/84	37/93	42/101	47/109	51/118	56/126	61/134	65/143	70/151	75/159	80/167	84/176
14	—	2/26	7/35	11/45	16/54	21/63	26/72	31/81	36/90	41/99	46/108	51/117	56/126	61/135	66/144	71/153	77/161	82/170	87/179	92/188
15	—	3/27	7/38	12/48	18/57	23/67	28/77	33/87	39/96	44/106	50/115	55/125	61/134	66/144	72/153	77/163	83/172	88/182	94/191	100/200
16	—	3/29	8/40	14/50	19/61	25/71	30/82	36/92	42/102	48/112	54/122	60/132	65/143	71/153	77/163	83/173	89/183	95/193	101/203	107/213
17	—	3/31	9/42	15/53	20/65	26/76	33/86	39/97	45/108	51/119	57/130	64/140	70/151	77/161	83/172	89/183	96/193	102/204	109/214	115/225
18	—	4/32	9/45	16/56	22/68	28/80	35/91	41/103	48/114	55/123	61/137	68/148	75/159	82/170	88/182	95/193	102/204	109/215	116/226	123/237
19	0/19	4/34	10/47	17/59	23/72	30/84	37/96	44/108	51/120	58/132	65/144	72/156	80/167	87/179	94/191	101/203	109/214	116/226	123/238	130/250
20	0/20	4/36	11/49	18/62	25/75	32/88	39/101	47/113	54/126	62/138	69/151	77/163	84/176	92/188	100/200	107/213	115/225	123/237	130/250	138/262

Find N, the number of pairs, in the leftmost column. Reject H_0, if the observed value exceeds the tabled value.

TABLE H
Critical Values for Spearman's r_S.

	Significance Level for a Directional Test at			
	.05	.025	.005	.001
	Significance Level for a Nondirectional Test at			
N	.10	.05	.01	.002
5	.900	1.000	—	—
6	.829	.886	1.000	—
7	.715	.786	.929	1.000
8	.620	.715	.881	.953
9	.600	.700	.834	.917
10	.564	.649	.794	.879
11	.537	.619	.764	.855
12	.504	.588	.735	.826
13	.484	.561	.704	.797
14	.464	.539	.680	.772
15	.447	.522	.658	.750
16	.430	.503	.636	.730
17	.415	.488	.618	.711
18	.402	.474	.600	.693
19	.392	.460	.585	.676
20	.381	.447	.570	.661
21	.371	.437	.556	.647
22	.361	.426	.544	.633
23	.353	.417	.532	.620
24	.345	.407	.521	.608
25	.337	.399	.511	.597
26	.331	.391	.501	.587
27	.325	.383	.493	.577
28	.319	.376	.484	.567
29	.312	.369	.475	.558
30	.307	.363	.467	.549

Source: Glasser, G. J., and Winter, R. F., "Critical Values of the Coefficient of Rank Correlation for Testing the Hypothesis of Independence," *Biometrika* 48 (1961):444. As used in R. B. McCall. 1980. *Fundamental statistics for psychology*, 3rd ed. New York: Harcourt Brace Jovanovich.

APPENDIX B

Answers to Even-numbered Questions

CHAPTER 1

2. (A) D (B) D (C) I (D) I (E) D

4. (A) Inferential
   (B) Sample; freshmen women
   (C) Use of toothpaste
   (D) Whiteness of teeth rating
   (E) Numerical

6. (A) Inferential
   (B) Samples; people who live in cities or rural areas
   (C) Cities or rural
   (D) Age at death
   (E) Numerical

8. (A) Descriptive
   (B) Population
   (C) Sex
   (D) Preferred major
   (E) Categorical

12. It would be meaningless to try to average categorized data. For example, what is the average of an apple and an orange?

14. Meaningfulness is the independent variable. Number of trigrams recalled is the dependent variable.

16. Reaction time, grams of caffeine

18. An independent variable is manipulated by the experimenter; a dependent variable is the one that is measured.

CHAPTER 2

2. (A)

Number of Aggressive Acts	f
1	1
2	2
3	2
4	0
5	3
6	0
7	2
8	0
9	1

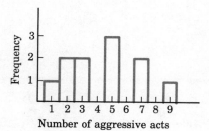

(C)

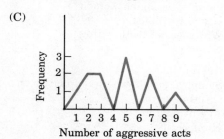

(D) 8

(E) Mode = 5; median = 5; mean = 4.45

4. (A) C   (B) C   (C) B   (D) Low

6. Measures of variability indicate the degree to which scores in a distribution differ.

8. The mean is best suited for statistical inference because it is the most stable estimate of central tendency among samples from a population.

10. As shown in Question 9, treatment effects may not be limited to measures of central tendency. More importantly, measures of variability are rough measures of the confidence we can have in our conclusions about means. In general, the greater the variance, the less the confidence in the mean.

12. (A) American League
 East: M = 52.857, standard deviation = 3.943, range = 11
 West: M = 48.429, standard deviation = 8.296, range = 24
 National League
 East: M = 51.667, standard deviation = 6.446, range = 17
 West: M = 51.167, standard deviation = 7.581, range = 23
 (B) Most variability: American League West;
 Least variability: American League East

14. The range is the difference between the largest and smallest score. The variance is the average squared deviation from the mean, and the standard deviation is the positive square root of the variance. The standard deviation is preferred over the variance, because it is expressed in the same units as the original measure. The range is not particularly useful because it is easily influenced by extreme scores.

16. (A) Range = 79, mode = 97, median = 86, mean = 70.81
 (B) Variance = 796.73, standard deviation = 28.23
 (C) $z(62)$ = $-.31$; $z(100)$ = 1.03; $z(25)$ = -1.62; $z(33)$ = -1.34;
 $z(55)$ = $-.56$
 (D) 42.58

CHAPTER 3

2. 0 heads = .25; 1 head = .5; 2 heads = .25

4. (A) .31 (B) .69

6. With these data alone, the best guess (inference) is .305.

8. (A) .79 (B) .67 (C) .54

10. No. The first gambler's probabilities sum only to .82, and the third gambler's probabilities sum to 1.11. Only the second gambler, whose odds sum to 1.00 could be correct.

12. (A) 43/60 (B) 17/60 (C) 2/3 (D) 1/3 (E) 4/5 (F) 32/43
 (G) 9/20 (H) 9/17

14. (A) If plain and peanut M&Ms are equally likely, P(M&M is red) = .191; P(peanut M&M is red) = .204.
 (B) P(M&M is green) = .089; P(plain M&M is green) = .037.
 (C) No. If color and type were independent, the probability of a given color would be the same for both types.

16. (A)

		Same-sex Toys	Opposite-sex Toys	
TV Watching	Less than 1 hour	.32	.08	.40
	4 hours	.48	.12	.60
		.80	.20	1.0

(B) Not related: the cell probabilities equal the product of the appropriate marginal probabilities.

18. (A) .648, .487   (B) 1223.7

20.

	B	Not B	
A	.20	.45	.65
Not A	.05	.30	.35
	.25	.75	

CHAPTER 4

2. (B) Sampling distribution

4. σ_μ increases as σ increases;
σ_μ increases as N decreases.

6. Standard error of the mean.

8. No. Sample means can be expected to vary. Either (or neither) result could be the true population mean.

10. Standard error of the mean. It gives information about the extent to which the means from samples of size N can be expected to vary.

12. Decreases, zero

14. (A) 3.79   (B) 5.59   (C) 237.62   (D) 92   (E) 900

16. $\mu = 6.5$; $\sigma = 1.708$

18. 6.24

20. (A) .267
     (B) 1.13
     (C) A sample mean of 8.5 would be more likely if $N = 50$ than if $N = 900$ (if we are correct to assume that the population mean is 82.5).

CHAPTER 5

2. It does present a problem unless the sample size is very large (e.g., N is 100 or greater). For very large N, the sampling distribution will be approximately normally distributed.

4. 68.26%, 95.44%, 99.74%

6. When N is very large.

8. (A) $t = -1.00$
(B) Yes

10.

df	t	p
1	$t > 12.706$	.025
2	$1.886 < t < 2.920$	.05
10	$t > 2.764$	.01
20	$t > 2.845$	.005
20	$1.325 < t < 2.086$	.075
29	$0 < t < 1.699$	.45

12. Degrees of freedom may be thought of as a measure assigned to a set of observations. It is a measure of the extent to which the values of the observations are free to vary and still satisfy some constraint.

14. $M = 1.813$, $t_7 = 1.396$. We cannot reject the null hypothesis at the .05 level.

16. The average increase in SAT scores was 2.5 points. $t_7 = 1.08$, which is not significant. We can conclude that attending college has no effect on SAT scores.

18. $t_{20} = .613$, we cannot reject the null hypothesis that the IQs of students in his class differ from the average.

20. (A) $t_9 = .33$
(B) No
(C) It is no different than fertilizer B

22. $t_7 = 19.44$; the mean underfill is significant.

CHAPTER 6

2. The tabled value is the critical value. It represents the minimum value a statistic must have in order to reject the null hypothesis at a specified level of significance.

4. A directional hypothesis specifies the direction of the effect and uses a one-tailed test. A nondirectional hypothesis predicts a difference, but the direction of the difference is not specified; it uses a two-tailed test.

6. Yes. Significance, for example, at the .05 level, states a *probability*. In the long run, 5 times out of 100 the experimenter would be making an error in rejecting H_0.

8. (A) Correct decision   (B) Correct decision   (C) Type I error   (D) Correct decision   (E) Type II error

10. (A) Decreases   (B) Decreases   (C) Decreases   (D) Increases

12. Type II, type I

14. (A) Hypothesis that CAI is effective

   (B) Hypothesis that CAI is not effective

16. Compare the results with findings of previous research, replicate the experiment, and statistical inference (rejecting H_0 at some level of confidence can be considered as concluding that the results are reliable).

CHAPTER 7

2. Variance due to extraneous variables can be controlled.

4. (A) $t = -1.47$, not significant   (B) $t = .64$, not significant
   (C) $t = 3.37, p < .05$ (D) $t = -2.86, p < .05$

6. (1) Sampling distribution of mean differences is symmetrical;   (2) Its mean is equal to $\mu_1 - \mu_2$;   (3) Its variance equals the sum of the squares of the standard errors of the respective means.

8. (1) Increased N;   (2) Factors used to control extraneous variance;
   (3) Use of one-tailed test (when the directional hypothesis is true).

10. $t_3 = 5.38$, which is greater than the critical value of 2.353 needed to reject H_0. The improvement in his bowling scores can be attributed to the new ball.

12. (A) $t = 1.916$, not significant;   (B) $t = 3.710$, reject H_0 and conclude the drug reduced blood pressure.

14. $t_8 = 3.418$, which is significant at the .05 level. Reject H_0 and conclude that players at Spike University jump higher.

16. $t = 1.63$, we cannot reject H_0, so there is no significant difference between the profit margins of urban and suburban stores.

18. $t = .17$. There is no evidence for a significant difference between the textbooks, so the principal should not authorize purchase of the new book.

CHAPTER 8

2. F

4. Two degrees of freedom values are used in the F statistic: $df_1 = N_1 - 1$ and $df_2 = N_2 - 1$, because two separate estimates of variance are involved.

6. (1) The distribution does not extend below zero;   (2) For small values of df, the distribution is positively skewed;   (3) As degrees of freedom increase, the distribution becomes symmetrical and bell-shaped.

8. Expected value ≈ 1, which is the same as the mean.

10.

α	df Numerator	df Denominator	Critical Value
.05	1	1	161
.05	∞	∞	1.00
.05	24	13	2.42
.01	24	2	3.78
.01	40	16	3.01
.01	5	80	3.25
.01	6	11	5.07

12. One-tailed test, used to test a directional experimental hypothesis: s_1^2 is expected to be larger than s_2^2.

14. We cannot predict anything about the sample variance, because sample means and variances are unrelated (provided the population is normally distributed).

16. A scatterplot shows the joint distribution of two variables in which pairs of scores are shown as points in two-dimensional space.

CHAPTER 9

2. An advantage of a within-subject design is that the between-subject variability is removed from the analysis. However, if scores among treatment conditions are not related, the within-subject design may

have less power than a between-subject design, because the degrees of freedom are less in a within-subject design.

4. Disagree. A within-subject design cannot be analyzed using a between-subject analysis of variance.

6. The analysis of variance requires three assumptions: (1) Subjects are sampled randomly; (2) The population is normally distributed; (3) The variances of samples are equivalent.

8.

Source	SS	df	MSS	F
Between	54	3	18	2.25
Within	64	8	8	
Total	118	11		

The critical value needed to reject H_0 for $F(3,8)$ at the .05 level is 4.07. We cannot reject H_0.

10.

Source	SS	df	MSS	F
Between	56	2	28	2.93
Within	86	9	9.56	
Total	142	11		

The critical value needed to reject H_0 for $F(2,9)$ at the .05 level is 4.26. We cannot reject H_0.

12.

Source	SS	df	MSS	F
Between	52.8	3	17.6	2.8
Within	100.4	16	6.28	
Total	153.2	19		

The critical value needed to reject H_0 for $F(3,16)$ at the .05 level is 3.24. We cannot reject H_0.

14.

Source	SS	df	MSS	F
Between	289.43	2	144.71	40.20
Within	90.00	25	3.60	
Total	379.43	27		

Since our computed F exceeds the critical value of $F(2,25)$ needed to reject H_0, we can conclude that significant differences exist among the means:

The difference between .2 mg and .8 mg is significant, $t_{25} = 5.20$.
The difference between .8 mg and 1.2 mg is significant, $t_{25} = 4.62$.
The difference between .2 mg and 1.2 mg is significant, $t_{25} = 8.96$.

16.

Source	SS	df	MSS	F
Between	145.80	2	72.9	7.53
Within	116.20	12	9.68	
Total	262.00	14		

The critical value needed to reject H_0 for $F(2,12)$ is 3.88, so we reject H_0.

18.

Source	SS	df	MSS	F
Between	8751.0	2	4375.5	21.95
Within	2989.5	15	199.3	
Total	11,740.5	17		

The critical value for F (2,15) at the .05 level is 3.68. Since our computed value of F is larger than the critical value, we can conclude that there are differences produced by the types of instruction. Contrary to the initial hypothesis, it appears that the memorize condition resulted in the fastest time to make the airplane, while the active group was slowest.

20.

Source	SS	df	MSS	F
Between	958.2	2	479.1	16.87
Within	766.6	27	28.4	
Total	1724.8	29		

(A) The critical value of F (2,27) at the .05 level is 3.35. We can therefore conclude that the means are different, and the orienting instructions had an effect on test performance.

(B) The teacher's prediction that the teach group would be superior to the study group was supported: $t_{27} = 4.53$, $p < .05$. The prediction that the study group would be superior to the control group was not supported, $t_{27} = .88$.

CHAPTER 10

2. Simultaneously; cells; levels

4. No. When an interaction is present, the main effects cannot be interpreted directly. In this example, one must specify the level of A in order to draw any conclusions about the effects of B.

6.

Source	SS	df	MSS	F
Groups	99.65	7		
Route (A)	58.46	3	19.49	4.629
Weather (B)	35.04	1	35.04	8.323
A × B	6.13	3	2.04	.485
Within group	67.33	16	4.21	
Total	166.98	23		

Both main effects are significant at the .05 level. We can conclude that Speedy runs faster in the rain than in sunny weather (means of 20.83 and 23.25, respectively). In addition, there are differences in running times among the different routes (means of 22.5, 24.33, 21.17, and 20.165 for routes 1–4, respectively). There is no interaction between routes and weather.

8.

Source	SS	df	MSS	F
Groups	85.76	7		
Rod (A)	6.57	3	2.19	.652
Line (B)	71.07	1	71.07	21.152
$A \times B$	8.12	3	2.71	.807
Within group	53.77	16	3.36	
Total	139.53	23		

The type of fishing line is significant: Examining the means, we see that line 2 is stronger than line 1. There are no differences among fishing rods, nor any interaction.

10.

Source	SS	df	MSS	F
Groups	623.6	8		
Detergent (A)	34.3	2	17.1	1.248
Water temperature (B)	442.7	2	221.4	16.161
$A \times B$	146.6	4	36.6	2.672
Within group	246.7	18	13.7	
Total	870.3	26		

In this experiment, only the water temperature was a significant factor, with cold water again yielding the highest "greasiness" ratings (mean of 55.33 compared to 46.33 and 47.22).

12. (A)

Source	SS	df	MSS	F
Groups	1330.95	3		
Teach (A)	2.45	1	2.45	.170
Test (B)	0.05	1	.05	.003
$A \times B$	1328.45	1	1328.45	92.41
Within group	230.0	16	14.375	
Total	1560.95	19		

Neither of the main effects is significant, but the interaction is highly significant.

(C) There is strong evidence for transfer appropriate processing in this experiment. Students whose instruction focused on concepts did better on a test of concepts than on computation; students who were taught computation did better on that test than on a test of concepts.

14.

Source	SS	df	MSS	F
Groups	4432.55	3		
Achievement level (A)	2354.45	1	2354.45	49.05
Instruction (B)	1462.05	1	1462.05	30.46
$A \times B$	616.05	1	616.05	12.83
Within group	768.00	16	48.00	
Total	5200.55	19		

(A) Yes. It appears that self-paced instruction is somewhat better for high-achieving students, but worse for low-achieving students, relative to traditional instruction.

(B) Yes, $t_{16} = 7.49$

(C) Yes, $t_{16} = 1.37$, not significant.

(D) The self-paced instruction is not significantly different from traditional instruction for high-achievement students, but it is much worse for low-achieving students. Therefore, the teacher would be better off continuing with the traditional instruction (at least for low-achieving students).

16.

Source	SS	df	MSS	F
Groups	8811.500	3		
Relaxation (A)	4648.167	1	4648.167	25.150
Drug (B)	4160.667	1	4160.667	22.510
$A \times B$	2.666	1	2.666	.014
Within group	3696.330	20	184.817	
Total	12,507.830	23		

There is a significant main effect for relaxation training and for drug, but no interaction. Drug Y appears superior to Drug X, and relaxation training is better than no training. The most promising treatment for the doctor to pursue is Drug Y given in combination with relaxation training.

18.

Source	SS	df	MSS	F
Groups	286.654	3		
Sex (A)	3.969	1	3.969	3.07
Age (B)	272.484	1	272.484	210.74
$A \times B$	10.201	1	10.201	7.889
Within group	46.546	36	1.293	
Total	333.200	39		

(A) There is a significant main effect of age, and an interaction between age and sex.

(B) In grade 3, boys and girls are about the same speed, $t_{36} = .747$, but by 7th grade, boys are faster, $t_{36} = 3.23$, $p < .05$.

CHAPTER 11

2. ρ^2 is a parameter, r^2 is the statistic that estimates ρ^2.

4. If two variables are correlated, it does not mean that changes in one are *responsible* for producing changes in the other. For example, there may be some other variable responsible for both.

6. (A) $y = 14{,}142.86 + 61{,}428.57x$

(B) $24{,}892.86$

8. (A) $y = 26.51 - 9.00x$

(B) 17.51

10. (B) $r = .431$. The scores are moderately and positively correlated: In general, the higher the score on the mechanical reasoning test, the higher the GPA.

(C) $y = 1.97 + .016x$

(D) 2.66

(E) 51.875

12. Yes, both share the same pattern of results, so the sign of the correlation coefficient should be the same for both.

14. $r = .294$; not significant at the .05 level.

16. $r = -.635$. As the duration of the marriage increases, the frequency of arguments decreases.

18. (A) $r = -.519$; this is significant at the .05 level.

(B) $t_{11} = -2.014$, which is also significant at the .05 level.

20. (A) $r = -.077$; there is no relationship between the test scores and number of errors.

(B) It is difficult to evaluate the effectiveness of his test, because he chose only the top scoring applicants. With this restricted range, one cannot tell if the test was useful or not. If he had randomly selected applicants regardless of their test score, a relationship might have been detected.

CHAPTER 12

2. Every observation is independent of the others.

4. Numerical measures assign numbers to subjects, whereas categorical measures assign categories to subjects.

6. Pearson's χ^2 tests the null hypothesis that the data in a $R \times C$ contingency table come from a population in which the proportion of events in the C categories is the same for all R categories. This is the same as testing for independence between the R and C variables.

8. $\chi^2 = 9.01$, retain H_0

10. $\chi^2 = 17.322$, reject H_0

12. $\chi^2 = .828$, retain H_0

14. $\chi^2 = 13.028$, $p < .05$. We can conclude that a relationship exists between birth order and academic achievement.

16. (A) Yes, there were significant differences in voting patterns; $\chi^2 = 307.16$; $p < .05$. (B) 978.

18. $\chi^2 = 3.06$, not significant. Men do not appear to pump their own gas more than women do.

CHAPTER 13

2. Tests based on counting involve determining the frequencies that events in categories occur. Tests based on ordering involve ranking measures in relative order on some dimension.

4.

Parametric	Nonparametric
t test (related)	sign test, Wilcoxon test
t test (independent)	Mann-Whitney U
within-subject analysis of variance	Friedman test
between-subject analysis of variance	Kruskal-Wallis
Pearson's r	Spearman's r_S

6. $W = 9$. For a two-tailed test, the computed value of W needed to be less than 3. We cannot reject the hypothesis that the rations are equivalent.

8. $H = 8$. Since the critical value for χ^2 is 5.99, we can reject H_0 and conclude that there is a significant difference somewhere among the groups.

10. $r_S = .189$, not significant; there is no evidence for a relationship between order and test score.

12. $U = 180$. We can reject H_0 and conclude that the women in this sample watch more television than the men.

14. $H = 4.53$, which does not exceed the critical value of $\chi^2 = 5.99$ needed to reject H_0. We cannot conclude that people of different income groups rated the administration differently.

16. $W = 15.5$. We needed a value smaller than 10 to reject the null hypothesis for a two-tailed test at the .05 level. We can conclude that the systems do not differ.

18. $U = 42$. We cannot reject H_0 at the .05 level. There is no difference between the number of questions asked by men and women students.

INDEX